ENVIRONMENTAL RESEARCH ADVANCES SERIES

BIOGEOGRAPHY

ENVIRONMENTAL RESEARCH ADVANCES SERIES

BIOGEOGRAPHY

MIHAILS GAILIS
AND
STEFANS KALNIŅŠ
EDITORS

Nova Science Publishers, Inc.
New York

NOTICE TO THE READER

The Publisher has taken reasonable care in the preparation of this book, but makes no expressed or implied warranty of any kind and assumes no responsibility for any errors or omissions. No liability is assumed for incidental or consequential damages in connection with or arising out of information contained in this book. The Publisher shall not be liable for any special, consequential, or exemplary damages resulting, in whole or in part, from the readers' use of, or reliance upon, this material.

Independent verification should be sought for any data, advice or recommendations contained in this book. In addition, no responsibility is assumed by the publisher for any injury and/or damage to persons or property arising from any methods, products, instructions, ideas or otherwise contained in this publication.

This publication is designed to provide accurate and authoritative information with regard to the subject matter covered herein. It is sold with the clear understanding that the Publisher is not engaged in rendering legal or any other professional services. If legal or any other expert assistance is required, the services of a competent person should be sought. FROM A DECLARATION OF PARTICIPANTS JOINTLY ADOPTED BY A COMMITTEE OF THE AMERICAN BAR ASSOCIATION AND A COMMITTEE OF PUBLISHERS.

LIBRARY OF CONGRESS CATALOGING-IN-PUBLICATION DATA
  Biogeography / Mihails Gailis and Stefans Kalnin', editors.
    p. cm.
  Includes index.
  ISBN 978-1-60741-494-0 (hardcover)
  1. Biogeography. I. Gailis, Mihails. II. Kalnin', Stefans.
  QH84.B544 2009
  578.09--dc22

                                                              2009034026

Published by Nova Science Publishers, Inc. ✝ New York

CONTENTS

PREFACE

Biogeography is the study of the geographic distribution of taxa and their attributes in space and time. Investigating biogeographic patterns and processes requires considerable amounts of data collected over large spatial and/or temporal scales. Evolutionary biogeography integrates distributional, phylogenetic, molecular and paleontological data in order to discover biogeographic patterns and assess the historical changes that have shaped them, following a step-wise approach. This book aims to understand the biogeography of the Drosophilidae in Africa and in other regions. This book also examines the development of preventive and educational components of conservation biogeography. The relevance of biogeographical evolutionary analyses for conservation science is discussed as well. In addition, the general methods that can be applied to prioritize areas for protection at regional global scales are also briefly commented on and exemplified with Mexico as a case study.

Chapter 1 - Evolutionary biogeography integrates distributional, phylogenetic, molecular and paleontological data in order to discover biogeographic patterns and assess the historical changes that have shaped them, following a step-wise approach. Although most of the authors involved in the theoretical development of biogeography or that apply specific methods usually conceive them as representing alternative "schools", they can be used to answer different questions. This review introduces the steps of this integrative approach and details the available methods. An evolutionary biogeographical analysis involves five steps: (1) recognition of biotic components (sets of spatio-temporally integrated taxa due to common history), through panbiogeography and methods used to identify areas of endemism; (2) contrastation of the biotic components and identification of the vicariant events that fragmented them, through cladistic biogeography and comparative phylogeography; (3) establishment of a hierarchic arrangement of the components in a biogeographic system of realms, regions, dominions, provinces and districts; (4) identification of cenocrons (sets of taxa with similar origins and ages), dated using intraspecific phylogeography, molecular clocks and fossils; and (5) formulation of a geobiotic scenario, that explains the evolution of the biotic components and cenocrons, integrating geological and tectonical information.

Chapter 2 - Three of the most recognized biogeographic paradigms on earth are the latitudinal gradients of species richness (the decreased richness in biological diversity from equatorial to polar regions), Bergmann's rule (the increase of body size in cold climates) and the unimodal pattern of diversity with depth (species richness peaking at intermediate depths). Understanding the causes of such patterns is still one of greatest contemporary challenges for biogeographers and ecologists. Moreover, though the generality and causal predictors of the first two paradigms have been fully debated in the terrestrial biome, their relevance in marine systems is still poorly understood. In this chapter, I review the present knowledge of these broad-scale biogeographic patterns in the marine systems and use cephalopod molluscs as a case study. Latitudinal gradients of species richness are present in coastal cephalopod fauna and both climate and area extent predict much of the diversity variation. Yet, in the open ocean, diversity does not decline monotonically with latitude and is positively correlated to the availability of oceanic resources. Therefore, a much stronger linkage between patterns of cephalopod diversity and bottom-up processes is found in the pelagic ecosystem. Additionally, cephalopod diversity does not show the classical hump-shaped response to depth. It declines sharply from sub-litoral and epipelagic zones to the slope and bathypelagic habitats and then steadily to abyssal depths. This suggests that higher thermal energy availability and productivity in shallow habitats promote diversification rates, and rejects hypotheses such as biome area, environmental stability and mid-domain effect. Climate also seems to play the most important role in structuring the latitudinal distribution of body size in these marine ectotherms. This evidence holds up to the concept of the "temperature-size rule" but does not support hypotheses for Bergmann's rule relating to resource availability, seasonality (or fasting endurance) and competition. Thus, it is evident that species-area-energy theories that have been formulated for the terrestrial biosphere also apply to the marine systems. In fact, although the phyletic composition and life-history characteristics of terrestrial and marine fauna are quite distinct, there is no reason to assume that the causal predictors behind these widespread biogeographic patterns will differ between these two realms.

Chapter 3 - The Drosophilidae is one of the most evolutionary successful muscomorphan families, with nearly 4,000 species showing astonishing morphological and ecological diversity. It is also unique in biological sciences thanks to the wealth of over 100 years of genetic and ecophysiological research, culminated in the sequencing of the complete genomes of 12 of its species. The aim of this Chapter was to trace the evolution of the Drosophilidae in the Afrotropical region in light of the accumulated faunistic and ecological observations and recent advances in the family phylogenetics and African paleoenvironmental studies. First, a reanalysis of the geographical distribution of 527 species belonging to 31 genera and two subfamilies revealed three main centers of endemism: West Africa (WA), East and South Africa (ESA) and the Insular Indian Ocean (IIO). Species richness shows a longitudinal cline in WA and a latitudinal cline in ESA, with Upper Guinea and South Africa being the most speciose, respectively. However, there are only four endemic genera in Africa, showing its old affinities with other continents. We have thus reviewed the origin and evolution of major drosophilid genera and species groups by reanalyzing recently published sequences of the nuclear gene *Amyrel* in the Schizophora, the Drosophilidae and the Afrotropical subgenus *Zaprionus*. Relative rate test showed no departure from the assumption of strict molecular clock, and thus *Amyrel* evolutionary rate was calibrated using the fossil record. The results show the family to have originated in the Early Eocene (50 MYA) and diversified in the Middle to Late Eocene. We also suggest that the Drosophilidae may be of Euramerican origin. Although Africa was an isolated continent from the Cretaceous to the Early Miocene, it may have been inhabited by early drosophiline radiations (*Chymomyza* and the *latifasciaeformis* group of *Scaptodrosophila*) in the Late Eocene. These taxa are associated with palm trees, a dominant feature of the African Eocene environment. Molecular dating of the basal genus *Lissocephala* whose species breed early in premature fig fruits (syconia) show the genus to start diversification in the Oligocene. More 'derived' lineages (*Zaprionus* and Old World *Sophophora*) breed later on mature and ripening syconia. This ecological succession may recapitulate a phylogenetic trend towards an increase in alcohol tolerance and the utilization of sweet resources. Neogene climatic and tectonic changes have proposed many speciation models of Afrotropical drosophilids: the ecotonal and refugial hypotheses. A reanalysis of *Amyrel* divergence and geographical distribution in three clades (*Drosophila melanogaster* subgroup, *D. montium* group and the genus *Zaprionus*) showed that no single mode of speciation has prevailed. Although in an analysis at both the family and continental scales only general patterns are relevant, we hope that our review will serve as a lightening rod for future integrative research aiming to understand the biogeography of the Drosophilidae in Africa and in other regions.

Chapter 4 - Conflicting opinions regarding the relative importance of dispersal versus vicariance in understanding Caribbean biogeography continue to stimulate lively debate, despite recent advances in regional tectonics, geology, palaeogeography and palaeontology. The Greater Antilles are young geographical features, probably mid-Miocene and Hispaniola has a rich terrestrial invertebrate fossil record in the form of Dominican Republic amber inclusions. Spiders are common in Dominican amber and all fossil species belong in extant families. The island is unique in terms of its known spider fauna, in that a similar number of families have described species recorded on Hispaniola from both fossils in amber and from the extant fauna. It is also the region of the world where the amber fauna is most similar to the Recent fauna, and Miocene Hispaniolan spider biodiversity was presumably similar to that at present.

Fossils form empirical data with both phylogenetic and temporal implications and are of paramount importance for understanding historical biogeography. They can play a decisive role in falsifying hypotheses proposed solely on the distributions of extant taxa. It is also possible to generate hypotheses for both fossil and extant faunas that are ultimately falsibiable through the discovery of new Hispaniolan taxa. The amber spider fauna has clear affiliations to the present South American fauna, so how did they colonize the island from the mainland: dispersal or vicariance? The high Miocene biodiversity (including taxa with poor dispersal capabilities) on a geologically young island would seem to add more weight in favour of the vicariance model for explaining on-island spider lineages, although dispersal cannot be excluded for some of the highly dispersive taxa. However, a case-by-case approach is required in order to determine the relative contributions of the different models. The future research direction for understanding the biogeographical patterns will lie in the application of cladistic and phylogenetic biogeographical approaches. It is interesting to note that preliminary obersvations based on pedipalp morphology suggest some fossil taxa are more derived than extant forms. The contribution that the Dominican (and Mexican) amber fossil fauna can make to our understanding of the origins of Hispaniolan biodiversity, and Caribbean biogeography in general, should not be underestimated.

Chapter 5 - The distribution and abundance of bird species along a rural-urban gradient placed in Comunidad Valenciana (Spain) was examined. At a regional scale (2800 km^2) I used data on presence-absence of breeding bird species in 10 x 10 km squares in order to determine the patterns of bird species' richness along the rural-urban gradient. Such squares were grouped into four sectors according to a decreasing degree of urbanization and they represent: Mediterranean transitional woodland shrub, mostly woody crops, mostly permanently irrigated land, and mostly urban areas. At a local scale (Valencia, 76 Km^2) I used the abundance of wintering and breeding bird species recorded in 118 squares of 700 x 700 m. Such squares were grouped into four classes depending on the cover of rural and urban landscapes: permanently irrigated land, mixed, mostly urban, and urban landscape. At regional scale, breeding bird richness did not show significant changes along the gradient, whereas data at the local scale revealed a decreasing number of bird species with increasing urbanization. Patterns at the regional scale were presumably related with: (i) the spatial variation of bird richness which increased from inland to the coastal fringe; (ii) a greater diversity of habitats because of development; (iii) a moderate level of urbanization; (iv) the use of a grain large enough to sample different habitats many of them sufficiently large to fulfil the requirements of many species. At the local scale, the following factors were apparently important: (i) a higher degree of urbanization; (ii) the smaller grain employed which sampled a less diverse set of habitats, and the extent of them was in many cases too small to fulfil habitat requirements of sensitive species; and (iii) the fact that urban parks were rarely suitable to open-field bird species of the surrounding habitat. The proportion of threatened species showed some variation along the gradient depending on the spatial scale and criterion employed, but overall the number of threatened species increased with increasing urbanization. The study area is a very developed region, where conservation strategies can not be designed without a full consideration of the intricate set of socio-economic processes which have driven most of the changes in the landscape. Currently, efforts should be orientated to conserve the biological communities and the ecological role of the agricultural land which still constitutes the landscape matrix at many spatial scales and administrative boundaries.

Chapter 6 - The extinct charophyte family Clavatoraceae was a significant component of the early Cretaceous lacustrine macrobenthos before the radiation of aquatic angiosperms. The most plesiomorphic and oldest representatives of this family occurred during the Late Jurassic and were scattered in a few localities in the central part of the Peri-Tethyan domain (Western Europe and Northern Africa) and on the margins of the North American Interior Seaway (Morrison Formation). In the beginning of the Early Cretaceous (Berriasian-Barremian), clavatoraceans were dominant in the Peri-Tethyan region and expanded eastwards, reaching the Chinese basins during the Valanginian and Hauterivian. During Barremian and Aptian times clavatoraceans reached their maximum diversity in the Peri-Tethyan region and their maximum palaeogeographic extension worldwide. Two species achieved a cosmopolitan distribution (Eurasia, North and South America and Northern and Eastern Africa), while four species were subcosmopolitan, mainly in the Northern Hemisphere. In contrast, the remaining clavatoracean species showed a marked endemic distribution in specific areas of the Tethyan region, which at that time was an archipelago of large islands enhancing allopatric speciation. During the Albian, the clavatoraceans began to decline. After a significant gap in the fossil record, Late Cretaceous clavatoraceans were either relict forms of previous cosmopolitan species in Northeast Asia or belonged to newly evolved species of endemic distribution in Southern Europe. The family Clavatoraceae became extinct near the Cretaceous-Tertiary boundary.

A historical biogeographic analysis of the clavatoraceans shows that the Tethyan region was a main focus of speciation for the family. A few species that originated in the Peri-Tethyan archipelago migrated elsewhere, sometimes reaching a worldwide distribution contemporaneous with the palaeogeographic extension of wetlands. Certain adaptations, such as the conjoint disposition of gametangia or the capability to colonise new biotopes provide keys to understanding the worldwide migration of particular species. The dispersal of most Tethyan species required substantial time spans - usually several million years – to achieve a cosmopolitan biogeographic range. The animal vectors of clavatoracean propagules are unknown, but may include ancestral birds and dinosaurs. The decline of the family Clavatoraceae in the Albian was marked by the extinction of many endemic species and by significant biogeographic range restrictions for cosmopolitan species. In the Latest Cretaceous, only two completely isolated species remained prior to extinction.

Chapter 7 - Investigating biogeographic patterns and processes requires considerable amounts of data collected over large spatial and/or temporal scales. Availability of such large datasets has recently increased thanks to the rapid developments of geographical information systems, satellite images, and data accessibility through the Internet. But beyond these technical advances, many biogeographic studies are now based on data collected by volunteers from the general public, so-called citizen scientists. The shared principles of these programs most likely to improve large-scale investigations have hardly been highlighted. In this chapter, we first browse existing citizen-science monitoring programs particularly useful for biogeography. We then highlight whether and how these data are valuable to address current challenges in biogeography and large-scale conservation targets. Using concrete examples, we further explain why these data should be particularly efficient to develop the preventive and educational component of conservation biogeography.

Chapter 8 - The relevance of biogeographical evolutionary analyses for conservation science is discussed and highlighted. The general methods that can be applied to prioritize areas for protection at regional global scales are briefly commented and exemplified with Mexico as a case study. We conclude that a biogeographical atlas, representing a synthesis of distributional patterns of taxa from a country or area, may help determine priorities for the selection of areas for conservation.

In: Biogeography
Editors: M. Gailis, S. Kalninš, pp. 1-63

ISBN: 978-1-60741-494-0
© 2010 Nova Science Publishers, Inc.

Chapter 1

EVOLUTIONARY BIOGEOGRAPHY: PRINCIPLES AND METHODS

Juan J. Morrone

Museo de Zoología "Alfonso L. Herrera", Departamento de Biología Evolutiva, Facultad de Ciencias, Universidad Nacional Autónoma de México (UNAM), Apartado Postal 70-399, 04510 Mexico D.F., Mexico.

ABSTRACT

Evolutionary biogeography integrates distributional, phylogenetic, molecular and paleontological data in order to discover biogeographic patterns and assess the historical changes that have shaped them, following a step-wise approach. Although most of the authors involved in the theoretical development of biogeography or that apply specific methods usually conceive them as representing alternative "schools", they can be used to answer different questions. This review introduces the steps of this integrative approach and details the available methods. An evolutionary biogeographical analysis involves five steps: (1) recognition of biotic components (sets of spatio-temporally integrated taxa due to common history), through panbiogeography and methods used to identify areas of endemism; (2) contrastation of the biotic components and identification of the vicariant events that fragmented them, through cladistic biogeography and comparative phylogeography; (3) establishment of a hierarchic arrangement of the components in a biogeographic system of realms, regions, dominions, provinces and districts; (4) identification of cenocrons (sets of taxa with similar origins and ages), dated using intraspecific phylogeography, molecular clocks and fossils; and (5) formulation of a geobiotic scenario, that explains the evolution of the biotic components and cenocrons, integrating geological and tectonical information.

INTRODUCTION

Biogeography is the study of the geographic distribution of taxa and their attributes in space and time (Hausdorf & Hennig 2007). In addition to recognizing distributional patterns of plants, animals and other organisms, biogeographers identify natural biotic units to provide a biogeographic regionalization of the Earth, postulate hypotheses about the processes that may have shaped such patterns, and based on the discovered patterns, help predict the consequences of global planetary changes and select areas for biodiversity conservation (Morrone 2004a). Biogeography is a rather peculiar discipline (Nelson 1978, 1985), occupying an intermediate area among geography, geology and biology, and being practiced by systematists, ecologists, paleontologists, anthropologists, naturalists, and geographers, among others. For this reason, biogeography is so heterogeneous in its principles and methods, lacking the conceptual unity of other sciences (Morrone 2009).

Biogeography of the last two decades has shown an extraordinary theoretical and methodological renovation. Morrone & Crisci (1995) considered that biogeography is passing through a revolution concerning its foundations, basic concepts, methods and relationships with other disciplines. Andersson (1996) detected a problem with the ontology of biogeography, because there is no consensus about which phenomena should be considered as "biogeographic". Crisci (2001) referred to external and internal forces that characterize this "revolution": the former include the paradigm of plate tectonics in earth sciences, cladistics as the basic language of biology and how biologists perceive biogeography; whereas the latter include the proliferation of methods and a seemingly endless philosophical debate. Other important developments in molecular biology, informatics, geographic information systems, ecology and geology should be added (Morrone 2009). Until recently, there has been little communication between biogeographers practicing different approaches, and too little integration of their concepts, even though the need to do so has been recognized for some time (Lomolino et al 2006).

Biogeographic Patterns

Biogeographic patterns are non-random, repetitive arrangements of organisms and clades in geographic space. The study of certain specific patterns constitutes the scope of particular biogeographic approaches, *e.g.*, specific richness patterns and distribution of life forms are studied in ecological biogeography, chorological patterns are studied in areography, structural and functional patterns of ecological systems are studied in macroecology, and biogeographic homology patterns are studied in evolutionary biogeography (Espinosa Organista et al 2002).

Biogeographic homology is the basic concept of evolutionary biogeography (Morrone 2004a). In its more general form, homology means equivalence of parts, and constitutes a sorting procedure used to establish meaningful comparisons within a hierarchical system (de Pinna 1991; Nelson 1994; Rieppel 2004; Williams 2004). Biogeographic homology allows us to identify biotic components, namely the sets of spatio-temporally integrated taxa that coexist in a given area. Homology is the relationship between the homologues (in biogeography, the biotic components) rather than the homologues themselves (Rieppel 1991; Nelson 1994; Williams 2004). Several authors have recognized two stages in the proposition of homologies, that have been named primary and secondary homology by de Pinna (1991). Primary homology, represents a conjecture on the correspondence between parts of different organisms and secondary homology represents a test of such conjecture by congruence with similar statements in the cladogram. In biogeography, both stages have been implicitly recognized by several authors (Morrone & Crisci 1990, 1995; Donoghue et al 2001; Hausdorf & Hennig 2003; Riddle & Hafner 2006).

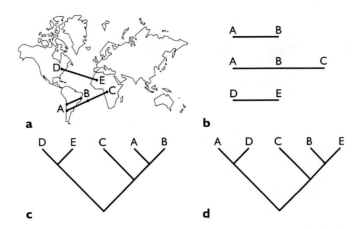

Figure 1. Biogeographic homology. a-b, Primary biogeographic homology, biotic components drawn as generalized tracks; c-d, secondary biogeographic homology (general area cladograms): c, the general area cladogram corroborates the hypothesis of primary biogeographic homology; d, the general area cladogram falsifies the hypothesis of primary biogeographic homology.

Primary biogeographic homology (Figure 1) is a conjecture on a common biogeographic history, which means that different taxa, even when having completely different means of dispersal, are spatio-temporally integrated in a biotic component (Morrone 2001, 2004a, 2009). A panbiogeographic analysis allows comparing individual tracks in order to detect generalized tracks. In addition to sorting distributions of the taxa analyzed into generalized tracks, smaller units or areas of endemism can be detected within them. Both areas of endemism and generalized tracks represent biotic components. Secondary biogeographic homology is the cladistic test of the formerly recognized biotic components. A cladistic biogeographic analysis allows one to compare area cladograms —obtained by replacing terminal taxa in taxon-area cladograms by the areas of endemism they inhabit— in order to obtain a general area cladogram.

Biogeographic Processes

Biogeographic processes (Figure 2) are those that shape the geographic distribution of taxa. There are three basic biogeographic processes: dispersal, vicariance, and extinction. Once patterns have been discovered, explanations on the processes that have shaped them are sought, and the hypotheses tested until robust theories become accepted.

Dispersal is the expansion of the distributional area of a taxon, covering all types of geographic translocation (Myers & Giller 1988). For classical dispersalists (*e.g.*, Darwin 1859; Wallace 1876; Matthew 1915), it meant the movement by active migration or by passive transfer of a species from its center of origin, usually crossing a preexisting barrier, and allowing it to colonize a new area and, eventually, differentiate into new taxa. More recent authors do not usually imply a precise center of origin but the ancestral area where the taxon evolved (Bremer 1995). It is useful to distinguish between the movement of an organism within its area of distribution, named dispersion (Platnick 1976), organismic dispersal (Wiley 1981) or intra-range dispersal (MacDonald 2003), from extra-range or biogeographic dispersal.

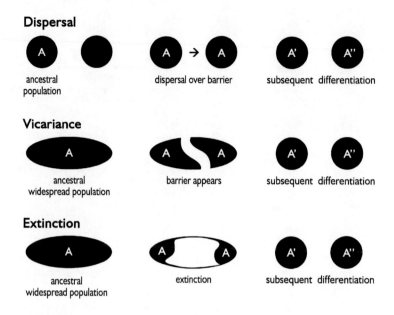

Figure 2. Three main biogeographic processes: dispersal, vicariance, and extinction.

There are four explanatory models of biogeographical dispersal (Pielou 1992; Morrone 2009):

(1) Long distance dispersal: random movement of organisms through barriers, that allows to the establishment of species in very distant areas. It was the most popular model in earlier dispersalism, and it has recently received some support to explain disjunct distributions on widely separated areas that apparently were never in contact.

(2) Diffusion or range expansion: gradual movement of populations crossing adjacent suitable habitats, during several generations.

(3) Secular migration: movement involving a short range distance that occurs so slowly that the species evolves in the meantime.

(4) Geodispersal or concerted dispersal: simultaneous movement of several taxa due to the effacement of a barrier, followed by the emergence of a new barrier that produces subsequent vicariance. As a result of geodispersal, biogeographic convergence occurs.

Vicariance is the appearance of a barrier that allows the fragmentation of the distribution of an ancestral species, after which the descendant species may evolve in isolation. The appearance of the barrier causes the disjunction, so they both have the same age. Area fragmentation is not the only way to cause vicariance. The process known as dynamic vicariance (Zunino & Zullini 2003) implies that climatic changes may act displacing a biotic component gradually in a certain direction, which finally finds a barrier that causes vicariance.

Extinction is the local extirpation or total disappearance of a taxon. It has the potential to obscure biogeographic patterns, because biotas may appear different simply because one region has experienced differential extinction (Lieberman 2003b, 2005). Although extinction is a fact, mechanisms explaining it usually do not concern biogeographers, because it does not form patterns. Different taxa may actually be biogeographically congruent but not appear so in a cladistic biogeographic analysis because of pruning due to extinction (Lieberman 2004).

It has been debated extensively whether dispersal or vicariance represent the most relevant process to explain biogeographic patterns (Nelson 1978; Platnick & Nelson 1978; Humphries & Parenti 1999). During the 19th century and the first decades of the 20th century, dispersalism emphasized dispersal through a stable geography from centers of origin to explain the distribution of organisms (Darwin 1859; Wallace 1876; Matthew 1915). During the second half of the 20th century, usually associated with the acceptance of plate tectonics, vicariance arose as a more appropriate explanation than dispersal (Croizat 1964; Croizat et al 1974). Intraspecific phylogeography (Avise 2000), however, shows that dispersal continues being relevant when explaining the distribution of organisms.

In the mid 19th century, Hooker (1844-60) discovered that, quite paradoxically, both dispersal and vicariance could explain the same disjunctions. For example, we may have a species inhabiting two disjunct areas, which could be due to a widespread ancestral distribution when both areas were united (vicariance explanation), or that it evolved in one area and then dispersed to the other (dispersal explanation). Can we choose between them? The solution to the vicariance-dispersal opposition does not consist in choosing a process or the other, but in adopting a different reasoning where vicariance includes dispersal, although the latter occurs before the geographic barrier appears (Croizat 1958, 1964; Savage 1982; Grehan 1991; Brooks & McLennan 2001; Morrone 2004a, 2009). According to this dispersal-vicariance model (Figure 3), geographic distributions evolve in two steps: (1) dispersal: when climatical and geographic factors are favorable, organisms expand actively their geographic distribution according to their dispersal capabilities or vagility, thus acquiring their ancestral distribution or primitive cosmopolitism; and (2) vicariance: when organisms have occupied all the available geographic or ecologic space, their distribution may stabilize. This allows the isolation of populations in different sectors of the area (subspecies, races or varieties), and the differentiation of new species through the appearance of geographic barriers.

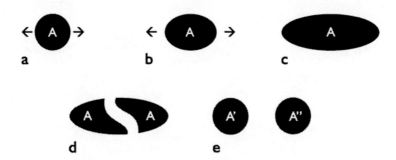

Figure 3. Stages of the dispersal-vicariance model. a-c, Dispersal; d-e, Vicariance.

Biotic Components and Cenocrons

It has been considered traditionally that areas of endemism are the basic biogeographic units (Nelson & Platnick 1981; Platnick 1991; Humphries & Parenti 1999); however,

Henderson (1991) and Andersson (1996) found that focusing on areas instead of biotas was reductionist, and panbiogeographers (Croizat 1958, 1964; Craw et al 1999) considered areas of endemism to be artificial units, preferring instead to recognize generalized tracks. Hausdorf (2002) proposed that biotic elements are more appropriate biogeographic units. Units proposed by other authors include lineages (Jeannel 1942; Ringuelet 1961), horofaunas and cenocrons (Reig 1962, 1981), chorotypes (Baroni-Urbani et al 1978), and dispersal or distributional patterns (Halffter 1987). All these concepts basically refer to two different entities: biotic components and cenocrons. Biotic components are sets of spatio-temporally integrated taxa that coexist in given areas, representing biogeographic units, from a synchronic or proximal perspective. Cenocrons are sets of taxa that share the same biogeographic history, constituting identifiable subsets within a biotic component by their common biotic origin and evolutionary history, from a diachronic perspective.

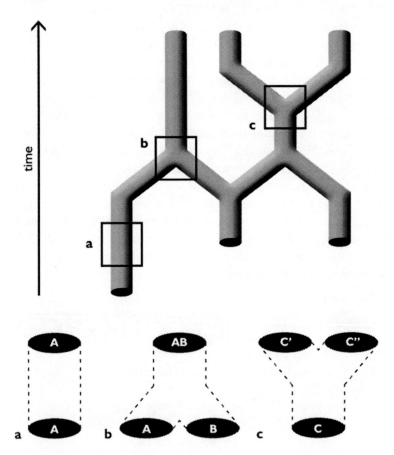

Figure 4. Model showing the stages of biotic evolution. a, Stable biotic component; b, biotic convergence, as a result of geodispersal; c, vicariance.

Biotic components (Figure 4) are basically like other biological entities perduring in time (Boniolo & Carrara 2004). Cenocrons allow one to represent how the convergence of biotic components occurs during biotic evolution. If during some time a biotic component evolves in relative isolation, taxa integrated within it will represent a biotic unit, within which it could be possible to track the cenocrons that have contributed to it. When there is biotic convergence, as a result of geodispersal, two or more cenocrons are combined into a single component. Vicariance splits a biotic component into two or more descendent components. Knowledge on fossils, intraspecific phylogeography and molecular clocks help identify cenocrons (Morrone 2009).

The concepts similar to biotic components (Real et al 1992; Morrone 2004a) are the lineages (Jeannel 1942), generalized tracks (Croizat 1958, 1964), horofaunas (Reig 1962, 1981), and areas of endemism (Nelson & Platnick 1981), whereas dispersal or distributional patterns (Halffter 1987) are similar to cenocrons. Biotic elements (Hausdorf 2002; Hausdorf & Hennig 2007) and chorotypes (Baroni-Urbani et al 1978) seem to correspond to biotic components, although they may help identify cenocrons, because they may allow identification of biotic units even when substantial dispersal affected the distributions of the taxa analyzed.

Evolutionary Biogeography

As a consequence of the frequent episodes of geodispersal, biotic evolution is rarely divergent, resulting in a reticulate rather than a branching structure (Upchurch & Hunn 2002; Riddle & Hafner 2006). To analyze this complexity, we should try to discover the instances of vicariance as well as those where biotic convergence has occurred. A step-wise approach (Figure 5) may allow one to identify particular questions, choose the most appropriate methods to answer them, and finally integrate them in a coherent framework. Most of the authors involved in the theoretical development of biogeography or that apply their methods usually conceive them as representing alternative "schools", but they can be used to answer different questions, which can be different steps of an evolutionary biogeographic analysis.

This step-wise approach comprises five steps, each corresponding to particular questions, methods, and techniques. Panbiogeography and methods for identifying areas of endemism are used to identify biotic components, which are the basic units of evolutionary biogeography. Cladistic biogeography uses phylogenetic data to test the historical relationships between these biotic components. Based on the results of the panbiogeographic and cladistic biogeographic analyses, a regionalization or biogeographic classification may be achieved. Intraspecific phylogeography, molecular clocks and fossils are incorporated to help identify the different cenocrons that become integrated in a biotic component. Finally, the geological and biological knowledge available can be integrated to construct a geobiotic scenario that may help explain the way the biotic components analyzed evolved.

This step-wise approach does not imply that every biogeographer must follow all the steps, but that anybody may articulate a specific biogeographic question and choose the most appropriate method to answer it. Given some time, as the different analyses accumulate, coherent theories may be formulated by integrating them. I defend this approach within the philosophical framework of integrative pluralism (Mitchell 2002). It does not imply an eclectic or "anything goes" approach, but that the different methods are compatible because they give partial solutions, when answering particular questions. Integrative pluralism with respect to methods coexists with the objective of integration in order to explain biotic evolution (Morrone 2009).

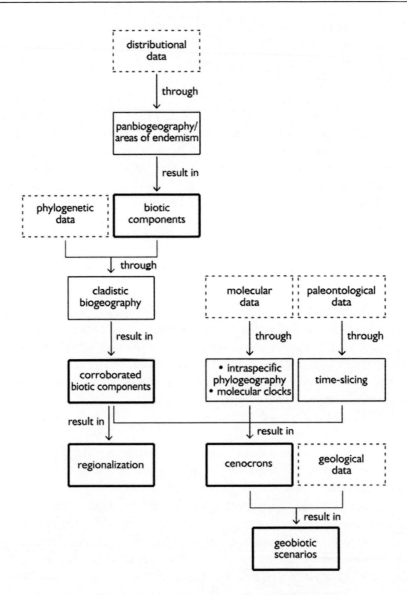

Figure 5. Flow chart with the five steps of an evolutionary biogeographic analysis.

IDENTIFICATION OF BIOTIC COMPONENTS

Biotic components are sets of spatiotemporally integrated taxa that coexist in given areas (Morrone 2009). The identification of biotic components, the basic biogeographic units, is the first stage of an evolutionary biogeographic analysis. There are two basic ways to represent biotic components: generalized tracks and areas of endemism. The former are studied by panbiogeography, whereas the latter are the units of cladistic biogeography. We may distinguish generalized tracks and areas of endemism by their scales (larger or smaller, respectively), although they both represent biotic components (Morrone 2001c, 2004a). The aim of panbiogeography is to recognize generalized tracks, whereas cladistic biogeography emphasizes the recognition of areas of endemism and their relationships as fundamental issues (Morrone & Crisci 1995; Nelson & Platnick 1981; Szumik et al. 2002, 2006).

Panbiogeography

Panbiogeography emphasizes the spatial or geographic dimension of biodiversity to allow a better understanding of evolutionary patterns and processes (Craw et al. 1999). Croizat (1958, 1964) formulated his approach in terms of three metaphors: "Earth and life evolve together," "space + time + form = the biological synthesis," and "life is the uppermost geological layer." Until the development of panbiogeography, biogeographers followed the fashionable geological ideas of their times. The lack of a method by which biogeography and geology could be integrated is illustrated by the cases in which former opponents of continental drift rapidly transferred their models to mobile models of Earth history.

Panbiogeography is based on four assumptions (Craw et al. 1999): (1) distributional patterns constitute an empirical database for biogeographic analyses; (2) distributional patterns provide information about where, when, and how taxa evolved; (3) the spatial and temporal component of these distributional patterns can be represented graphically; and (4) testable hypotheses about historical relationships between the evolution of distributions and Earth history can be derived from geographic correlations between distribution graphs and geological or geomorphic features.

A panbiogeographic analysis comprises three basic steps (Figure 6): (1) construct individual tracks for two or more different taxa; (2) obtain generalized tracks based on the comparison of the individual tracks; and (3) identify nodes in the areas where two or more generalized tracks intersect.

Individual tracks. Individual tracks are defined as the primary spatial coordinates of a species or supraspecific taxa (Crisci et al. 2003). Operationally, an individual track is a line graph drawn on a map that connects the different localities or distributional areas of a taxon according to their geographic proximity. From the topological viewpoint, an individual track is a minimum-spanning tree that for n localities contains n - 1 connections (Page 1987). When a track is drawn, the criterion for connecting the different localities of a species is simple. When any locality is chosen, the nearest locality to it is found, and they are connected by a line; then, this pair of localities is connected with the nearest locality to any of them; the nearest locality to any of the three is united, and so on. The result is an unrooted cladogram, where the sum of the segments connecting the localities is minimal, following a sort of geographic parsimony (Morrone 2009).

Individual tracks can be oriented by formulating a hypothesis on the sequence of the disjunctions implied in it (Figure 7). The most common way to orient a track is designating a baseline, which represents a spatial correlation between the individual track and a geographic or geological feature. On a global scale, ocean or marine basins can be used as baselines, whereas on smaller scales, we may use some evident geological features such as rivers and mountain chains. If we are orienting a strictly terrestrial track, the situation is more complex because we have to decide whether a mountain range is more relevant than a river or any other geological feature.

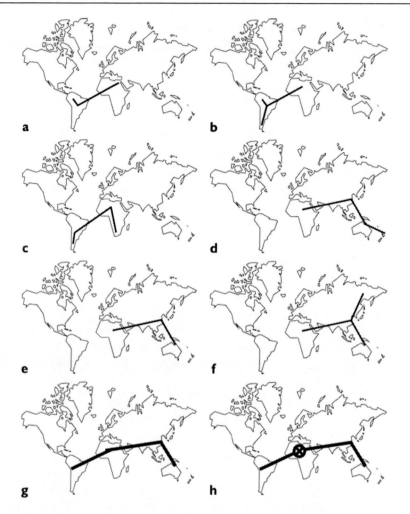

Figure 6. Steps of a panbiogeographic analysis. a-f, Obtaining individual tracks; g, identifying generalized tracks; h, identifying nodes.

Page (1987) suggested the possibility of using cladistic information for orienting individual tracks, and his suggestion was accepted by Craw (1988) and Grehan (1991). McDowall (1978) discussed the possibility of testing generalized tracks, noting the problem of using phylogenetic criteria to construct the individual tracks on which they are based. Platnick & Nelson (1988) noticed that the application of this criterion can be analogous to that of Hennig's progression rule. If the results of panbiogeographic analyses represent hypotheses of primary biogeographic homology, which we will falsify in a cladistic biogeographic analysis (Morrone 2001c), it seems problematic to orient the tracks by means of phylogenetic information. This implies that the phylogenetic hypotheses are part of both the panbiogeographic and the cladistic biogeographic analyses, falling in a circular sequence of reasoning. Therefore, I find it inappropriate to use this criterion (Morrone 2009).

Another criterion for orienting individual tracks is the location of main massings, which are defined as the greatest concentration of biological diversity in the range of the taxon, such as number of species or genetic diversity. In general, main massings represent areas of numerical, genetic, or morphological diversity of a group (Page 1987), which may be identified by a grid analysis (Craw et al. 1999). Platnick & Nelson (1988) and Humphries & Seberg (1989) complained that Croizat referred to main massings as "dispersal centers," "places of origin," "centers of emergence," "ancestral centers of radiation," and "centers of origin," and even Page (1990b) admitted that the concept of main massing was "horribly vague." If a track is oriented from the main massing toward the periphery, the inference involved would be similar to that from dispersal biogeographers (Crisci et al. 2003), so this criterion might also be inappropriate.

Of the available criteria for orienting tracks, the less problematic is the baseline. When the analyses are undertaken on continental scale, however, the use of geological or tectonic characteristics is somewhat more difficult to carry out (Morrone 2004a). For this reason, most of the published panbiogeographic analyses do not orient the individual tracks.

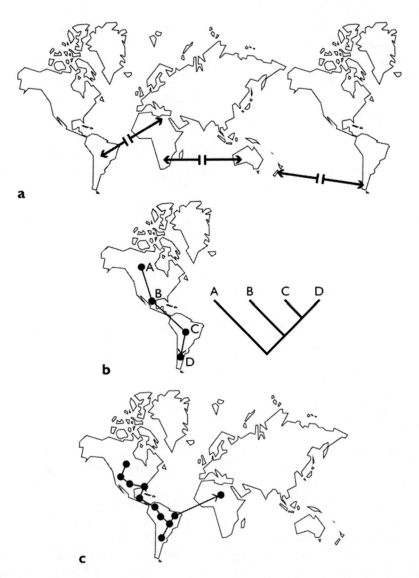

Figure 7. Criteria for orienting individual tracks. a, Baselines; b, phylogenetic information; c, main massing.

Generalized tracks. Generalized or standard tracks result from the significant superposition of different individual tracks (Zunino & Zullini 2003). They indicate the preexistence of ancestral biotic components that became fragmented by geological or tectonic events (Craw 1988). When we compare oriented individual tracks, we can determine that they belong to the same generalized track when they agree in both their structure and their direction (Craw 1988).

The generalized track is the most important concept in panbiogeography. However, McDowall (1978) noted that there are still some questions that research should address: How many individual tracks must coincide in order for a generalized track to be identified? How good must the coincidence be for an individual track to be considered part of a generalized track? How does one interpret and weigh the patterns indicated by noncongruent or conflicting generalized tracks?

From an epistemological perspective, identifying generalized tracks is an inductive process. Although inductivism seems to be a crude and primitive way to conceive scientific knowledge (Reig 1981), it can play a basic part of pattern seeking, as recognized by Rosen (1988a, c). Generalized tracks and areas of endemism represent valid conjectures, the most basic hypothetical statements, which may be falsified by cladistic biogeographic analyses (Morrone 2001c, 2009).

Nodes. They are complex areas where two or more generalized tracks superimpose. They are usually interpreted as tectonic and biotic convergence zones, areas of ancient geography around which evolution has taken place (Heads 2004). The recognition of nodes is one of the most important contributions of panbiogeography. Croizat and his followers have based many of their critiques to cladistic biogeography on its inability to distinguish these types of complex areas because of their attachment to the implicit hierarchy in cladograms. This would not be totally correct because compound areas would behave as species of hybrid origin, showing conflicting relationships with different "paternal" areas. Nodes are particularly interesting from the evolutionary biogeographic viewpoint because they allow us to speculate on the existence of compound or complex areas (Morrone 2009). Nodes may represent the location of endemism, high diversity, distributional boundaries, disjunction, anomalous absence of taxa, incongruence and convergence of characters, and unusual hybrids, among other features (Heads 2004).

Areas of Endemism

They are defined as areas of nonrandom distributional congruence between different taxa (Morrone 1994). How do we recognize an area of endemism? Müller (1973) suggested a protocol for working out "dispersal centers," which has been applied to identify areas of endemism (Morrone 1994; Morrone et al. 1994). It consists basically of plotting the ranges of species on a map and finding the areas of congruence between several species. This approach assumes that the species' ranges are small compared with the region itself, that the limits of the ranges are known with certainty, and that the validity of the species is not in dispute. According to Linder (2001), areas of endemism should meet four criteria: they must have at least two endemic species; the ranges of the species endemic to them should be maximally congruent; they should be narrower than the whole study area, so that several areas are located; and they should be mutually exclusive.

Several issues concerning areas of endemism should be addressed. Crisp et al. (1995) suggested that the alternative procedures for identifying areas of endemism were controversial, especially questioning whether the hierarchical model of parsimony analysis of endemicity (PAE) was adequate for that purpose. Humphries & Parenti (1999) argued that including species that are ecologically very different can argue for a historical rather than ecological explanation for the areas of endemism identified. Linder (2001) proposed three optimality criteria to help choose the best estimate of the areas of endemism: the number of areas identified, the proportion of the species restricted to the areas of endemism, and the congruence of the distributions of the species restricted to the areas of endemism. Roig-Juñent et al. (2002) enumerated some problems with the identification of areas of endemism: lack of distributional data, bias toward locality data, and subjectivity in drawing the exact limits of the areas of endemism.

Methods

Six methods can be applied to identify biotic components (Morrone 2009). I will deal herein with the minimum spanning-tree method (Croizat 1964), track compatibility (Craw 1988), PAE (Rosen 1988b), and endemicity analysis (Szumik et al. 2002). Connectivity and incidence matrices (Page 1987), which have never been applied empirically, and nested areas of endemism analysis (Deo & DeSalle 2006), an adaptation of the nested clade analysis (Templeton 1998), are not dealt with herein.

Minimum-spanning tree method. Croizat did not focus on explicit methods (Grehan 1991; Humphries & Parenti 1999), but one may consider the minimum-spanning tree method as the first formalization of panbiogeography (Craw 1988; Page 1987). It consists of delineating on maps the individual tracks of different taxa and then superimposing them in order to find the generalized tracks. Nodes are identified in the areas where two or more generalized tracks superimpose. With few individual tracks, the minimum-spanning tree method is easy to apply.

The algorithm comprises the following steps (Morrone 2004b): (1) construct individual tracks for different taxa, connecting the localities where they are distributed by a minimum-spanning tree; (2) if possible, orient the individual tracks with baselines; (3) recognize similar tracks (in oriented tracks, they should have the same direction), which will be considered as part of the same generalized track; (4) recognize nodes in the areas where two or more generalized tracks superimpose; and (5) indicate on a map the generalized tracks, baselines, and nodes. Software Trazos2004 (Rojas Parra 2007) is available for this method.

Track compatibility. Craw (1988, 1989a) formalized a quantitative method based on character compatibility (Lequesne 1982). Individual tracks are coded in an area × track matrix that is analyzed for track compatibility, where two individual tracks are compatible or congruent with each other if one is a subset of the other or they are the same in a pairwise comparison. The largest set of compatible tracks is called a clique, and it is used to construct the generalized track. The individual tracks are coded and input into an area × track matrix, where the presence of a track in an area is indicated with "1" and its absence with "0". The matrix is analyzed in search of compatibility between the tracks. The set of compatible tracks represents a clique that is used to construct the generalized track connecting the areas, which is drawn on a map.

The algorithm consists of the following steps (Craw 1989a; Espinosa Organista et al. 2002; Morrone 2004b; Morrone et al. 1996): (1) construct an r x c matrix, where r (rows) represent the localities or distributional areas and c (columns) represent the individual track, where each entry is "1" or "0," depending on whether the track is present or absent; (2) use a compatibility program to find the largest clique of compatible tracks; (3) map out the largest cliques as generalized tracks connecting the localities or areas; (4) use a statistical test to evaluate the percentage of randomly generated matrices where the largest clique size is as large as or larger than the largest clique in the real data in order to provide a statistical test of the level at which the largest clique attains significance; (5) identify baselines (if possible) for the generalized tracks; and (6) represent the generalized tracks, nodes, baselines, and main massings on a map. Software available include CLIQUE of package PHYLIP (Felsenstein 1993), CLINCH (Fiala 1984), SECANT 2.2 (Salisbury 1999), and TNT (Goloboff et al., http://www.zmuc.dk/public/phylogeny/TT/).

Parsimony analysis of endemicity. PAE (Figure 8) was formulated originally by Rosen (1985) and fully developed by Rosen (1988b) and Rosen & Smith (1988). It constructs cladograms based on the parsimony analysis of a presence-absence data matrix of species and supraspecific taxa (Cecca 2002; Cracraft 1991; Escalante & Morrone 2003; Morrone 1994, 1998; Myers 1991; Nihei 2006; Porzecanski & Cracraft 2005; Posadas & Miranda-Esquivel 1999; Rosen 1988b; Rosen & Smith 1988; Trejo-Torres 2003).

Crisci et al. (2003) distinguished three variants of PAE according to the units analyzed: localities, areas of endemism, and grid cells. There are several other units, such as hydrological basins (Aguilar-Aguilar et al. 2003), real and virtual islands (Luna-Vega et al. 1999, 2001; Trejo-Torres & Ackerman 2001), transects (García-Trejo & Navarro 2004; León-Paniagua et al. 2004; Navarro et al. 2004; Trejo-Torres & Ackerman 2002), communities (Ribichich 2005), and political entities (Cué-Bär et al. 2006). García-Barros (2003) proposed a more appropriate classification based on the objectives of the analysis: to infer historical relationships between areas, to identify areas of endemism, and to classify areas (as a phenetic association method).

In order to root the PAE cladograms, a hypothetical area with all "0" is added to the matrix. However, some authors (Cano & Gurrea 2003; Ribichich 2005) have used an area coded with all "1." This alternative rooting groups areas according to shared absences, which would imply depletion through time starting from a cosmopolitan biota (Cecca 2002).

PAE may be used for panbiogeographic analyses, where the clades obtained are considered generalized tracks (Craw et al. 1999; Luna-Vega et al. 2000; Morrone & Márquez 2001). Luna-Vega et al. (2000) and García-Barros et al. (2002) proposed that when the most parsimonious cladograms have been obtained, it is possible to remove or exclude the taxa supporting the different clades and analyze the reduced matrix to search for alternative clades supported by other taxa. This procedure has been named parsimony analysis of endemicity with progressive character elimination (PAE-PCE) (García-Barros 2003; García-Barros et al. 2002).

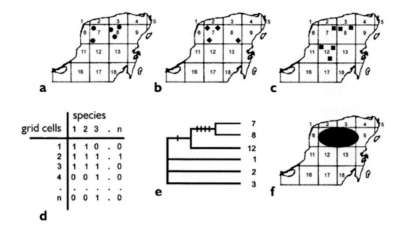

Figure 8. Parsimony analysis of endemicity (PAE). a-c, Distributional maps of three species, with grid cells superimposed; d, data matrix of grid cells x species; e, cladogram of grid cells; f, area of endemism obtained.

PAE has received some criticism. Linder & Mann (1998) criticized Morrone's (1994) approach for identifying areas of endemism with PAE because grid cells can be used only as presence-absence data, and undercollecting may result in grid cells being omitted. Some authors suggested that PAE is not a valid historical method because it does not take into account the phylogenetic relationships of the taxa analyzed (García-Barros et al. 2002; Humphries 1989, 2000; Santos 2005). According to Rosen (1988b; see also Nihei 2006; Trejo-Torres 2003; Trejo-Torres & Ackerman 2002), there are two possible interpretations for PAE cladograms: static and dynamic. The former assumes that cladograms constitute an alternative to phenetic classification methods, whereas according to the latter, cladograms are hypotheses on the historical or ecological relationships of the areas analyzed. If we interpret the external area with all "0" as an area lacking suitable conditions for the taxa to survive therein (ecological interpretation), relationships will indicate ecological affinities. If we interpret the external area as a geologically ancient area, where none of the taxa has yet evolved (historical interpretation), relationships will indicate biotic interchanges or vicariance events. Most of the authors who have used PAE explored historical interpretations of the detected patterns, usually from a vicariance viewpoint; for ecological interpretations, see Trejo-Torres & Ackerman (2002), Trejo-Torres (2003), and Ribichich (2005).

Enghoff (2000) considered PAE an extreme "assumption 0" approach because only the widespread taxa provide evidence of area relationships. Morrone & Márquez (2001) considered PAE an incomplete implementation of Brooks parsimony analysis (BPA). Szumik et al. (2002) criticized the use of PAE for identifying areas of endemism because an explicit optimality criterion is used a posteriori to select areas of endemism found by what they considered less appropriate means. Brooks & Van Veller (2003) criticized the use of PAE as a cladistic biogeographic method, which is erroneous because it has a different objective. Parenti & Humphries (2004) suggested that PAE adopts protocols directly from phylogenetic systematics and violates some of the basic assumptions of cladistic biogeography.

Nihei (2006) presented a revision of PAE, including a discussion of its history and applications. He suggested that most of the criticisms dealt with its method rather than its theory and that they usually resulted from the confusion between the dynamic and static approaches. Nihei warned biogeographers applying PAE to be aware of the problems and limitations of both dynamic and static PAE and to evaluate new variations of PAE.

PAE-PCE consists of the following steps (Craw 1989b; Crisci et al. 2003; Lomolino et al. 2006; Morrone 1994, 2004b; Posadas & Miranda-Esquivel 1999): (1) construct an r x c matrix, where r (rows) represents the units analyzed (e.g., localities, distributional areas, grid cells) and c (columns) represents the taxa, where each entry is "1" or "0," depending on whether the taxon is present or absent in the locality, and a hypothetical area coded with all "0" is added to the matrix in order to root the resulting cladograms; (2) analyze the matrix with a parsimony algorithm, and if more than one cladogram is found, calculate the strict consensus cladogram; (3) connect on a map the area relationships supported by two or more taxa as generalized tracks or areas of endemism; (4) remove the taxa supporting the previous generalized tracks or areas of endemism; and (5) repeat steps 2-4 until no more taxa support any clade. Software available include Hennig86 (Farris 1988), PHYLIP (Felsenstein 1993), NONA (Goloboff 1998), PAUP (Swofford 2003), Pee-Wee (Goloboff, http://www.zmuc.dk/public/phylogeny/Nona-PeeWee/), and TNT (Goloboff et al., http://www.zmuc.dk/public/phylogeny/TNT/). For reading and editing data files and cladograms: Winclada (Nixon 1999), compatible with NONA, Pee-Wee, and Hennig86; and MacClade (Maddison & Maddison, http://macclade.org/macclade.html), compatible with PAUP.

Endemicity analysis. Szumik et al. (2002) proposed a method that takes into consideration the spatial position of the species in order to identify the set of grid cells that represent an optimal area of endemism according to a score based on the number of species endemic to it (Szumik & Roig-Juñent 2005). In order to assign the values of endemicity to the sets of grid cells evaluated, Szumik et al. (2002) suggested four criteria. Szumik & Goloboff (2004) developed an endemicity value that gives weight to each species, considering its adjustment to the evaluated area. The degree of adjustment between the distributional area of each species and the area of endemism under evaluation depends on the relationship between the number of grid cells where it is found and the total number of grid cells. Additionally, the endemicity value increases with the number of grid cells where the presence of the species is assumed or inferred and decreases with the number of grid cells outside the area of endemism where the species is observed or assumed to be present (Szumik & Goloboff 2004).

The algorithm consists of the following steps (Szumik et al. 2002, 2006; Szumik & Goloboff 2004; Szumik & Roig-Juñent 2005): (1) plot species localities on a map with a grid; (2) assign values of endemicity to all possible sets of grid cells, counting the species that may be considered endemic to them according to the four criteria defined by Szumik et al. (2002); (3) choose the sets of grid cells with the highest endemicity scores; and (4) draw the sets of grid cells on a map as areas of endemism. Software available include NDM and VNDM (Goloboff 2004; Szumik et al. 2006).

Evaluation of the methods. There are no studies evaluating the performance of all available methods. Linder (2001) compared PAE using equally weighted data, PAE using his weighting procedure, and UPGMA using the Jaccard coefficient. He found that PAE with equally weighted data performed worst under the criterion of optimality and that weighted PAE outperformed the other two methods in the quality of the results. In addition, the majority of the areas of endemism identified in the consensus cladogram from the latter analysis had at least two endemic species, thus making it unnecessary to check each area of endemism. Moline & Linder (2006) compared PAE and endemicity analysis, finding that both identified adequate areas of endemism, differing in the number of areas identified, the proportion of species endemic to them, the congruence of the distributions of species restricted to the areas, and the performance score calculated.

TESTING RELATIONSHIPS BETWEEN BIOTIC COMPONENTS

Because the generalized tracks that result from panbiogeographic analyses are unrooted, they connect geographic areas but do not specify a precise sequence of fragmentation. For example, given a generalized track joining Australia, New Zealand, and Chile, which of the three areas first separated from the others? In order to determine this sequence, phylogenetic data must be incorporated (Morrone 2009).

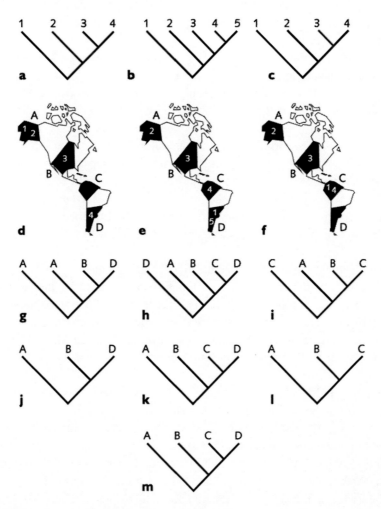

Figure 9. Steps of a cladistic biogeographic analysis. a-c, Three taxonomic cladograms; d-f, maps showing the distribution of the species of the three taxa analyzed; g-i, taxon-area cladograms; j-l, resolved area cladograms; m, general area cladogram.

Cladistic biogeography (Figure 9) assumes a correspondence between the phylogenetic relationships of the taxa and the relationships between the areas they inhabit (Nelson & Platnick 1980, 1981; Platnick & Nelson 1978). It uses information on the cladistic relationships between the taxa and their geographic distribution to form hypotheses on the relationships between areas. If several taxa show the same pattern, such congruence is evidence of common history (Crisci & Morrone 1992; Crisci et al. 2003; Ebach & Humphries 2002; Enghoff 1996; Humphries & Parenti 1999; Humphries et al. 1988; Morrone 1997; Morrone & Crisci 1995; Sanmartín & Ronquist 2002; Wiley 1988a; Zunino & Zullini 2003). Cladistic biogeography originated from the joining of three largely independent research programs: Hennig's (1950) phylogenetic systematics, Croizat's (1958, 1964) panbiogeography, and Wegener's (1929) continental drift. To them, Nelson & Platnick (1981) added the deductive-hypothetical method of Popper (1959, 1963).

We can raise an analogy between systematics and biogeography. In systematics we study taxa, classifying them by their shared characters, whereas in biogeography we study areas, classifying them by their shared taxa. This implies equivalence between taxa (systematics) and areas (biogeography).

Cladistic biogeography poses three questions (Nelson & Platnick 1978): (1) is endemism geographically nonrandom, and, if so, what are the areas of endemism?; (2) given a list of areas of endemism, are the interrelationships of their endemic taxa geographically nonrandom, and, if so, what patterns are formed by their interrelationships?; and (3) given one or more patterns of interrelationships, as represented by one or more general area cladograms, does the pattern correlate with the geological history?

A cladistic biogeographic analysis consists of three basic steps: (1) construction of taxon-area cladograms, from the taxonomic cladograms of two or more different taxa, by replacing their terminal taxa with the areas they inhabit; (2) obtaining resolved area cladograms from the taxon-area cladograms (when demanded by the method applied); and (3) obtaining a general area cladogram, based on the information contained in the resolved area cladograms.

Taxon-area cladograms. Taxon-area cladograms are obtained by replacing the name of each terminal taxon in the cladograms of the taxa analyzed with the area where it is distributed. For example, if a taxon (1 (2 (3, 4))) has species 1 distributed in North America, species 2 in Africa, species 3 in South America, and species 4 in Australia, by replacing the four species in the cladogram with the areas where they are distributed, we obtain the taxon-area cladogram (North America (Africa (South America, Australia))).

Resolved area cladograms. The construction of taxon-area cladograms is simple when each taxon is endemic to a single area and each area has only one taxon, but it is more complex when taxonomic cladograms include widespread taxa, redundant distributions, and missing areas. In these cases, some methods require that taxon-area cladograms be turned into resolved area cladograms, which are also known as fundamental area cladograms (Nelson & Platnick 1981), area cladograms (Page 1990a), and areagrams (Ebach et al. 2005a).

When any of the terminal taxa of a taxon-area cladogram inhabits two or more of the studied areas, it is a widespread taxon (Nelson & Platnick 1981) or a mast (for "multiple areas on a single terminal") (Ebach et al. 2005a). For example, if a taxon (1 (2, 3)) has species 1 distributed in both North America and Africa, species 2 in South America, and species 3 in Australia, by replacing the three species in the cladogram by the areas where they are distributed, we obtain the taxon-area cladogram (North America-Africa (South America, Australia)). As a result of the widespread taxon, North America and Africa appear together in the taxon-area cladogram.

Under assumption 0 (Zandee & Roos 1987), the areas inhabited by a widespread taxon are considered as a monophyletic group in the resolved area cladogram, meaning that the taxon is treated as a synapomorphy of the areas. Under assumption 1 (Nelson & Platnick 1981), the widespread taxon is not considered as a synapomorphy in constructing the resolved area cladograms, and the areas inhabited by it can constitute monophyletic or paraphyletic groups in the resolved area cladograms. Under assumption 2 (Nelson & Platnick 1981), only one of the occurrences is considered as evidence, whereas the other can "float" in the resolved area cladograms, therefore constituting the areas involved monophyletic, paraphyletic, or polyphyletic groups. The three assumptions show an inclusion relationship because topologies obtained under assumption 0 are included within those obtained under assumption 1, and those obtained under assumption 1 are included within those from assumption 2. Some authors prefer assumption 2, especially to deal with widespread taxa (Humphries 1989, 1992; Humphries et al. 1988; Morrone & Carpenter 1994; Nelson & Platnick 1981), considering that widespread taxa are a source of ambiguity, because a future analysis can show that a widespread taxon really represents two or more different taxa, not necessarily related, and inhabiting different areas; a taxon may have a widespread distribution due to dispersal; and a taxon may have a wide distribution because it did not respond with speciation to a vicariance event. Other authors accept the informative value of widespread taxa, thus preferring assumption 0 (Brooks 1990; Enghoff 1996; Wiley 1988a; Zandee & Roos 1987). Enghoff (1995) and van Veller et al. (1999) considered assumption 2 to be less informative because it offers more solutions than assumptions 0 or 1. Zandee & Roos (1987), Wiley (1988a), Enghoff (1996), and van Veller et al. (1999, 2000) argued that assumptions 1 and 2 distort the phylogenetic relationships between the terminal taxa of the taxon-area cladogram. However, Page (1989a, 1990a) indicated clearly that assumptions 1 and 2 are interpretations about relationships between areas, not between taxa. The main criticism of assumption 0 is that it is too restrictive, not considering the possibility of dispersal to explain the distributions of widespread taxa (Page 1989a, 1990a).

Van Soest (1996) and van Soest & Hajdu (1997) proposed an alternative treatment for widespread taxa called "no-assumption coding." These authors reasoned that because taxa have different means of dispersal and are affected differentially by the environment, when comparing different taxon-area cladograms, usually there will be areas of greater and smaller size partially superposed. If the smaller areas are chosen as units of the analysis, then the greater ones will generate widespread taxa, and later manipulation with assumptions 0, 1, and 2 will imply the assumption of some of the processes previously indicated. Additionally, current biogeographic methods allow areas to occur on a single position in the general area cladogram, whereas the history of the biota may indicate various positions. To remedy these limitations, van Soest (1996) proposed a different coding of widespread taxa, where if two or more taxon-area cladograms have taxa widespread in the same areas, this combination can be treated as a single area. The comparison between the different area cladograms will determine whether the large area is united in a general area cladogram with its constituent areas (demonstrating that in fact it is a single area) or it constitutes a paraphyletic or polyphyletic group (indicating that the individual areas are independent).

Sanmartín & Ronquist (2002) proposed another treatment for widespread taxa, within the approach of the event-based methods. For these authors, the problem of a widespread taxon is solved by treating it as an unsolved polytomy, consisting of a unit for each of the involved areas. In order to optimize the ancestral distribution of the widespread taxon, three options are applied. These options allow different sets of possible ancestral distributions, each with different associated costs. The option "recent" assumes that the wide distribution is recent, due to dispersal. The option "ancient" assumes that the wide distribution is ancestral, due to vicariance and extinction. The option "free" assumes that the wide distribution is unsolved, allowing any combination of biogeographic processes and any resolution of the polytomy.

Ebach et al. (2005a) suggested that assumptions 1 and 2 may inadvertently use paralogy and widespread taxa or masts and yield spurious results. They proposed that the "transparent method," along with paralogy-free subtree analysis, is appropriate to deal with this problem. This is done by treating all taxon-area cladograms as individual points that may be part of a common pattern (the general area cladogram). Area cladograms with masts are viewed in terms of proximal relationships, and masts are resolved so each area is represented once. For example, if we have the taxon-area cladogram AB (C (D, E)), resolving mast AB implies that two area cladograms are obtained: A (C (D, E)) and B (C (D, E)), which together result in AB (C (D, E)). Ebach et al. (2005a) suggested that the transparent method should be implemented before the paralogy-free subtree analysis.

Redundant distributions occur when an area appears more than once in a taxon-area cladogram because in this area, two or more terminal species are distributed. In the taxon (1 (2 (3 (4, 5)))), if species 1 and 5 are distributed in North America when the species are replaced by the areas, this area will appear twice in the taxon-area cladogram. If the species constitute a monophyletic group, obtaining a resolved area cladogram is simple. There is no special treatment for redundant distributions under assumption 0, although Kluge (1988) proposed a weighting scheme in which a smaller weight is given to the components involving redundant distributions. Under assumption 1, it is interpreted that the redundant distributions are due to duplicated patterns followed by extinction, whereas assumption 2 adds the possibility that sympatry may be due to dispersal (Enghoff 1996; Page 1990a). Most of the authors prefer assumption 2 to treat redundant distributions (Enghoff 1996; Morrone & Carpenter 1994; Nelson & Platnick 1981; Page 1990a).

When no terminal taxon is distributed in one of the areas analyzed, this area will not be represented in the taxon-area cladogram. In the taxon (1 (2, 3)), if no species inhabits Africa (one of the study areas) when the areas are replaced by the species of the cladogram, this area will not appear in the taxon-area cladogram. Missing areas, which are caused by extinction or insufficient studies, are treated as noninformative. They are coded with "?" so that they can be placed in all possible positions in the resolved area cladograms. Also, it is possible to treat them as primitively absent by coding them with "0" (Kluge 1988).

Some authors have determined that the assumptions are not mutually exclusive because it would be possible to deal with widespread taxa under one assumption and with redundant distributions under another (Enghoff 1996; Morrone & Crisci 1995; Page 1990a). Van Veller et al. (1999, 2000, 2001) argued that in order to obtain valid general area cladograms, two requirements should be fulfilled: (1) resolved area cladograms from the different taxa analyzed should be obtained under the same assumption because the common pattern for all taxa would have to be explained by the same process; and (2) the sets of resolved area cladograms obtained under the three assumptions should be successively inclusive, namely, those obtained under assumption 0 would be included within those obtained under assumption 1, and the latter would have to be found within those obtained under assumption 2. The reason is that the processes are additive (each assumption incorporates or includes those of the preceding assumption). The first requirement is not valid because different taxa can be affected by different processes, although they may show a common pattern. In relation to the second requirement, Ebach & Humphries (2002) suggested that it is violated by the use of assumption 0 when introducing artificial internal nodes for widespread taxa.

General area cladograms. On the basis of the information from the different resolved area cladograms, a general area cladogram is derived. It represents a hypothesis on the biogeographic history of the taxa analyzed and the areas where they are distributed. The general area cladogram that results from the analysis may be falsified with a geological area cladogram or geogram (Swenson et al. 2001), based on geological or tectonic data (Morrone & Carpenter 1994). Another way to evaluate general area cladograms is through the calculation of items of error (Morrone & Carpenter 1994; Nelson & Platnick 1981), dtermining the terminal number of nodes and areas that are necessary to add to the taxon-area cladogram so that it agrees with the general area cladogram, that is, to map one cladogram onto the other to determine their congruence. The smaller the number of nodes and terminal areas that must be added, the more parsimonious will be the general area cladogram analyzed, and for that reason it will be chosen. This can be carried out manually or with the program Component version 1.5 (Page 1989b).

From an epistemological point of view, general area cladograms represent testable hypotheses in the framework of Popper's (1959, 1963) hypothetico-deductive method (Nelson & Platnick 1981; Platnick & Nelson 1978). However, some authors have suggested that cladograms are not general hypotheses in Popper's sense (Andersson 1996). An important aspect of general area cladograms is their retrodictive power (Morrone 1997). When we have obtained a general area cladogram, we may use it to carry out predictions or retrodictions related to taxa still not analyzed (which are expected to agree with the general pattern), with geological or tectonic hypotheses, or the relative ages of biotas (when a molecular clock is available for some of the studied taxa) (Morrone 2009).

Methods

There are fourteen cladistic biogeographic methods (Crisci et al. 2003; Humphries & Parenti 1999; Morrone 2004a; Morrone & Crisci 1995). I will deal herein with component analysis (Nelson & Platnick 1981), Brooks parsimony analysis (Wiley 1987), three area statement analysis (Nelson & Ladiges 1991a, 1991c), tree reconciliation analysis (Page 1990a), paralogy-free subtree analysis (Nelson & Ladiges 1996), dispersal-vicariance analysis (Ronquist 1997a), and area cladistics (Ebach & Edgecombe 2001). Seven other methods, namely, reduced consensus cladogram (Rosen 1978), ancestral species maps (Wiley 1980, 1981), quantitative phylogenetic biogeography (Mickevich 1981), component compatibility (Zandee & Roos 1987), quantification of component analysis (Humphries et al. 1988), vicariance event analysis (Hovenkamp 1997), and phylogenetic analysis for comparing trees (Wojcicki & Brooks 2005), which have been applied occasionally, are not dealt with here.

For details on these latter methods, see Humphries & Parenti (1999), Goyenechea et al. (2001), and Morrone (2004a).

A recent biogeographic development that merits special comment here is comparative phylogeography. It is intended to compare phylogeographic structure exhibited by sympatric species to determine whether they exhibit congruent patterns, geographically structured by vicariance events (Abogast & Kenagy 2001; Cunningham & Collins 1998; Riddle & Hafner 2006; Taberlet et al. 1998; Zink 2002). Incongruent patterns may indicate that the species colonized the area more recently, whereas congruent patterns may suggest a longer history of association of the different species (Zink 1996). This approach is similar to cladistic biogeography (Lanteri & Confalonieri 2003; Lieberman 2004; Morrone 2004a; Riddle & Hafner 2006; Santos 2007), so I will not deal with it separately. Some authors have considered BPA to be an appropriate method for comparative phylogeographic studies (Taberlet et al. 1998). Others have not used any formal method for comparison (e.g. Costa 2003; Morales-Barros et al. 2006; Palma et al. 2005; Riddle et al. 2000; Weisrock & Janzen 2000). However, the analysis with any cladistic biogeographic method is possible (Bermingham & Martin 1998; Lapointe & Rissler 2005).

Component analysis. This method (Figure 10) was proposed by Nelson & Platnick (1981). It solves the problems derived from redundant distributions, widespread taxa, and missing areas using assumptions 0, 1, and 2 and then finds the general area cladogram through the intersection of the sets of resolved area cladograms (Biondi 1998; Enghoff 1996; Morrone 1997; Morrone & Carpenter 1994; Nelson 1984; Nelson & Platnick 1981; Page 1988, 1990a; Humphries & Parenti 1999). Enghoff (1996) suggested treating widespread taxa under assumption 0 and redundant distributions under assumption 2. If it is not possible to find a general area cladogram, it may be possible to find a cladogram shared for some of the sets (Crisci et al. 1991). If more than one general area cladogram are obtained, it is possible to build a consensus cladogram (Morrone & Carpenter 1994).

Zandee & Roos (1987) criticized the use of consensus techniques to obtain the general area cladogram, but Page (1990a) clarified that consensus trees are not the only way to obtain a general area cladogram. Alternatively, component analysis has been criticized for its preference for assumptions 1 and 2 instead of assumption 0, which Wiley (1987, 1988a, 1988b) considered more parsimonious. Lieberman (2004) argued that the presence of artificial incongruence due to extinct clades may lead to problems with component analysis because it is not designated to deal with any sort of incongruence. However, I argue that this represents a problem for all cladistic biogeographic methods.

The algorithm consists of the following steps (Biswas & Pawar 2006; Humphries & Parenti 1999; Morrone 1997; Nelson & Platnick 1981; Page 1990a): (1) obtain the taxonomic cladograms of the taxa distributed in the areas analyzed; (2) replace the terminal taxa from the taxonomic cladograms with the areas inhabited by them to obtain taxon-area cladograms; (3) derive resolved area cladograms from each taxon-area cladogram, applying assumptions 0, 1, or 2; and (4) intersect the sets of resolved area cladograms obtained for each taxon-area cladogram in order to find the general area cladogram. If no general area cladogram results from the intersection, check whether an area cladogram is shared by at least some of the sets. If more than one general area cladogram is obtained, build a consensus cladogram. Software available include Component version 1.5 (Page 1989b).

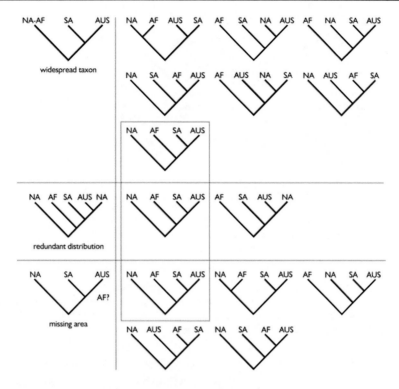

Figure 10. Component analysis. Left column, taxon-area cladograms with: a, a widespread taxon; b, a taxon with a redundant distribution; c, a taxon with a missing area. Right column, resolved area cladograms, showing the result of the intersection of the three sets of resolved area cladograms. AF, Africa; AUS, Australia; NA, North America; SA, South America.

Brooks parsimony analysis. BPA (Figure 11) was proposed by Wiley (1987, 1988a, b) and posteriorly modified by Brooks (1990). It is based on the ideas developed initially by Brooks (1981, 1985) for historical ecology. It is a parsimony analysis of taxon-area cladograms that are codified as two-state variables and analyzed as characters (Biondi 1998; Brooks 2004; van Veller et al. 2000). In order to apply BPA, a data matrix is constructed on the basis of taxon-area cladograms, and it is analyzed with a parsimony algorithm. Brooks (1990) and Brooks & McLennan (1991) proposed another strategy for dealing with parallelisms (dispersal events) that represent falsifications of the null hypothesis. It is named secondary BPA and consists of duplicating the involved area and dealing with each of the resulting areas separately. The analysis of the data matrix allows one to determine whether it was really a unique area or whether they were different areas incorrectly treated as a single one (Lomolino et al. 2006).

An alternative implementation of BPA (Kluge 1988) differs in three aspects. It considers missing areas to be uninformative, coding them with "0." It considers widespread taxa, caused either by dispersal or by not having responded to vicariance, to be irrelevant and therefore does not take them into account. Because for redundant distributions it is impossible to determine which distribution is irrelevant (by being due to dispersal) and which one is not, Kluge (1988) suggested that they be eliminated one at a time, with the resulting columns weighted in proportion to their number (e.g., if there are two redundant distributions, each of the columns will weigh 0.5, and if there are three, 0.33).

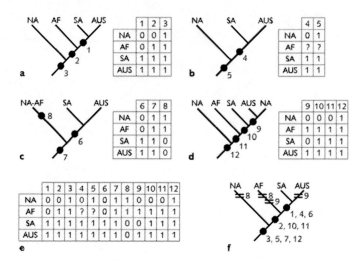

Figure 11. Primary Brooks parsimony analysis (BPA). a-d, taxon-area cladograms and matrices derived from them: a, trivial case; b, taxon with a missing area; c, widespread taxon; d, taxon with a redundant distribution; e, data matrix with all the information; f, general area cladogram obtained. AF, Africa; AUS, Australia; NA, North America; SA, South America.

Lieberman (1997, 2000, 2003a, 2004) proposed another modification of BPA, named modified BPA, intended to interpret geodispersal within a cladistic biogeographic framework. The biogeographic analysis is divided into two separate analyses: one to retrieve congruent episodes of vicariance and another to retrieve congruent episodes of geodispersal (Lieberman 2004). The vicariance analysis produces a cladogram that makes predictions about the relative sequence of vicariance events that fragmented biotas. The geodispersal analysis produces a cladogram that provides information about the relative sequence of vicariance events that joined the biotas. Both cladograms provide complementary information about the biotic history and can be placed in a geological framework. The procedure implies optimizing the ancestral states in the area cladograms in order to estimate whether distributions implied expansions or contractions of the ancestral areas and building the vicariance and geodispersal data matrices, following the same procedure as BPA. The best-supported patterns of vicariance and geodispersal emerge from the parsimony analysis of these matrices. If both cladograms are similar, the same geological processes may have produced vicariance and geodispersal at different times (Lieberman 2004) (e.g., cyclical sea level rise and fall). If the cladograms are very different, they may imply that vicariance and geodispersal have been caused by noncyclical processes (e.g., continental collisions).

BPA has received some criticism (Cracraft 1988; Ebach & Edgecombe 2001; Ebach & Humphries 2002; Ebach et al. 2003; Enghoff 2000; Miranda Esquivel et al. 2003; Nelson & Ladiges 1991c; Page 1993a; Parenti 2007; Platnick 1988; Ronquist & Nylin 1990; Siddall 2005; Siddall & Perkins 2003). Cracraft (1988) wrote that BPA relies on a questionable analogy to methods in systematics, so it has the potential to obscure the history of a biota rather than reveal it. For some authors, it tends to overestimate dispersal and extinction events (Dowling 2002). Van Welzen (1992) and Enghoff (2000) stated that BPA sometimes groups areas based on absent taxa. Ebach & Edgecombe (2001) noted that when taxa are mapped on the general area cladogram, anomalous reconstructions may appear as descendants dispersing along with their ancestors, thus necessitating a posteriori interpretations. On one hand, Ebach et al. (2003) found that BPA sometimes gives spurious results. On the other hand, it has been argued that the parsimony principle should be used for analyzing the data, not for interpreting the results (Page 1989a). Miranda Esquivel et al. (2003) concluded that the events and duplication of areas in secondary BPA are ad hoc, so this method introduces a scheme of verification, not of falsification. Siddall & Perkins (2003) compared the performance of BPA and tree reconciliation obtained with TreeMap, finding that sometimes BPA gives less parsimonious results. Siddall (2005) concluded that BPA lacks an optimality criterion and the coherence of a research program because published descriptions of the method are self-contradictory. Furthermore, he suggested that rules for a posteriori duplication of entities in secondary BPA are not specified clearly and that both primary and secondary BPA arrive at solutions that may defy logical or temporally consistent interpretation. Parenti (2007) criticized the codification strategy used to deal with geodispersal in modified BPA because it specifies a direction that makes it an extension of Hennig's progression rule.

Brooks et al. (2001, 2003) responded to the criticisms suggesting that all the authors have applied the method incorrectly because they did not take into account the modifications that were performed after the original formulation. These authors argued that primary BPA finds the most parsimonious general area cladogram, indicating through homoplasy how the null hypothesis of vicariance may be falsified. Secondary BPA integrates the incongruent elements, choosing the general area cladogram that postulates the smallest number of duplicated areas, each one of which represents a falsification of the null hypothesis. Brooks et al. (2003) responded to Siddall & Perkins (2003) by arguing that TreeMap cannot treat widespread taxa properly and that BPA is a better method.

The algorithm of primary BPA consists of the following steps (Dowling 2002; Wiley 1987, 1988a, b): (1) obtain the taxonomic cladograms of the taxa distributed in the areas analyzed; (2) replace the terminal species in the taxonomic cladograms with the areas inhabited by them to obtain taxon-area cladograms; (3) label the components and the widespread terminal species (assumption 0) in the taxon-area cladograms; (4) construct a data matrix in which areas are the rows and components and widespread terminal species are the columns, coding "1" if the area is present and "0" if it is absent, use "?" for missing areas, and add a row with all "0" to root the cladogram; (5) analyze the data matrix with a parsimony algorithm in order to obtain the general area cladogram; and (6) optimize the components in the general area cladogram to identify vicariance events (synapomorphies), dispersal events (parallelisms), and extinctions (reversals). Software available include Hennig86 (Farris 1988), PHYLIP (Felsenstein 1993), NONA (Goloboff 1998), PAUP (Swofford 2003), Pee-Wee (Goloboff, http://www.zmuc.dk/public/phylogeny/Nona-PeeWee/), and TNT (Goloboff et al., http://www.zmuc.dk/public/phylogeny/TNT/). For reading and editing data files and cladograms: Winclada (Nixon 1999), compatible with NONA, Pee-Wee, and Hennig86; and MacClade (Maddison & Maddison, http://macclade.org/macclade.html), compatible with PAUP.

Three area statement analysis. Nelson & Ladiges (1991a, c) developed this method (Figure 12), which is based on the three item statement analysis (Marques 2005; Nelson & Ladiges 1993; Nelson & Platnick 1991; Williams & Humphries 2003). In contrast to component coding, applied in BPA, in three-item coding each node is considered a relationship between branches, where some branches are related more closely than other branches. Each separate relationship is expressed minimally as a three-item statement (Williams and Humphries 2003). All area statements are input in a data matrix and analyzed with a parsimony or compatibility algorithm. The three-item statement analysis has received some criticism (Farris 2000; Farris & Kluge 1998; Harvey 1992; Kluge 1993), and a few authors have defended it (de Pinna 1996; Marques 2005; Scotland 2000; Siebert & Williams 1998). The main reasons for controversy have to do with the relationship of this approach to cladistics, observations and homology, data transformation, parsimony, and synapomorphies (Marques 2005).

The algorithm consists of the following steps (Nelson & Ladiges 1991a, c, 1993): (1) obtain the taxonomic cladograms of the taxa distributed in the areas analyzed; (2) replace the terminal taxa from the taxonomic cladograms with the areas inhabited by them to obtain taxon-area cladograms; (3) construct a data matrix in which areas are the rows, and components and widespread terminal species are the columns, coding the relationships as three area statements, adding a row with all "0" to root the cladogram; and (4) analyze the data matrix with a parsimony or compatibility algorithm in order to obtain the general area cladogram. Software available include TAS (Nelson & Ladiges 1991b), which implements assumptions 0 and 1, and the matrix obtained may be analyzed with any parsimony or compatibility algorithm.

Tree reconciliation analysis. Tree reconciliation analysis (Figure 13) was developed independently in molecular systematics, parasitology, and biogeography as a way to describe historical associations between genes and organisms (Goodman et al. 1979), parasites and hosts (Mitter & Brooks 1983), and taxa and areas (Page 1990a, 1993b; Page & Charleston 1998), respectively. It was formalized by Page (1994a, b) as a general method that maximizes the amount of codivergence or shared history between area cladograms from different taxa, minimizing losses (due to extinctions or lack of collection) and duplications (independent vicariance events) when different area cladograms are combined to obtain a general area cladogram (Morrone & Crisci 1995; van Veller et al. 2000).

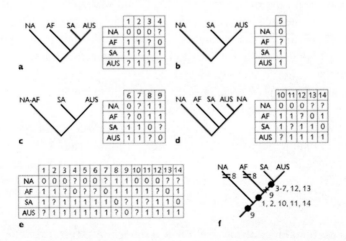

Figure 12. Three-area Statement Analysis. a-d, Taxon-area cladograms and matrices derived from them: a, trivial case; b, taxon with a missing area; c, widespread taxon; d, taxon with a redundant distribution; e, data matrix with all the information; f, general area cladogram obtained. AF, Africa; AUS, Australia; NA, North America; SA, South America.

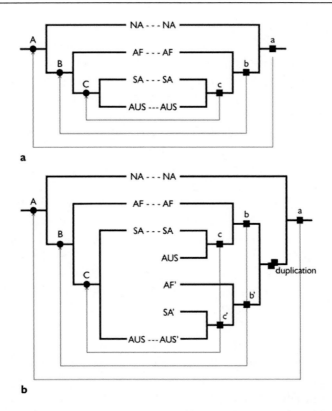

Figure 13. Reconciled tree analysis. a, Example with total codivergence; b, example where there is no total codivergence and we need to duplicate a component (b) to reconcile both cladograms. AF, Africa; AUS, Australia; NA, North America; SA, South America.

When there is correspondence between the taxa and the components of the area cladograms that are compared, they are reconciled easily. In most cases, there is no complete correspondence between the cladogram topologies, so in order to reconcile them, some components must be duplicated. In the example, component b was duplicated, giving rise to an identical component b' that contains terminal taxa Africa', South America', and Australia'. In biogeography, codivergence between areas and organisms is equivalent to vicariance, duplications are equivalent to speciation independent of the vicariance events, horizontal transference is equivalent to dispersal, and losses are equivalent to extinction events.

This method has been criticized for not considering host switching or dispersal (Page & Charleston 1998). Charleston (1998) developed a solution using mathematical structures called "jungles" that contain all possible partial orderings in which the associate cladogram may be tracked in the host cladogram, considering codivergence, duplication, sorting, and host-switching events and all the existing known associations. When the costs are calculated for each of these events, it is possible to find the subgraphs that correspond to the least costly reconstructions of the association.

Huelsenbeck et al. (2000) stated that in inferring host-switching events, it is assumed that the host and parasite cladograms are estimated without error. They suggested that a Bayesian estimation can be used in models where host-switching events are assumed to occur at a constant rate over the entire evolutionary history of the association. This method provides information on the probability that an event of host switching is associated with a particular pair of branches and reduces the possibility that a particular phylogenetic hypothesis may be overturned if a reexamination of the group results in a different cladogram.

Crisci et al. (2003) discussed a problem of the tree reconciliation analysis and dispersal-vicariance analysis: that they violate metricity when considering that duplications are less probable than vicariance events. Violation of metricity assumes in the calculations the existence of a nonmetric space whose geometric properties are difficult to explore. Dowling (2002) enumerated some criticisms to the program TreeMap; the most important is that it may overestimate duplications and underestimate dispersal.

The algorithm consists of the following steps (Crisci et al. 2003): (1) obtain the taxonomic cladograms of the taxa distributed in the areas analyzed; (2) replace the terminal taxa from the taxonomic cladograms with the areas inhabited by them to obtain taxon-area cladograms; (3) superimpose each component of a cladogram on the components of the other cladograms; (4) assume maximum cospeciation and no dispersal, attributing the differences between the cladograms to duplication or extinction events; and (5) choose the reconstruction that implies the maximum cospeciation and minimum number of losses and duplications. Software available include Component version 2.0 (Page 1993a), TreeMap (Page 1994c), and TreeFitter version 1.3 (Ronquist 2002).

Paralogy-free subtree analysis. Paralogous areas are those that conflict with duplications of themselves. Nelson & Ladiges (1996, 2001) suggested that because of geographic paralogy the components may provide biogeographic information but not be directly informative. This means that we may have contradictory relationships due to sympatric speciation, lack of response to vicariance events, incorrect definition of areas, and other explanations, which can lead to erroneous interpretations (Nelson & Ladiges 2001). Paralogy-free subtrees simplify the cladistic biogeographic analysis, so that geographic data need not be associated with paralogous nodes, preventing artifactual results, if not completely then at least to a significant degree (Nelson & Ladiges 2003; Parenti 2007). Nelson & Ladiges (1996) developed an algorithm that constructs paralogy-free subtrees, starting off at the most terminal groups of the cladogram. The procedure reduces complex cladograms to paralogy-free subtrees, meaning that geographic data are associated only with informative nodes, and areas duplicated or redundant in the descendants of each node do not exist. These are the only data relevant for cladistic biogeography. When obtained, paralogy-free subtrees are represented in a component or a three-item matrix and analyzed with a parsimony algorithm. Before paralogy-free subtrees are obtained, the transparent method (Ebach et al. 2005a) may be implemented to resolve widespread taxa.

The algorithm consists of the following steps (Nelson & Ladiges 1996): (1) obtain the taxonomic cladograms of the taxa distributed in the areas analyzed; (2) replace the terminal taxa from the taxonomic cladograms with the areas inhabited by them to obtain taxon-area cladograms; (3) resolve widespread taxa with the transparent method; (4) identify the paralogy-free subtrees starting at each terminal node and progressing to the base of each taxon-area cladogram, and when a node leads to one or more terminal taxa that are geographically widespread and part of that distribution overlaps with that of another taxon or taxa, reduce the widespread distribution to the nonoverlapping geographic element; (5) represent the nodes of all the paralogy-free subtrees in a component or three-item matrix; and (6) analyze the data matrix with a parsimony algorithm to obtain the general area cladograms.

Software available include TASS (Nelson and Ladiges 1995) and Nelson05 (Cao & Ducasse 2005).

Dispersal-vicariance analysis. Proposed by Ronquist (1997a), based on the ideas developed by Ronquist & Nylin (1990) to analyze host-parasite relationships, this method reconstructs ancestral distributions from one given phylogenetic hypothesis, without assuming a particular process a priori, taking into account vicariance, dispersal, and extinction (Miranda Esquivel et al. 2003). It reconstructs the biogeographic history of individual taxa, but it can also be used to find the general relationships of an area, especially when these relationships do not conform to a hierarchical pattern, that is, when there are reticulate relationships due to biogeographic convergence. The biogeographic reconstruction is based on a cost matrix, which is constructed according to the following premises (Crisci et al. 2003): (1) vicariance events have a null cost of 0 (speciation is assumed to be by vicariance); (2) duplication events receive a null cost of 0 (this implies sympatric speciation); (3) dispersal events receive a cost of 1 per area unit added to a distribution; and (4) extinction events receive a cost of 1 per area unit deleted from a distribution.

Ronquist (1997b) developed a modification of this method, called constrained dispersal-vicariance analysis, which distinguishes between random dispersals (those that imply that the taxon passes through a barrier) and predictable dispersals (those that occur when a barrier disappears). In constrained DIVA, the cost matrix is constructed according to the following rules: (1) vicariance events receive a benefit value of -1; (2) duplication events receive a null cost of 0; (3) extinction events receive a cost of 1; (4) random dispersal events receive a cost of 1; and (5) predictable dispersal events receive a benefit value of -1.

The algorithm consists of the following steps (Biswas & Pawar 2006; Ronquist 1997a): (1) obtain the taxonomic cladograms of the taxa distributed in the areas analyzed; (2) replace the terminal taxa from the taxonomic cladograms with the areas inhabited by them, to obtain taxon-area cladograms; (3) construct a cost matrix assigning values to vicariance, duplication, dispersal, and extinction events; and (4) obtain the general area cladogram that optimizes these values. Software available include DIVA version 1.2 (Ronquist 1996).

Area cladistics. Area cladistics (Ebach 2003; Ebach & Edgecombe 2001; Ebach & Humphries 2002) is derived from component and three area statement analyses, although it uses a different approach for resolving problems of the taxon-area cladograms. Area cladistics begins by replacing the names of the terminal taxa of two or more taxon-area cladograms and deriving areagrams and then resolves paralogy using the transparent method (Ebach et al. 2005a). Areagrams of different taxa inhabiting the same areas are combined, and patterns are searched. When geographic congruence between the different areagrams is found, we may interpret the biotic history. Biotic congruence is evidence of vicariance, and we can infer that biotic components inhabiting sister areas were once part of the same biotic component, which posteriorly diverged.

Williams & Ebach (2004) categorized cladistic biogeographic methods as either transformational or taxic. The latter, which includes area cladistics, is no more than proximal relationships of two areas in relation to a third (area homology). A combination of area homologies forms a general areagram or general area cladogram, which represents geographic homology consisting of area clades that share a common history. Williams & Ebach (2004) argued that the alternatives (e.g., identification of dispersal routes, centers of origin, or even vicariance events) are themselves artifacts of the transformational perspective.

The algorithm consists of the following steps (Ebach 2003; Ebach & Humphries 2002): (1) obtain the taxonomic cladograms of the taxa distributed in the areas analyzed; (2) replace the terminal taxa from the taxonomic cladograms with the areas inhabited by them to obtain taxon-area cladograms; (3) resolve widespread taxa with the transparent method; (4) represent the nodes of the taxon-area cladograms in either a component or a three-item matrix; and (5) analyze the data matrix with a parsimony or compatibility algorithm to obtain the general area cladogram. Software available include 3item (Ebach et al. 2005b) and Nelson05 (Cao & Ducasse 2005).

Evaluation and classification of the methods. Morrone & Carpenter (1994) compared empirically four cladistic biogeographic methods: component analysis, BPA, three area statement analysis, and tree reconciliation analysis. They analyzed ten different data sets from the literature, calculating the items of error necessary to reconcile the original taxon-area cladograms with the different general area cladograms obtained. They concluded that none of the methods was consistently better than the others, producing results that might be considered the least ambiguous; the main sources of ambiguity were dispersal and the existence of duplicated lineages combined with extinction. Apparently, the different methods are affected differentially by them; for example, BPA is more affected by dispersal, whereas component analysis is more affected by redundant distributions (Morrone & Crisci 1995). Another comparative analysis (Biondi 1998) basically agreed with these conclusions. After these contributions, several new methods have been described, so we lack a complete comparative study.

Van Veller et al. (2000) and van Veller & Brooks (2001) compared component compatibility, BPA, component analysis, tree reconciliation analysis, and three area statement analysis to evaluate the implementation of assumptions 0, 1, and 2 in agreement with both requirements previously indicated. In relation to the requirement that the solved sets of cladograms of areas are successively inclusive, van Veller et al. (2000) detected violations by component analysis, tree reconciliation analysis, and three area statement analysis when there are redundant distributions or a combination of widespread taxa and redundant distributions. Van Veller & Brooks (2001) concluded that when vicariance and extinction are the most probable explanations, BPA, component analysis, and tree reconciliation analysis would give the same general area cladogram, but when dispersal is the most reasonable explanation, secondary BPA represents each dispersal event as a falsification of the null hypothesis, whereas component analysis and tree reconciliation analysis remove a priori data, or duplicate lineages, and postulate extinctions a posteriori to avoid falsification.

Some authors (Crisci 2001; Crisci et al. 2003; Ronquist & Nylin 1990; Sanmartín & Ronquist 2002) have considered the existence of two types of cladistic biogeographic methods:

1. Pattern-based methods: those that search for general patterns of relationships between areas, without initial assumptions about particular biogeographic processes. These methods would belong to cladistic biogeography in the strict sense (e.g., Crisci 2001; Crisci et al. 2003). They include component analysis, BPA, component compatibility, and paralogy-free subtree analysis.

2. Event-based methods: those that derive explicit models of particular biogeographic processes. These methods are excluded from cladistic biogeography by Crisci et al. (2003) & Crisci (2001). They include tree reconciliation analysis, vicariance events analysis, and dispersal-vicariance analysis.

Van Veller et al. (2000; see also Biswas & Pawar 2006; Brooks & van Veller 2003; Ebach 2001; Morrone 2005; van Veller & Brooks 2001; van Veller et al. 2003; Wojcicki & Brooks 2005; Zandee & Roos 1987) consider the existence of two groups of methods, whose purpose is to implement different research programs:

1. A posteriori methods: those that deal with dispersal, extinction, and duplicated lineages after the parsimony analysis of a data matrix based on the unmodified taxon-area cladograms. They are intended to implement "vicariance biogeography" sensu Zandee & Roos (1987) or "phylogenetic biogeography" sensu van Veller et al. (2003). [The name *phylogenetic biogeography* is inappropriate because for decades it has been the name of Hennig's (1950) and Brundin's (1966) approach.] A posteriori methods include BPA and component compatibility.
2. A priori methods: those that allow modification of the area relationships in the taxon-area cladograms to deal with dispersal, extinctions, or duplicated lineages in order to obtain resolved area cladograms and provide the maximum fit to a general area cladogram. They are intended to implement "cladistic biogeography" sensu Zandee & Roos (1987) and van Veller et al. (2003). A priori methods include component analysis, tree reconciliation analysis, and three area statement analysis.

Ebach & Humphries (2002) stated that a posteriori methods correspond to a generation paradigm, whereas a priori methods correspond to a discovery paradigm. Generation paradigm methods are based on previous beliefs rather than facts. Additionally, these authors criticized the generation methods for immunizing their results to avoid falsification, whereas discovery methods are superior because they allow the free exploration of the data. Ebach et al. (2003) also noticed both uses of the term *phylogenetic biogeography,* considering that Van Veller et al.'s (2003) use corresponded to Hennig's (1950) *parasitological method,* which is inappropriate for implementing cladistic biogeography. Brooks (2004) suggested that, in addition to using different methods, cladistic and phylogenetic biogeographies are research programs justified by different ontologies. Cladistic biogeographic methods modify the original data in order to provide maximum fit to the null hypothesis of vicariance, amounting to an a priori parsimony criterion (ontology of simplicity). Phylogenetic biogeography falsifies the null hypothesis of a vicariance explanation when the data do not support it, amounting to an a posteriori parsimony criterion. Data that appear to conflict with the null hypothesis indicate that it is flawed, determining a posteriori the minimum number of falsifications that are necessary to explain all the data (ontology of complexity).

Humphries & Ebach (2004) postulated that the dichotomy between cladistic and phylogenetic biogeographies has to do with the interpretation of cladograms as homology-based hierarchies or as phylogenetic trees, respectively. This interpretation can be extended to other approaches that also treat cladograms as phylogenetic trees: ancestral area analysis and intraspecific phylogeography. In this respect, it would be pertinent to refer to O'Hara's (1988) distinction between chronicle and history. A chronicle is a description of a series of events, arranged in chronological order, and not accompanied by any causal statements or explanations, whereas a history contains statements about causal connections. Cladistic biogeography estimates a chronicle, and phylogenetic biogeography estimates a history (Morrone 2009). Another way to distinguish both groups of methods is to apply the distinction between the philosophical positions of reciprocal illumination and total evidence (Rieppel 2004). A posteriori methods work under the principle of reciprocal illumination (Hennig 1950), searching for consilience or congruence between data from different taxa. A priori methods work under the principle of total evidence (Kluge 1989), relying on the largest set of data analyzed simultaneously.

I am unconvinced of the existence of two different research programs (Ebach & Morrone 2005; Morrone 2005, 2009). Although the distinction between the two groups of methods is valid, I suggest that they still can be considered to implement cladistic biogeography. The question of which method to apply is not easy to answer; all have their supporters and their critics. A practical approach is to apply more than one and then compare the differences between the results obtained.

REGIONALIZATION

One of the most striking facts about the geographic distributions of taxa is that they have limits. Because these limits are repeated for different taxa, they allow the recognition of biotic components. Biotic components are nested within other larger components, so they can be ordered hierarchically in a system of realms, regions, dominions, provinces, and districts. Given the historical and logical primacy of classification over process explanations (Rieppel 1991, 2004), this stage of the analysis takes place before cenocrons are elucidated and a geobiotic scenario is proposed (Morrone 2009).

Viloria (2005) found a lack of uniformity and consensus in the criteria that have been used for regionalization. Analyzing Venezuela as a case study, he detected a mixture of classificatory systems based on distinct criteria, and a lack of rigor in the equivalences or synonymies between these regionalizations. He concluded that two different classificatory systems should be recognized: one based on areas of endemism and other on biotic elements. In order to have a single, standardized system of regionalization, another important thing is a unified nomenclature, paralleling the different codes used in systematics (Viloria 2005). An international group of specialists on different fossil taxa discussed the principles of classification and nomenclature of marine paleobiogeographic classification in order to reach a consensus on the nomenclature of biogeographic areas (Westermann 2000). A similar effort is taking place for neobiogeographic classification (Ebach et al., 2008).

Realms, Regions, and Transition Zones

Biogeographic regionalization implies the recognition of successively nested areas. These areas should be natural; that is, they should correspond to biotic components. Classically, the following five categories have been used: *realm, region, dominion, province,* and *district*

(Morrone 2009). Whenever more categories are necessary, the prefix *sub-* may be added to them (e.g., *subregion, subprovince*).

Sometimes it is more difficult to determine the exact boundaries of two realms or regions. As a result, authors have described transition zones (Darlington 1957; Halffter 1987; Ruggiero & Ezcurra 2003), which represent events of biotic hybridization, promoted by historical and ecological changes that allowed the mixture of different biotic components. Transition zones may have depauperate biotas, but in some cases they harbor a particularly high biodiversity. From the evolutionary viewpoint, transition zones deserve special attention because they represent areas of intense biotic interaction (Morrone 2009). The analysis of transition zones by ecological biogeographers is mostly quantitative, whereas for evolutionary biogeographers it is qualitative. In panbiogeographic analyses, transition zones are indicated by the presence of nodes (Escalante et al. 2004), whereas in cladistic biogeographic analyses, putative transition zones give conflicting results because they appear to be sister areas to different areas. Ruggiero & Ezcurra (2003) analyzed the compatibility between historical and ecological analyses in South America, finding consistent similarities, which point to the possibility of reciprocal illumination (in the sense of Hennig 1950) between evolutionary and ecological biogeography (Morrone 2009).

Regionalization of the World

Modern biogeographic classification began with de Candolle (1820, 1838) for plants (phytogeography) and Wallace (1876) for animals (zoogeography). De Candolle (1820) recognized twenty botanical regions: northern Asia, Europe, and America; southern Europe and north of the Mediterranean; Siberia; Mediterranean area; eastern Europe to the Black and Caspian seas; India; China, Indochina, and Japan; Australia; south Africa; east Africa; tropical west Africa; Canary Islands; northern United States; northwest coast of North America; the Antilles; Mexico; tropical America; Chile; southern Brazil and Argentina; and Tierra del Fuego. De Candolle (1838) added another twenty regions, making a total of forty. They have remained the basis for some more modern treatments (e.g., Good 1974; Takhtajan 1969). Wallace's (1876) system of six zoogeographic regions, built on a previous work by Sclater (1858), is probably the most generally known biogeographic global terrestrial regionalization (Lankester 1905). It consists of six regions: Nearctic, Neotropical, Palearctic, Ethiopian, Oriental, and Australian. These large zoogeographic regions were divided into subregions, which basically correspond to de Candolle's (1820) regions.

Alongside the development of these systems were efforts to construct ecogeographic systems (Allen 1871; Udvardy 1969). They are based on the assumption that adaptations to natural surroundings generally confine species in definite areas. Thus, ecogeographic areas or biomes can be delineated by correlating plant and animal distributions with climatic, edaphic, and other environmental factors (Cox & Moore 1998).

Craw & Page (1988) considered Wallace's system defective because it was based on a belief in the permanence of continents and oceans since the time of origin of modern life. They proposed instead a biogeographic system in which areas now widely separated by oceans and sea basins are related to one another by generalized tracks. De Candollean and Wallacean regions are not parts of this system but boundaries where different biotic components interrelate in space and time. In fact, Craw & Page (1988) concluded that the natural biogeographic regions for terrestrial and freshwater taxa are not present-day landmasses but the world's major ocean basins. Parenti (1991) reviewed the relevance of ocean basin evolution to the distribution of freshwater fishes and terrestrial organisms, modifying slightly Craw & Page's (1988) regions to include the Tethys Sea, which she considered important in understanding the composite Indo-Australian area.

Cox (2001) examined the floral kingdoms or realms from de Candolle, Engler, and Takhtajan and the zoogeographic regions from Sclater and Wallace and analyzed their differences. He reviewed levels of endemism of Takhtajan's system, concluding that the Cape realm should be treated as a region of the African realm and that the Antarctic realm should be divided and transferred to the Neotropical and Australian realms. He also considered the names *Neotropical, Nearctic,* and *Palearctic,* used for both floral realms and faunal regions, to be cumbersome and unnecessary, replacing them with the names *South American, North American,* and *Eurasian,* respectively. Morrone (2002) stated that a single biogeographic scheme for all organisms, to serve as a general reference system, would be a desirable goal. Based on several panbiogeographic and cladistic biogeographic papers that have shown that some of the units recognized in traditional phytogeographic and zoogeographic systems do not represent natural units, he presented a general system of biogeographic realms and regions, intending to incorporate their conclusions. This biogeographic regionalization (Figure 14) is as follows:

1. Holarctic realm: Europe, Asia north of the Himalayan mountains, northern Africa, North America (excluding southern Florida), and Greenland. From a paleogeographic viewpoint, it corresponds to the paleocontinent of Laurasia. It comprises two regions: Nearctic (New World, namely Canada, most of the United States, and northern Mexico) and Palearctic (Old World, namely Eurasia and Africa north of the Sahara).

2. Holotropical realm: basically the tropical areas of the world, between 30° south latitude and 30° north latitude, corresponding to the eastern portion of the Gondwana paleocontinent (Crisci et al. 1993). It comprises four regions: Neotropical (tropical South America, Central America, south central Mexico, the West Indies, and southern Florida), Afrotropical or Ethiopian (central Africa, the Arabian peninsula, Madagascar, and the West Indian Ocean islands), Oriental (India, Himalaya, Burma, Malaysia, Indonesia, the Philippines, and the Pacific islands); and Australian Tropical (northwestern Australia).

3. Austral realm: southern temperate areas in South America, South Africa, Australasia, and Antarctica that correspond to the western portion of the paleocontinent of Gondwana (Crisci et al. 1993; Moreira-Muñoz 2007). It comprises six regions: Andean (southern South America below 30° south latitude, extending through the Andean highlands north of this latitude, to the Puna and North Andean Paramo), Antarctic (Antarctica), Cape or Afrotemperate (South Africa), Neoguinean (New Guinea plus New Caledonia), Australian Temperate (southeastern Australia), and Neozealandic (New Zealand).

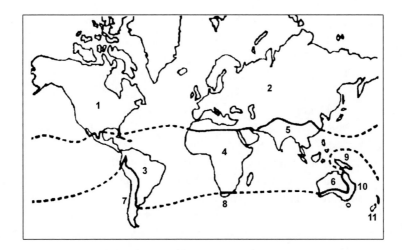

Figure 14. Biogeographic regionalizations of the World (Morrone 2002). 1-2, Holarctic realm; 1, Nearctic region; 2, Palearctic region; 3-6, Holotropical realm, 3, Neotropical region; 4, Afrotropical region; 5, Oriental region; 6, Australian Tropical region; 7-11, Austral realm; 7, Andean region; 8, Cape region; 9, Neoguinean region; 10, Australian Temperate region; 11, Neozelandic region.

Marine regionalization began with Ortmann (1896). In the twentieth century, important contributions included Ekman (1935), Briggs (1995), and Pierrot-Bults et al. (1986). Briggs (1995) recognized twenty-three marine biogeographic regions: Indo-West Pacific, Eastern Pacific, Western Atlantic, Eastern Atlantic, Southern Australian, Northern New Zealand, Western South America, Eastern South America, Southern Africa, Mediterranean-Atlantic, Carolina, California, Japan, Tasmanian, Southern New Zealand, Antipodean, Subantarctic, Magellan, Eastern Pacific Boreal, Western Atlantic Boreal, Eastern Atlantic Boreal, Antarctic, and Arctic.

IDENTIFICATION OF CENOCRONS

Dispersal explanations have resided traditionally on narrative frameworks, lacking a general theory to explain distributional patterns, so they have been rejected by panbiogeographers and cladistic biogeographers as *ad hoc* explanations. After establishing biogeographic homology patterns, however, dispersal explanations can help establish when the cenocrons assembled in the identified components, incorporating a time perspective to the study of biotic evolution (Morrone 2009).

Time-slicing

Hunn & Upchurch (2001) emphasized the relevance of time in evolutionary biogeography, because data on the temporal distribution may provide important constraints in biogeographic analyses, helping reinforce or overturn specific hypotheses. The incorporation of temporal information requires assigning time values to cladogenetic events (Crisci et al 2003; Morrone 2009).

Upchurch & Hunn (2002) considered that the extent to which temporal data are incorporated into biogeography depends on the researcher's choice regarding sources of spatial data and analytical methodologies. It has been suggested that the temporal ranges of organisms are important because distribution patterns seem to 'decay' through time as new ones are superimposed (Grande 1985; Hunn & Upchurch 2001; Upchurch & Hunn 2002; Upchurch et al 2002). Donoghue & Moore (2003) postulated that cladistic biogeographic methods are susceptible to the confounding effects of pseudo-incongruence and pseudo-congruence, if they do not incorporate information on the absolute timing of the diversification of the lineages. Pseudo-incongruence means that different area cladograms may show conflict when the taxa evolved at the same time, but diversified in response to different events. Pseudo-congruence means that different area cladograms may show the same area relationships, although the taxa diversified at different times, presumably under different underlying causes. Riddle & Hafner (2006), however, have argued that time alone might not necessarily cause us to resort to an explanation of pseudo-congruence.

Cladistic biogeographers avoid using temporal data because of the risk of incorporating ideas of unobserved processes in the elucidation of biogeographic patterns. This would imply unverifiable assumptions, with the risk of falling back on narrative scenarios. The need for considering time in biogeography, however, becomes clearer in cases of biogeographic convergence (Morrone 2009). The terms convergence and divergence have been proposed by Hallam (1974) to distinguish two extreme biogeographic patterns. Widespread taxa and redundancy identify biogeographic convergence, whereas vicariance is the most common interpretation of divergence patterns. Convergence can be the result of area coalescence (due to the elimination of geographic barriers). Analyses of biogeographic convergence are unlikely to show congruence.

Events of biogeographic convergence produce a sort of overprinting of past biogeographic histories by more recent patterns (e.g. reticulated area histories), which lowers the chances of establishing area relationships through congruence. The solution to problems posed by instances of biogeographic convergence is time-slicing (Grande 1985; Upchurch & Hunn 2002). While assessments of faunal similarity are usually undertaken with faunas of successive geological ages, traditional cladistic biogeography has only used data on organism relationships and spatial distributions on a single time plane (usually the present). Time-slicing may reconcile the use of time and a synchronic approach. Ideally, paleobiogeographers should be able to use a synchronic approach for each time slice they identify. This is difficult because of the limits imposed by geological constraints (e.g. insufficient precision or resolution of chronological correlations, incompleteness of the fossil record, etc.).

In order to apply time-slicing, three methods are available. Two of them, parsimony analysis of endemicity (PAE) and area cladistics, have been already dealt. The third one, temporally partitioned component analysis (TPCA), is dealt herein.

Temporally partitioned component analysis (TPCA). Upchurch & Hunn (2002) proposed this method, also known as chronobiogeography, to incorporate explicitly temporal data into a cladistic biogeographic analysis (see also Hunn & Upchurch 2001 and Upchurch et al 2002). After geodispersal, due to area coalescence events (biogeographic convergence), "the histories of areas and biotas will have a reticulated rather than a branching structure, raising the question as to how well cladistic biogeographic techniques will be able to accurately analyze and depict a reticulate system" (Upchurch & Hunn 2002: 280). The starting point of TPCA is the existence of taxon-area cladograms for the taxonomic groups on which the analysis is based. Although the introduction of time-slicing may tend to uncover reticulate histories, it is interesting to note that, ideally, synchronic relations would be found for each time slice on the basis of phylogenetic relations.

The algorithm comprises the following steps (Upchurch & Hunn 2002): (1) prune or temporally partition the taxon cladograms by deleting all taxa that did not exist at a particular designated time-slice; (2) find optimal area cladograms for each particular time-slice, by determining which area relationships provide the best (under some designated optimality criterion) explanation for the spatial distributions observed in the taxon cladogram; and (3) use a randomization test to determine whether the degree of fit between area and taxon cladogram for each time-slice is greater than would be expected by chance.

Intraspecific Phylogeography

Intraspecific phylogeography studies the principles and processes governing the geographic distribution of genealogical lineages, especially those within and among closely related species, based on molecular data (Avise et al 1987; Avise 2000; Lanteri & Confalonieri 2003; Lomolino et al 2006). The maternal, non-recombinant mode of inheritance of mitochondrial DNA (mtDNA) and the rapid evolution of mtDNA sequences make it possible to obtain haplotypes (combinations of alleles at multiple linked loci), which can be used to obtain intraspecific phylogenetic hypotheses (Crisci et al 2003). Once the population genetic structure has been assessed based on mtDNA, it is possible to obtain a network or cladogram of haplotypes, which allows to analyze historical patterns and the processes that have shaped them, e.g. dispersal, vicariance, range expansion and colonization (Templeton 2004). This knowledge can offer insight into when relatively recent cenocrons incorporated to a biotic component.

The results of a phylogeographic analysis are usually represented as a parsimony network connecting the studied haplotypes (Figure 15), where the number of restriction sites differences indicates the relative distance between haplotypes and groups of haplotypes. When mapped, disjunctions and sympatry between the haplotypes may give us clues on their evolutionary histories. Highly divergent groups inhabiting disjunct areas can indicate independent evolutionary histories for a relatively long period of time, usually due to vicariance, whereas lack of geographic structure may indicate recent dispersal.

Intraspecific phylogeography is conceptually positioned between macroevolution and microevolution (Riddle & Hafner 2004; Lomolino et al 2006). With its focus on historical processes, it contextualizes and balances the ecogeographic perspective that tend to emphasize natural selection's role in microevolution (Avise 2000). In addition, it provides the promising basis for a bridge between ecological and evolutionary biogeography (Morrone 2009).

In animal species, mtDNA has proven to be useful for phylogeographic studies, but in plants it evolves quickly with respect to gene order but slowly in nucleotide sequence, so it is of limited utility (Crisci et al 2003). Chloroplast DNA (cpDNA) has shown to be structured geographically in several plant species, it is also transmitted maternally and has exhibited considerable intraspecific variation (Soltis et al 1992; Avise 2000, 2004).

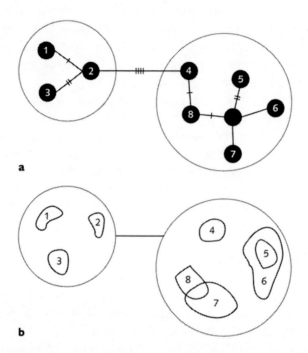

Figure 15. Representation of a phylogeographic hypothesis. a, Parsimony network connecting the
haplotypes, where the "dashes" indicate restriction site differences between haplotypes and groups of
haplotypes; b, geographic distribution of the haplotypes.

An explicit genealogical approach to intraspecific phylogeography implies extending the
lineage sorting theory to sister populations (Avise 2000). In terms of maternal genealogy,
there are three possible situations: (1) polyphyly, where some but not all extant matrilines in
one population join with some but not all extant matrilines in the other to form a clade; (2)
paraphyly, where all matrilines within one population from a clade nested within the broader
matrilineal history of the other population; and (3) reciprocal monophyly, when all extant
matrilines within each sister population are closer genealogically to each other than to any
matriline in the other. These three situations frequently characterize the same pair of sister
populations at different time depths.

Three phylogeographic hypotheses and four corollaries have been formulated by Avise
et al (1987):

1. Most species are composed of geographic populations whose members occupy
    recognizable matrilineal branches of an extended intraspecific pedigree. Populations
    of most species display significant phylogeographic structure supported by mtDNA
    data.
2. Species with limited or "shallow" phylogeographic population structure have life
    histories conducive to dispersal and have occupied ranges free of firm, long-standing
    impediments to gene flow. Non-subdivided, high-dispersal species may have limited
    phylogeographic structure.
3. Intraspecific monophyletic groups distinguished by large genealogical gaps usually
    arise from long term extrinsic biogeographic barriers to gene flow. Major

phylogeographic units within a species reflect long term historical barriers to gene flow. This hypothesis has four corollaries representing four aspects of genealogical agreement (Figure 16):

a.   Agreement across sequence characters within a gene: every deep phylogenetic split in the intraspecific gene tree is supported concordantly by multiple diagnostic characters, e.g. nucleotides or restriction sites, within the mitochondrial genome. If this is not the case, such matrilineal splits would not be evident in the analysis, nor would receive significant support.

b.   Agreement in significant genealogical partitions across multiple genes within a species: empirical examples show general agreement between deep phylogeographic topologies in multiple gene trees (such as mitochondrial and nuclear) within the species analyzed. These deep branch separations characterize the same sets of geographic populations.

c.   Agreement in the geography of gene-tree partitions across multiple codistributed species: several sympatric species with comparable natural histories or habitat requirements proved to be phylogeographically structured in similar fashion. In particular, divergent branches in the intraspecific gene trees might map consistently to the same geographic regions.

d.   Agreement of gene-tree partitions with spatial boundaries between traditionally recognized biogeographic units: an emerging generality from phylogeographic analyses is that deeply separated phylogroups at the intraspecific level are confined to biogeographic provinces or districts as recognize by systematic biogeography.

Five different phylogeographic patterns can be characterized for mtDNA gene cladograms (Avise 2000):

Category I (Deep gene tree, major lineages allopatric): It is characterized by the presence of spatially circumscribed haplotypes, separated by relatively large mutational distances. This pattern appears commonly in phylogeographic patterns of mtDNA. A long-term extrinsic barrier to genetic exchange is the most commonly invoked explanation.

Category II (Deep gene tree, major lineages broadly sympatric): It is characterized by pronounced phylogenetic gaps between some branches in a gene tree, with main lineages codistributed over a wide area. It could arise in a species where some anciently separated lineages might have been retained by chance, whereas many intermediate lineages were lost over time by gradual lineage sorting.

Category III (Shallow gene tree, lineages allopatric): Most or all haplotypes are related closely, yet are localized geographically. Contemporary gene flow has been low enough in relation to population size to have permitted lineage sorting and random drift to promote genetic divergence among populations that were in historical contact recently.

Category IV (Shallow gene tree, lineages sympatric): It is expected for high-gene-flow species of small effective size whose populations have not been sundered by long-term barriers.

Category V (Shallow gene tree, major distributions varied): It is intermediate between categories III and IV, and involves common lineages that are widespread plus closely related lineages that are confined to one or a few nearby localities. It implies low contemporary gene flow between populations that are connected tightly in history. Common haplotypes are often plesiomorphic and rare haplotypes are the presumed apomorphic conditions.

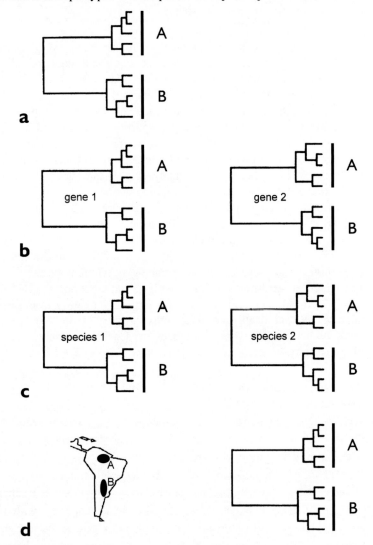

Figure 16. Schematic representation of four different aspects of genealogical agreement. a, Agreement across sequence characters within a gene; b, agreement in significant genealogical partitions across two different genes within a species; c, agreement in the geography of gene-tree partitions across two codistributed species; d, agreement of gene-tree partitions with spatial boundaries between two biotic components.

Phylogeographic patterns exhibiting large genetic gaps between phylogroups (groups of closely related haplotypes separated from other phylogroups by relatively large genetic gaps), such as those represented in categories I and II, are amenable to phylogenetic analyses using methods (Lomolino et al 2006). This can be done simply treating each phylogroup as a terminal unit in the cladistic analysis. When variable haplotypes are closely related and groups of populations are not clustered within the clearly reciprocally monophyletic phylogroups, such as those in categories III, IV and V, one can still present an unrooted cladogram to summarize haplotype relationships.

A phylogeographic analysis comprises the following steps (Templeton et al 1987, 1992; Lanteri & Confalonieri 2003): (1) construct a data matrix, where the haplotypes of a species or closely related species represent the terminal units, and gene sequences from mitochondrial DNA (animals) or chloroplast DNA (plants) represent the characters, and obtain a haplotype phylogram or network; (2) translate the phylogram into a network of nested clades, where each successive hierarchical level is considered to be more ancient than the subordinate clades or phylogroups; (3) represent the phylogroups on a map, and calculate clade distance (average distance of each haplotype in the particular clade from the center of its geographic distribution, measured as the great circle distance, to represent the distribution of the phylogroups), nested clade distance (average distance of the center of distribution for this haplotype from the center of distribution for the haplotype within which it is nested), and average distance for the terminal phylogroups in order to identify whether they are structured geographically or not; (4) use a statistical test to evaluate whether haplotypes of the nested clades are distributed randomly. Software available include TCS 1.18 (Clement et al 2000) and GEODIS (Posada et al 2000).

Molecular Clocks

Cladograms based on molecular data may be used as raw data in cladistic biogeography and intraspecific phylogeography. In addition, the assumption that the rate of molecular evolution is approximately constant over time for proteins in all lineages, allows inferring a clock-like accumulation of molecular changes (Zuckerland & Pauling 1965; Bromham & Woolfit 2004). The "ticks" of the molecular clock, which correspond to mutations, do not occur at regular intervals, but rather at random points in time (Gillespie 1991). This time is measured in arbitrary units and calibrated in millions of years by reference fossils or geological data (Sanderson 1998; Magallón 2004; Benton & Donoghue 2007), giving minimum estimates of the age of a clade, which in turn may help elucidate the relative minimum ages of the cenocron to which it belongs. Knowledge of relative minimum ages of divergence may help decide whether a specific dispersal or vicariance event better explains the patterns observed. If clock calibrations provide estimates smaller than those proposed by vicariance events, dispersal may be a better explanation.

The calibration of a molecular clock requires that we find two extant species for which the date of speciation can be determined from the fossil record, to establish the time since the speciation event. Then, we compare the DNA sequences of the same gene of both species and count the number of nucleotide substitutions. If all the substitutions are assumed to have arisen subsequently to the speciation event, the rate of DNA evolution for the gene under study is obtained by dividing the number of DNA differences between both species by the time since speciation. Assuming a constant mutation rate, we can extrapolate the approximate dates of speciation for other species, for which no fossil dates are available (Crisci et al 2003). In order to test the molecular clock hypotheses there are three available tests: the likelihood ratio test, the dispersion index and the relative rate test (Page & Holmes 1998).

As analyses from several taxa began to accumulate in the 1970s, it became apparent that the molecular clock is not always a good model for the process of molecular evolution

(Rutschmann 2006). If the null hypothesis of a constant rate is rejected or if we have evidence suggesting that rates vary, we may use methods that correct for rate heterogeneity (Rambaut & Bromham 1998) or that estimate divergence times by incorporating rate heterogeneity (Kishino et al 2001; Sanderson 2002). Some problems associated with molecular methods include the stochastic nature of molecular substitution, the assumption of rate constancy among lineages when such constancy is absent, and the link between substitution rate and elapsed time on the branches of a cladogram (Magallón 2004). Calibration made by reference to geological events runs the risk of circular reasoning, because the clock is used to test biogeographic hypotheses which involve an event potentially caused by a geological process (Crisci et al 2003). Magallón (2004) and Benton & Donoghue (2007) clarified the relationship of the fossil record and molecular dating methods (Figure 17), the former documenting first appearances of morphological features and the latter dating splits of molecular lineages.

There are several molecular dating methods, grouped into three main classes: methods that use a molecular clock and one global rate of substitution, methods that correct for rate heterogeneity, and methods that try to incorporate rate heterogeneity. Each method has it own algorithm. For a revision, see Rutschmann (2006). Software available incluye DNAMLK of package PHYLIP (Felsenstein 1993), BASEML of package PAML (Yang 1997), QDATE (Rambaut & Bromham 1998), PHYBAYES (Aris-Brosou & Yang 2001), RHINO (Rambaut 2001), maximum likelihood clock optimization method of PAUP (Swofford 2003), PATH (Britton et al 2002), BEAST (Drummond & Rambaut 2003), MULTIDIVTIME (Thorne & Kishino 2002), TREEEDIT (Rambaut & Charleston 2002), and R8s (Sanderson 2003).

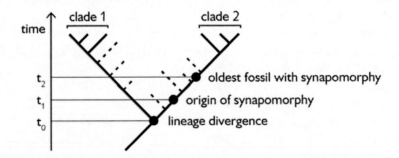

Figure 17. Relationship of lineage divergence, origin of a synapomorphy and occurrence of the oldest fossil with the synapomorphy, according to Magallón (2004). There is a temporal gap between the divergence of a taxon and its sister taxon (t_0), the origin of the synapomorphy (t_1) and the occurrence of the oldest fossil bearing such synapomorphy (t_2).

CONSTRUCTION OF A GEOBIOTIC SCENARIO

Once we have identified the biotic components and cenocrons, we may be able to construct a geobiotic scenario. By accounting biological data (means of dispersal, etc.) as well as non-biological data (past continental configurations, etc.) we can integrate a plausible scenario to help explain the episodes of vicariance/ biotic divergence and dispersal/ biotic convergence that shaped biotic evolution (Morrone 2009).

Geographic Features

Panbiogeographers and cladistic biogeographers have shown interest in geology, geophysics and plate tectonics (Heads 1989; Michaux 1989; Craw et al 1999). Geology and biogeography have a causal relationship, they are the independent and dependent variables, respectively (Michaux 1989). This does not imply that geological hypotheses necessarily validate biogeographic hypotheses. Geologists may not necessarily have interpreted geological history of the area adequately enough to justify validation. In fact, the relationship between geology and biogeography should be based on its capacity of "reciprocal illumination" (in the sense of Hennig 1950). In order to make this relation more fruitful, it would be important to develop a common language, which allows interconnecting the biological and geological systems. This is because evolution in space/ time of species, whole communities and biotas is a unique geobiotic phenomenon. Tracks and area cladograms constitute appropriate instruments to develop such common language (Morrone 2009).

Biogeographers have classified geographic features in terms of their impact on dispersal and vicariance (Simpson 1953; Rapoport 1975; Cox & Moore 1998; MacDonald 2003). The most important are barriers (geographic features that hinder dispersal) and corridors (geographic features that facilitate dispersal). Barriers are easily identified with geographic elements as mountains, rivers, seas, etc. In the marine environment, in addition to land barriers, there can be more subtle barriers, represented by changes in physico-chemical properties (Cecca 2002). Corridors include a variety of habitats that a large amount of the organisms found at either end of them find little difficulty in traversing them (Cox & Moore 1998). These terms are relative, because, for example, a cordillera may act as a barrier for certain species but be a corridor for others. Instead of barriers, cladistic biogeographers usually refer to vicariance events.

In some instances, physical or biological conditions make it easier or more difficult for certain species to cross a certain barrier. Features that are not equally favorable for dispersal of all species are called filters. For example, before the rise of the Isthmus of Panama, there was a chain of small islands (stepping stones) set upon a relatively shallow sea of about 150 m in depth occupying its place. These islands acted facilitating dispersal of some species but acted as a barrier for other species.

There are some areas completely surrounded by totally different environments, like islands, caves or high mountain peaks, where chances of dispersal are very low for most of the taxa. They are known as sweepstake routes, and differ from filters in kind, not merely in degree, for almost all the species that traverse them cannot survive (Cox & Moore 1998).

Plate Tectonics

When dealing with long-term changes in the biotic distributional patterns, continental drift may be a relevant factor (Briggs 1987; Cox & Moore 1998). The splitting and collision of land masses not only affect distributional patterns directly, but also new mountains, oceans or land barriers change the climatic patterns upon the land masses.

Continental drift was originally proposed by Wegener (1912), and found an enormous opposition. Plate tectonics was a mechanism that explained continental drift and made it a credible theory. Seafloor spreading is believed to be caused by great convection currents that bring material to the surface from the hot interior of the Earth, inducing the movement of the tectonic plates. These constitute the moving units at the surface of the Earth, and may contain continental masses or may consist of ocean floor. The movement of the plates had great relevance for the organisms on Earth. The movement of the continents relative to the poles and the Equator caused climatic changes. Additionally, shallow epicontinental seas covered parts of the continents or formed seas within them during the Jurassic and Cretaceous periods, forming barriers to dispersal. The splitting of continents also altered the patterns of water circulation in the oceans. Furthermore, the appearance of new mountains as a result of continental drift had also dramatic consequences for biotic distributional patterns.

A major feature of the late Paleozoic and Mesozoic was the supercontinent of Gondwana. It included the land areas that later became South America, Africa, Madagascar, Antarctica, Australia, New Zealand, and India. Its northern edge was broken up into a series of minor land masses or terranes, which moved northwards and joined the southern edge of Eurasia to form southern Europe, Tibet and two separate portions of China (Cox & Moore 1998). By the Silurian (435-410 Mya), there were three continents: Euramerica (North America plus Eurasia), Siberia and Gondwana. In the late Carboniferous to early Permian (300-270 Mya) Euramerica joined Gondwana, and in the late Permian (260 Mya) Siberia joined this land mass, forming the world continent of Pangaea. Pangea soon became divided into two land masses: Laurasia in the north and Gondwana in the south. From the Jurassic to the Cretaceous, Laurasia was penetrated by epicontinental seas and Gondwana started to break up into separate continents. India separated from the rest of Gondwana in the early Cretaceous. In the late Cretaceous, Europe and Asia were separated by the Obisk Sea, the former was connected to eastern North America (the Euramerican land mass) and the latter was connected to western North America (the Asiamerican land mass).

Sanmartín & Ronquist (2004) provided a recent synthesis of the ideas concerning the fragmentation of Gondwana. This supercontinent started to break up in the Jurassic (165-150 Mya), when rifting began between India and Australia-east Antarctica. Shortly after, the Madagascar-India block, which was adjacent to Somalia, broke away from Africa and began moving southeast, attaining its present position in the Early Cretaceous (121 Mya). India separated from Madagascar in the late Cretaceous (88-84 Mya), with the opening of the Mascarene Basin, and began drifting northward, eventually to collide with Asia *ca.* 50 Mya. South America began to separate from Africa in the Early Cretaceous (135 Mya), with the opening of the South Atlantic Ocean at the latitude of Argentina and Chile. Northern South America and Africa remained connected until the mid-late Cretaceous (110-95 Mya), when a transform fault opened between Brazil and Guinea. As a result, Africa started drifting northeast and collided with Eurasia in the Paleocene (60 Mya), whereas southern South America drifted southeast into contact with Antarctica. New Zealand, Australia, South America and Antarctica remained connected until the Late Cretaceous: east Antarctica was adjacent to southern Australia, whereas New Zealand and southern South America were in contact with west Antarctica. About 80 Mya, the Tasmantis block (New Zealand plus New Caledonia) broke away from west Antarctica and moved northwest, opening the Tasman Sea. New Zealand and New Caledonia were finally separated in the mid-Tertiary (40-30 Mya), when the Norfolk Ridge foundered, opening the New Caledonian Basin. Australia and South America remained in contact across Antarctica until the Eocene. Australia began to separate from Antarctica in the Late Cretaceous (90 Mya), but both continents remained in contact along Tasmania, and complete separation did not occur until the late Eocene (35 Mya) with the opening of the South Tasman Sea. Southern South America and Antarctica remained in contact through the Antarctic peninsula until the Oligocene (20-28 Mya). Following its separation from Antarctica, Australia began to drift rapidly toward Asia. New Guinea was then joined to the northern margin of the Australian plate. The collision of the Australian and Pacific plates in the Oligocene (30 Mya) initiated the tectonic uplift of New Guinea, but by the Early Miocene much of southern New Guinea was again submerged. Subsequent episodes of uplift in the Miocene, after the collision of the Australian and Asian plates, led to the accretion of numerous terranes to the northern margin of New Guinea. The link between North and South America, the Isthmus of Panama, was formed in the late Pliocene (2 Mya).

There is still considerable discussion about some aspects of tectonics. Theories postulating a lost Pacifica continent (Nur & Ben-Avraham 1980) or an expanding Earth (Shields 1996; McCarthy 2003, 2007) have been proposed to explain certain "anomalies" of Wegener's theory, but they have not gained support by geophysicists (Humphries & Ebach 2004).

Any of these theories implies a major role for vicariance in isolating populations of plant and animal species (Cox & Moore 1998). Continents split, and their fragments carry away its cargo of living organisms ("Noah's arks") and buried fossils ("Viking funeral ships") (McKenna 1973). Hallam (1974) characterized two main historical phenomena due to plate tectonics: convergence and divergence. Biogeographic convergence indicates intermixing of biotas due to continental shifting: two continents approach each other and their biotas merge. Biogeographic divergence occurs when two continents move apart or two land masses are isolated by a seaway.

REFERENCES

Abogast, B. S. & Kenagy, G. J. (2001). Comparative phylogeography as an integrative approach to historical biogeography. *Journal of Biogeography, 28*, 819-825.

Aguilar-Aguilar, R., Contreras-Medina, R. & Salgado Maldonado, G. (2003). Parsimony analysis of endemicity (PAE) of Mexican hydrological basins based on helminth parasites of freshwater fishes. *Journal of Bioeography, 30*, 1861-1872.

Allen, J. A. (1871). On the mammals and winter birds of East Florida. *Bulletin of the Museum of Comparative Zoology, 2*, 161-450.

Andersson, L. (1996). An ontological dilema: Epistemology and methodology of historical biogeography. *Journal of Biogeography, 23*, 269-277.

Aris-Brosou, S. & Yang, Z. (2001). *PHYBAYES: A program for phylogenetic analyses in a Bayesian framework.* London: Department of Biology, University College London.

Avise, J. C. (2000). *Phylogeography: The history and formation of species.* Cambridge, Massachusetts: Harvard University Press.

Avise, J. C. (2004). What is the field of biogeography, and where is it going? *Taxon, 53(4)*, 893-898.

Avise, J. C., Arnold, J., Ball, R. M., Bermingham, E., Lamb, T., Neigel, J. E., Reeb, C. A & Saunders, N. C. (1987). Intraspecific phylogeography: The mitochondrial DNA bridge between population genetics and systematics. *Annual Review of Ecology and Systematics, 18*, 489-522.

Baroni-Urbani, C., Ruffo, S. & Vigna Taglianti, A. (1978). Materiali per una biogeografia italiana fondata su alcuni generi di coleotteri cicindelidi, carabidi e crisomelidi. *Estrato dalle Memorie della Societa Entomologica Italiana, 56,* 35-92.

Benton, M. J. & Donoghue, P. C. J. (2007). Paleontological evidence to date the tree of life. *Molecular Biology and Evolution, 24,* 26-53.

Bermingham, E. & Martin, A. P. (1998). Comparative mtDNA phylogeography of Neotropical fishes: Testing shared history to infer the evolutionary landscape of lower Central America. *Molecular Ecology, 7,* 499-517.

Biondi, M. (1998). Comparison of some methods for a cladistically founded biogeographical analysis. *Memorie del Museo Civico di Storia Naturale di Verona, 2. serie, Sezione Scienze della Vita, 13,* 9-31.

Biswas, S. & Pawar, S. S. (2006). Phylogenetic tests of distribution patterns in South Asia: Towards an integrative approach. *Journal of Biosciences, 31(1),* 95-113.

Boniolo, G. & Carrara, M. (2004). On biological identity. *Biology and Philosophy, 19,* 443-457.

Bremer, K. (1995). Ancestral areas: Optimization and probability. *Systematic Biology, 44,* 255-259.

Briggs, J. C. (1987). *Biogeography and plate tectonics.* Amsterdam: Developments in Palaeontology and Stratigraphy 10, Elsevier Science Publishers.

Briggs, J. C. (1995). *Global biogeography.* Amsterdam: Developments in Palaeontology and Stratigraphy 14, Elsevier Science Publishers.

Britton, T., Oxelman, B., Vinnersten, A. & Bremer, K. (2002). Phylogenetic dating with confidence intervals using mean path lengths. *Molecular Phylogenetics and Evolution, 24,* 58-65.

Bromham, L. & Woolfit, M. (2004). Explosive radiations and the reliability of molecular clocks: Island endemic radiations as a test case. *Systematic Biology, 53(5),* 758-766.

Brooks, D. R. (1981). Hennig's parasitological method: A proposed solution. *Systematic Zoology, 30,* 229-249.

Brooks, D. R. (1985). Historical ecology: A new approach to studying the evolution of ecological associations. *Annals of the Missouri Botanical Garden, 72,* 60-680.

Brooks, D. R. (1990). Parsimony analysis in historical biogeography and coevolution: Methodological and theoretical update. *Systematic Zoology, 39,* 14-30.

Brooks, D. R. (2004). Reticulations in historical biogeography: The triumph of time over space in evolution. In M. V. Lomolino, Heaney, L. R., editors. *Frontiers of biogeography: New directions in the geography of nature.* Sunderland, Massachusetts: Sinauer Associates Inc.; pp. 125-144.

Brooks, D. R., Dowling, P. G., Van Veller, M. G. P. & Hoberg, E. P. (2003). Ending a decade of deception: A valiant failure, a not-so valiant failure, and a success story. *Cladistics, 20,* 32-46.

Brooks, D. R. & McLennan, D. A. (1991). *Phylogeny, ecology and behavior: A research program in comparative biology.* Chicago: University of Chicago Press.

Brooks, D. R. & McLennan, D. A. (2001). A comparison of a discovery-based and an event-based method of historical biogeography. *Journal of Biogeography, 28,* 757-767.

Brooks, D. R. & Van Veller, M. G. P. (2003). Critique of parsimony analysis of endemicity as a method of historical biogeography. *Journal of Biogeography, 30,* 819-825.

Brooks, D. R., Van Veller, M. G. P. & McLennan, D. A. (2001). How to do BPA, really. *Journal of Biogeography, 28*, 345-358.

Brundin, L. (1966). Transantarctic relationships and their significance as evidenced by midges. *Kungliga Svenska Vetenskaps Akademien Handlingar (series 4), 11*, 1-472.

Cano, J. M. & Gurrea, P. (2003). La distribución de las zigenas (Lepidoptera, Zygaenidae) ibéricas: Una consecuencia del efecto península. *Graellsia, 59(2-3)*, 273-285.

Cao, N. & Ducasse, J. (2005). *Nelson05: A Program for Cladistics and Biogeography.* Paris: Software available from the authors.

Cecca, F. (2002). *Palaeobiogeography of marine fossil invertebrates: Concepts and methods.* London and New York: Taylor and Francis.

Charleston, M. A. (1998). Jungles: A new solution to the host/ parasite phylogeny reconciliation problem. *Mathematical Biosciences, 149*, 191-223.

Clement, M., Posada, D. & Crandall, K. A. (2000). TCS: A computer program to estimate gene genealogies. *Molecular Ecology, 9*, 1557-1659.

Costa, L. P. (2003). The historical bridge between the Amazon and the Atlantic forest of Brazil: A study of molecular phylogeography with small mammals. *Journal of Biogeography, 30*, 71-86.

Cox, C. B. (2001). The biogeographic regions reconsidered. *Journal of Biogeography, 28*, 511-523.

Cox, C. B. & Moore, P. D. (1998). *Biogeography: An ecological and evolutionary approach.* Oxford: Blackwell Science.

Cracraft, J. (1988). Deep-history biogeography: Retrieving the historical pattern of evolving continental biotas. *Systematic Zoology, 37(4)*, 221-236.

Cracraft, J. (1991). Patterns of diversification within continental biotas: Hierarchical congruence among the areas of endemism of Australian vertebrates. *Australian Systematic Botany, 4*, 211-227.

Craw, R. C. (1988). Continuing the synthesis between panbiogeography, phylogenetic systematics and geology as illustrated by empirical studies on the biogeography of New Zealand and the Chatham Islands. *Systematic Zoology, 37(4)*, 291-310.

Craw, R. C. (1989a). New Zealand biogeography: A panbiogeographic approach. *New Zealand Journal of Zoology, 16*, 527-547.

Craw, R. C. 1989b. Quantitative panbiogeography: Introduction to methods. *New Zealand Journal of Zoology, 16*, 485-494.

Craw, R. C., Grehan, J. R. & Heads, M. J. (1999). *Panbiogeography: Tracking the history of life.* New York and Oxford: Oxford Biogeography series 11.

Craw, R. C. & Page, R. (1988). Panbiogeography: Method and metaphor in the new biogeography. In Ho, M. -W., Fox, S. W., editors. *Evolutionary processes and metaphors*. Chichester: John Wiley and Sons; pp. 163-189.

Crisci, J. V. (2001). The voice of historical biogeography. *Journal of Biogeography, 28*, 157-168.

Crisci, J. V., Cigliano, M. M., Morrone, J. J. & Roig-Juñent, S. (1991). Historical biogeography of southern South America. *Systematic Zoology, 40*, 152-171.

Crisci, J. V., De la Fuente, M. S., Lanteri, A. A., Morrone, J. J., Ortiz Jaureguizar, E., Pascual, R. & Prado, J. L. (1993). Patagonia, Gondwana Occidental (GW) y Oriental (GE), un modelo de biogeografía histórica. *Ameghiniana, 30*, 104.

Crisci, J. V., Katinas, K. & Posadas, P. (2003). *Historical biogeography: An introduction.* Cambridge, Massachusetts: Harvard University Press.

Crisp, M. D., H. P. Linder & P. H. Weston (1995). Cladistic biogeography of plants in Australia and New Guinea: Congruent pattern revealed two endemic tropical tracks. *Systematic Zoology, 44(4),* 457-473.

Croizat, L. (1958). *Panbiogeography. Vols. 1 and 2.* Caracas: Published by the author.

Croizat, L. (1964). *Space, time, form: The biological synthesis.* Caracas: Published by the author.

Croizat, L., Nelson, G. & Rosen, D. E. (1974). Centers of origin and related concepts. *Systematic Zoology, 23,* 265-287.

Cunningham, C. W. & Collins, T. M. (1998). Beyond area relationships: Extinction and recolonization in molecular marine biogeography. In DeSalle, R., Schierwater, B., editors. *Molecular approaches to ecology and evolution.* Basel: Birkhäuser; pp. 297-321.

Darlington, P. J. Jr. (1957). *Zoogeography: The geographical distribution of animals.* New York: John Wiley and Sons.

Darwin, C. R. (1859). *The origin of species by means of natural selection, or, the preservation of favoured races in the struggle for life.* London: John Murray.

De Candolle, A. P. (1820). Géographie botanique. In *Dictionnaire des Sciences Naturelles.* Strasbourg and Paris; pp. 359-422.

De Candolle, A. P. (1838). *Statistique de la famille des Composées.* Paris and Strasbourg: Treuttel and Würtz.

Deo, A. J. & DeSalle, R. (2006). Nested areas of endemism analysis. *Journal of Biogeography, 33,* 1511-1526.

Donoghue, M. J., Bell, C. D. & Li, J. (2001). Phylogenetic patterns in northern hemisphere plant geography. *Int. J. Plant Sci., 162,* S41-S52.

Donoghue, M. J. & Mooore, B. R. (2003). Toward an integrative historical biogeography. *Integrative and Comparative Biology, 43,* 261-270.

Dowling, A. P. G. (2002). Testing the accuracy of TreeMap and Brooks parsimony analyses of coevolutionary patterns using artificial associations. *Cladistics, 18,* 416-435.

Drummond, A. J. & Rambaut, A. (2003). *BEAST version 1.0.* Available at http://evolve. zoo.ox.ac.uk/beast/.

Ebach, M. C. (2001). Extrapolating cladistic biogeography: A brief comment on van Veller *et al.* (1999, 2000, 2001). *Cladistics, 17(4),* 383-388.

Ebach, M. C. (2003). Area cladistics. *Biologist, 50,* 169-172.

Ebach, M. C. & Edgecombe, G. D. (2001). Cladistic biogeography: Component-based methods and paleontological application. In Adrain, J. M., Edgecombe, G. D., Lieberman, B. S., editors. *Fossils, phylogeny, and form: An analytical approach.* New York: Kluwer/ Plenum; pp. 235-289.

Ebach, M. C. & Humphries, C. J. (2002). Cladistic biogeography and the art of discovery. *Journal of Biogeography, 29,* 427-444.

Ebach, M. C. & Humphries, C. J. (2003). Ontology of biogeography. *Journal of Biogeography, 30,* 959-962.

Ebach, M. C., Humphries, C. J, Newman, R. A., Williams, D. & Walsh, S. A. (2005a). Assumption 2: Opaque to intuition? *Journal of Biogeography, 32,* 781-787.

Ebach, M. & Morrone, J. J. (2005). Forum on historical biogeography: What is cladistic biogeography? *Journal of Biogeography, 32,* 2179-2183.

Ebach, M. C., Morrone, J. J., Parenti, L. R. & Viloria, Á. L. (2008). International Code of Area Nomenclature. *Journal of Biogeography, 35*, 1153-1157.

Ebach, M. C., Newman, R. A., Humphries, C. J. & Williams, D. M. (2005b). *3item version 2.0: Three-item analysis for cladistics and area cladistics.* Oxford: Published by the authors.

Ekman, S. (1935). *Tiergeographie des Meeres.* Leipzig: Akademische Verlagsgesellschaft.

Enghoff, H. (1995). Historical biogeography of the Holarctic: Area relationships, ancestral areas, and dispersal of non-marine animals. *Cladistics, 11(3),* 223-263.

Enghoff, H. (1996). Widespread taxa, sympatry, dispersal, and an algorithm for resolved area cladograms. *Cladistics, 12*, 349-364.

Enghoff, H. (2000). Reversals as branch support in biogeographical parsimony analysis. *Vie et Milieu, 50(4),* 255-260.

Escalante, T., Espinosa, D. & Morrone, J. J. (2003). Using parsimony analysis of endemicity to analyze the distribution of Mexican land mammals. *Southwestern Naturalist, 48(4),* 563-578.

Escalante, T. & Morrone, J. J. (2003).¿Para qué sirve el análisis de parsimonia de endemismos? In Morrone, J. J., Llorente Bousquets, J., editors. *Una perspectiva latinoamericana de la biogeografía.* Mexico, D.F.: Las Prensas de Ciencias, UNAM; pp. 167-172.

Espinosa Organista, D., Morrone, J. J., Llorente Bousquets, J. & Flores Villela, O. (2002). *Análisis de patrones biogeográficos históricos.* México, D.F.: Las Prensas de Ciencias, UNAM.

Farris, J. S. (1988). *Hennig86 reference. Version 1.5.* Port Jefferson, New York: Published by the author.

Farris, J. S. (2000). Diagnostic efficiency of three-taxon analysis. *Cladistics, 16*, 403-410.

Felsenstein, J. (1993). *Phylogenetic inference package (PHYLIP): Version 3.5.* Seatle: University of Washington.

Fiala, K. (1984). *CLINCH: Cladistic Inference using Character Compatibility 6.2.* Seatle: University of Washington.

García-Barros, E. (2003). Mariposas diurnas endémicas de la región Paleártica Occidental: Patrones de distribución y su análisis mediante parsimonia (Lepidoptera, Papilionoidea). *Graellsia, 59*, 233-258.

García-Barros, E., Gurrea, P., Luciáñez, M. J., Cano, J. M., Munguira, M. L., Moreno, J. C., Sainz, H., Sanza, M. J. & Simón, J. C. (2002). Parsimony analysis of endemicity and its application to animal and plant geographical distributions in the Ibero-Balearic region (western Mediterranean). *Journal of Biogeography, 29*, 109-124.

Gillespie, J. H. (1991). *The causes of molecular evolution.* New York: Oxford University Press.

Goloboff, P. (1998). *NONA ver. 2.0.* Available at http://www.cladistics.com/about_nona.htm.

Goloboff, P. (2004). *NDM/ VNDM programs ver. 1.5.* Available at www.zmuc.dk - /public/phylogeny/Endemism/.

Good, R. (1974). *The geography of the flowering plants.* London: Longman Group Limited.

Goodman, M., Czelusniak, J., Moore, G. W., Romero-Herrera, A. E. & Matsuda, G. (1979). Fitting the gene lineage into its species lineage: A parsimony strategy illustrated by cladograms constructed from globin sequences. *Systematic Zoology, 28*, 132-168.

Goyenechea, I, Flores Villela, O. & Morrone, J. J. (2001). Introducción a los fundamentos y métodos de la biogeografía cladística. In Llorente Bousquets, J., Morrone, J. J., editors. *Introducción a la biogeografía en Latinoamérica: Conceptos, teorías, métodos y aplicaciones*. Mexico, D.F.: Las Prensas de Ciencias, UNAM; pp. 225-232.

Grande, L. (1985). The use of paleontology in systematics and biogeography, and a time control refinement for historical biogeography. *Paleobiology, 11*, 234-243.

Grehan, J. R. (1991). Panbiogeography 1981-91: Development of an Earth/life synthesis. *Progr. Phys. Geogr., 15,* 331-363.

Halffter, G. (1987). Biogeography of the montane entomofauna of Mexico and Central America. *Annual Review of Entomology, 32,* 95-114.

Hallam, A. (1974). Changing patterns of provinciality and diversity of fossil animals in relation to plate tectonics. *Journal of Biogeography, 1,* 213-225.

Harvey, A. W. (1992). Three-taxon statements: More precisely, an abuse of parsimony? *Cladistics, 8,* 345-354.

Hausdorf, B. (2002). Units in biogeography. *Systematic Biology, 51(4),* 648-652.

Hausdorf, B. & Hennig, C. (2003). Biotic element analysis in biogeography. *Systematic Biology, 52(5),* 717-723.

Hausdorf, B. & Hennig, C. (2004). Does vicariance shape biotas? Biogeographical tests of the vicariance model in the north-west European land snail fauna. *Journal of Biogeography, 31,* 1751-1757.

Hausdorf, B. & Hennig, C. (2007). Biotic element analysis and vicariance biogeography. In Ebach, M. C., Tangney, R. S., editors. *Biogeography in a changing world*. Boca Raton: The Systematics Association Special Volume Series 70, CRC Press; pp. 95-115.

Heads, M. J. (1989). Integrating earth and life sciences in New Zealand natural history: The parallel arcs model. *New Zealand Journal of Zoology, 16,* 549-585.

Heads, M. J. (1994). A biogeographic review of *Parahebe* (Scrophulariaceae). *Botanical Journal of the Linnean Society of London, 115,* 65-89.

Heads, M. J. (2004). What is a node? *Journal of Biogeography, 31,* 1883-1891.

Henderson, I. M. (1991). Biogeography without area? *Australian Systematic Botany, 4,* 59-71.

Hennig, W. (1950). *Grundzüge einer Theorie der phylogenetischen Systematik.* Berlin: Deutscher Zentralverlag.

Hooker, J. D. (1844-60). *The botany of the Antarctic Voyage of H. M. Discovery ships Erebus and Terror in the years 1839-1843. I. Flora Antarctica (1844-47).* London.

Hovenkamp, P. (1997). Vicariance events, not areas, should be used in biogeographic analysis. *Cladistics, 13,* 67-79.

Hovenkamp, P. (2001). A direct method for the analysis of vicariance patterns. *Cladistics, 17,* 260-265.

Huelsenbeck, J. P., Rannala, B. & Larget, B. (2000). A Bayesian framework for the analysis of cospeciation. *Evolution, 54,* 352-364.

Humphries, C. J. (1989). Any advance on assumption 2? *Journal of Biogeography, 16,* 101-102.

Humphries, C. J. (1992). Cladistic biogeography. In Forey, P. L., Humphries, C. J., Kitching, I. J., Scotland, R. W., Siebert, D. J., Williams, D. M., editors. *Cladistics: A practical course in systematics*. Oxford: The Systematics Association Publication 10, Oxford Science Publications, Clarendon Press; pp. 137-159.

Humphries, C. J. (2000). Form, space and time; which comes first? *Journal of Biogeography, 27,* 11-15.

Humphries, C. J. & Ebach, M. C. (2004). Biogeography on a dynamic Earth. In Lomolino, M. V., Heaney, L. R., editors. *Frontiers of biogeography: New directions in the geography of nature.* Sunderland, Massachusetts: Sinauer Associates Inc.; pp. 67-86.

Humphries, C. J. & Parenti, L. R. (1999). *Cladistic biogeography- Second edition: Interpreting patterns of plant and animal distributions.* Oxford: Oxford University Press.

Humphries, C. J. & O. Seberg. (1989). Graphs and generalized tracks: Some comments on method. *Systematic Zoology, 38,* 69-76.

Hunn, C. A. & Upchurch, P. (2001). The importance of time/space in diagnosing the causality of phylogenetic events: Towards a "chronobiogeographical paradigm". *Systematic Biology, 50,* 391-407.

Jeannel, R. (1942). *La genese des faunes terrestres: Élements de biogéographie.* Paris: Presses Universitaires de France.

Kishino, H., Thorne, J. L. & Bruno, W. J. (2001). Performance of a divergence time estimation method under a probabilistic model of rate evolution. *Molecular Biology and Evolution, 18(3),* 352-361.

Kluge, A. G. (1988). Parsimony in vicariance biogeography: A quantitative method and a greater Antillean example. *Systematic Zoology, 37,* 315-328.

Kluge, A. G. (1989). A concern for evidence, and a phylogenetic hypothesis of relationships among Epicrates (Boidae, Serpentes). *Systematic Zoology, 38,* 7-25.

Kluge, A. G. (1993). Three-taxon transformation in phylogenetic inference: Ambiguity and distortion as regards explanatory power. *Cladistics, 9,* 246-259.

Lankester, E. R. (1905). *Extinct animals.* London: Constable.

Lanteri, A. A. & Confalonieri, V. A. (2003). Filogeografía: Objetivos, métodos y ejemplos. In Morrone, J. J., Llorente Bousquets, J., editors. *Una perspectiva latinoamericana de la biogeografía.* Mexico, D.F.: Las Prensas de Ciencias, UNAM; pp. 185-193.

Lapointe, F. J. & Rissler, L. J. (2005). Congruence, consensus, and the comparative phylogeography of codistributed species in California. *The American Naturalist, 166(2),* 290-299.

León-Paniagua L., García, E., Arroyo-Cabrales, J. & Castañeda-Rico, S. (2004). Patrones biogeográficos de la mastofauna. In Luna, I., Morrone, J. J., Espinosa, D., editors. *Biodiversidad de la Sierra Madre Oriental.* Mexico D.F.: Las Prensas de Ciencias, UNAM; pp. 469-486.

Lequesne, W. J. (1982). Compatibility analysis and its applications. *Zoological Journal of the Linnean Society, 74,* 267-275.

Lieberman, B. S. (1997). Early Cambrian paleogeography and tectonic history: A biogeographic approach. *Geology, 25,* 1039-1042.

Lieberman, B. S. (2000). *Paleobiogeography: Using fossils to study global change, plate tectonics and evolution.* New York: Kluwer Academic Press.

Lieberman, B. S. (2003a). Unifying theory and methodology in biogeography. *Evolutionary Biology, 33,* 1-25.

Lieberman, B. S. (2003b). Paleobiogeography: The relevance of fossils to biogeography. *Annual Review of Ecology and Systematics, 34,* 51-69.

Lieberman, B. S. (2004). Range expansion, extinction, and biogeographic congruence : A deep time perspective. In Lomolino, M. V., Heaney, L. R., edsitors. *Frontiers of biogeography: New directions in the geography of nature.* Sunderland, Massachusetts: Sinauer Associates Inc.; pp. 111-124.

Lieberman, B. S. (2005). Geobiology and paleobiogeography: Tracking the coevolution of the Earth and its biota. *Palaeobiogeography, Palaeoclimatology, Palaeoecology, 219,* 23-33.

Lieberman, B. S. & Eldredge, N. (1996). Trilobite biogeography in the Middle Devonian: Geological processes and analytical methods. *Paleobiology, 22,* 66-79.

Linder, H. P. (2001). On areas of endemism, with an example of the African Restionaceae. *Systematic Biology, 50(6),* 892-912.

Linder, H. P. & Mann, D. M. (1998). The phylogeny and biogeography of *Thamnochortus* (Restionaceae). *Botanical Journal of the Linnean Society of London, 128,* 319-357.

Lomolino, M. V., Riddle, B. R. & Brown, J. H. (2006). *Biogeography: Third edition.* Sunderland, Massachusetts: Sinauer Associates, Inc.

Luna-Vega, I., Alcántara, O., Espinosa Organista, D. & Morrone, J. J. (1999). Historical relationships of the Mexican cloud forests: A preliminary vicariance model applying parsimony analysis of endemicity to vascular plant taxa. *Journal of Biogeography, 26,* 1299-1305.

Luna-Vega, I., Alcántara, O., Morrone, J. J. & Espinosa Organista, D. (2000). Track analysis and conservation priorities in the cloud forests of Hidalgo, Mexico. *Divers. Distrib., 6,* 137-143.

Luna-Vega, I., Morrone, J. J., Alcántara Ayala, O. & Espinosa Organista, D. (2001). Biogeographical affinities among Neotropical cloud forests. *Plant Systematics and Evolution, 228,* 229-239.

MacDonald, G. M. (2003). *Biogeography: Space, time, and life.* New York: John Wiley and Sons.

Magallón, S. A. (2004). Dating lineages: Molecular and paleontological approaches to the temporal framework of clades. *International Journal of Plant Sciences, 165(4 Suppl.),* S7-S21.

Marques, A. C. (2005). Three-taxon statement analysis and its relation with primary data: Implications for cladistics and biogeography. In Llorente Bousquets, J., Morrone, J. J., editors. *Regionalización biogeográfica en Iberoamérica y tópicos afines: Primeras Jornadas Biogeográficas de la Red Iberoamericana de Biogeografía y Entomología Sistemática (RIBES XII.I-CYTED).* Mexico, D. F.: Las Prensas de Ciencias, UNAM; pp. 171-180.

Matthew, W. D. (1915). Climate and evolution. *Annals of the New York Academy of Sciences, 24,* 171-318.

McCarthy, D. (2003). The trans-Pacific zipper effect: Disjunct sister taxa and matching geological outlines that link the Pacific margins. *Journal of Biogeography, 30,* 1545-1561.

McCarthy, D. (2007). Are plate tectonic explanations for trans-Pacific disjunctions plausible? Empirical tests of radical dispersalist theories. In Ebach, M. C., Tangney, R. S., editors. *Biogeography in a changing world.* Boca Raton: The Systematics Association Special Volume Series 70, CRC Press; pp. 177-198.

McDowall, R. M. (1978). Generalized tracks and dispersal in biogeography. *Systematic Zoology, 27(1),* 88-104.

McKenna, M. C. (1973). Sweepstakes, filters, corridors, Noah's ark and beached Viking funeral ships in paleogeography. In Tarling, D. H., Runcorn, S. K., editors. *Implications of continental drift to the earth sciences, Vol. I.* London and New York: Academic Press; pp. 295-308.

Michaux, B. (1989). Generalized tracks and geology. *Systematic Zoology, 38(4),* 390-398.

Mickevich, M. F. (1981). Quantitative phylogenetic biogeography. In Funk, V. A., Brooks, D. R., editors. *Advances in cladistics: Proceedings of the First Meeting of the Willi Hennig Society.* New York: New York Botanical Garden; pp. 202-222.

Miranda Esquivel, D. R., Donato, D. & Posadas, P. (2003). La dispersión ha muerto, larga vida a la dispersion. In Morrone, J. J., Llorente Bousquets, J., editors. *Una perspectiva latinoamericana de la biogeografía.* Mexico, D.F.: Las Prensas de Ciencias, UNAM; pp. 179-184.

Mitchell, S. D. (2002). Integrative pluralism. *Biology and Philosophy, 17,* 55-70.

Mitter, C. & Brooks, D. R. (1983). Phylogenetic aspects of coevolution. In Futuyma, D. J., Slatkin, M., editors. *Coevolution.* Sunderland, Massachusetts: Sinauer; pp. 65-98.

Moline, P. M. & Linder, H. P. (2006). Input data, analytical methods and biogeography of *Elegia* (Restionaceae). *Journal of Biogeography, 33,* 47-62.

Morales-Barros, N., Silva, J. A. B., Miyaki, C. Y. & Morgante, J. S. (2006). Comparative phylogeography of the Atlantic forest endemic sloth (*Bradypus torquatus*) and the widespread three-toed sloth (*Bradypus variegatus*) (Bradypodidae, Xenarthra). *Genetica, 126,* 189-198.

Moreira-Muñoz, A. (2007). The Austral floristic realm revisited. *Journal of Biogeography, 34,* 1649-1660.

Morrone, J. J. (1994). On the identification of areas of endemism. *Systematic Biology, 43,* 438-441.

Morrone, J. J. (1995). Asociaciones históricas en biología comparada. *Ciencia, 46,* 229-235.

Morrone, J. J. (1997). Biogeografía cladística: Conceptos básicos. *Arbor, 158,* 373-388.

Morrone, J. J. (1998). On Udvardy's Insulantarctica province: A test from the weevils (Coleoptera: Curculionoidea). *Journal of Biogeography, 25,* 947-955.

Morrone, J. J. (2001). Homology, biogeography and areas of endemism. *Diversity and Distributions, 7,* 297-300.

Morrone, J. J. (2002). Biogeographic regions under track and cladistic scrutiny. *Journal of Biogeography, 29,* 149-152.

Morrone, J. J. (2004a). *Homología biogeográfica: Las coordenadas espaciales de la vida.* Mexico, D.F.: Cuadernos del Instituto de Biología 37, Instituto de Biología, UNAM.

Morrone, J. J. (2004b). Panbiogeografía, componentes bióticos y zonas de transición. *Revista Brasileira di Entomologia, 48,* 149-162.

Morrone, J. J. (2005). Cladistic biogeography: Identity and place. *Journal of Biogeography, 32,* 1281-1284.

Morron, J. J. (2009). *Evolutionary biogeography: An integrative approach with case studies.* Columbia Unversity Press, New York.

Morrone, J. J. & Carpenter, J. M. (1994). In search of a method for cladistic biogeography: An empirical comparison of component analysis, Brooks parsimony analysis, and three-area statements. *Cladistics, 10,* 99-153.

Morrone, J. J. & Crisci, J. V. (1995). Historical biogeography: Introduction to methods. *Annual Review of Ecology and Systematics, 26,* 373-401.

Morrone, J. J., Espinosa Organista, D. & Llorente Bousquets, J. (1996). *Manual de biogeografía histórica.* Mexico, D. F.: Universidad Nacional Autónoma de México.

Morrone, J. J. & Márquez, J. (2001). Halffter's Mexican Transition Zone, beetle generalised tracks, and geographical homology. *Journal of Biogeography, 28*, 635-650.

Müller, P. (1973). *The dispersal centres of terrestrial vertebrates in the Neotropical realm: A study in the evolution of the Neotropical biota and its native landscapes.* The Hague: Junk.

Myers, A. A. (1991). How did Hawaii accumulate its biota?: A test from the Amphipoda. *Global Ecol. Biogeogr. Lett., 1,* 24-29.

Myers, A. A. & Giller, P. S. (1988). Biogeographic patterns. In Myers, A. A., Giller, P. S., editors. *Analytical biogeography: An integrated approach to the study of animal and plant distributions.* London and New York: Chapman and Hall; pp. 15-21.

Nelson, G. (1978). From Candolle to Croizat: Comments on the history of biogeography. *Journal of the History of Biology, 11,* 269-305.

Nelson, G. (1984). Cladistics and biogeography. In Duncan, T., Stuessy, T. F., editors. *Cladistics: Perspectives on the reconstruction of evolutionary history.* New York: Columbia University Press; pp. 273-293.

Nelson, G. (1985). A decade of challenge the future of biogeography. *Journal of the History of Earth Sciences Society, 4,* 187-196.

Nelson, G. (1994). Homology and systematics. In *Homology: The hierarchical basis of comparative biology.* B. K. Hall, editor. San Diego: Academic Press; pp. 101-149.

Nelson, G. & Ladiges, P. Y. (1990). Biodiversity and biogeography. *Journal of Biogeography, 17,* 559-560.

Nelson, G. & Ladiges, P. Y. (1991a). Standard assumptions for biogeographic analysis. *Australian Systematic Botany, 4,* 41-58.

Nelson, G. & Ladiges, P. Y. (1991b). Three-area statements: Standard assumptions for biogeographic analysis. *Systematic Zoology, 40,* 470-485.

Nelson, G. & Ladiges, P. Y. (1991c). *TAS (MSDos computer program).* New York and Melbourne: Published by the authors.

Nelson, G. & Ladiges, P. Y. (1993). Missing data and three-item analysis. *Systematic Zoology, 40,* 470-485.

Nelson, G. & Ladiges, P. Y. (1995). *TASS.* New York and Melbourne: Published by the authors.

Nelson, G. & Ladiges, P. Y. (1996). Paralogy in cladistic biogeography and analysis of paralogy-free subtrees. *American Museum Novitates, 3167,* 1-58.

Nelson, G. & Platnick, N. I. (1978). The perils of plesiomorphy: Widespread taxa, dispersal, and phenetic biogeography. *Systematic Zoology, 27(4),* 474-477.

Nelson, G. & Platnick, N. I. (1980). A vicariance approach to historical biogeography. *Bioscience, 30(5),* 339-343.

Nelson, G. & Platnick, N. I. (1981). *Systematics and biogeography: Cladistics and vicariance.* New York: Columbia University Press.

Nelson, G. & Platnick, N. I. (1991). Three taxon statements: A more precise use of parsimony? *Cladistics, 7,* 351-366.

Nihei, S. S. (2006). Misconceptions about parsimony analysis of endemicity. *Journal of Biogeography, 33,* 2099-2106.

Nixon, K. C. (1999). WinClada ver. 1.0000. Ithaca, New York: Published by the author. Available athttp://www.cladistics.com/about_winc.htm.

Nur, A. & Ben-Avraham, Z. (1980). Lost Pacifica continent: A mobilistic speculation. In D. E. Rosen, Nelson, G., editors. *Vicariance biogeography: A critique*. New York: Columbia University Press; pp. 341-358.

O'Hara, R. J. (1988). Homage to Clio, or, toward an historical philosophy for evolutionary biology. *Systematic Zoology, 37(2)*, 142-155.

Ortmann, A. E. (1896). *Grundzuge der marinen Tiergeographie*. Jena.

Page, R. D. M. (1987). Graphs and generalized tracks: Quantifying Croizat's panbiogeography. *Systematic Zoology, 36*, 1-17.

Page, R. D. M. (1988). Quantitative cladistic biogeography: Constructing and comparing area cladograms. *Systematic Zoology, 37*, 254-270.

Page, R. D. M. (1989a). Component *user's manual. Release 1.5*. Auckland: Published by the author.

Page, R. D. M. (1989b). Comments on component-compatibility in historical biogeography. *New Zealand Journal of Zoology, 16*, 471-483.

Page, R. D. M. (1990a). Component analysis: A valiant failure? *Cladistics, 6(2)*, 119-136.

Page, R. D. M. (1990b). Temporal congruence and cladistic analysis of biogeography and cospeciation. *Systematic Zoology, 39(3)*, 205-226.

Page, R. D. M. (1993a). Component *user's manual. Release 2.0*. London: The Natural History Museum.

Page, R. D. M. (1993b). Genes, organisms, and areas: The problem of multiple lineages. *Systematic Biology, 42*, 77-84.

Page, R. D. M. (1994a). Maps between trees and cladistic analysis of historical associations among genes, organisms, and areas. *Systematic Biology, 43*, 58-77.

Page, R. D. M. (1994b). Parallel phylogenies: Reconstructing the history of host-parasite assemblages. *Cladistics, 10*, 155-173.

Page, R. D. M. (1994c). TreeMap. *Release 3.1*. Oxford: University of Oxford.

Page, R. D. M. & Charleston, M. A. (1998). Trees within trees: Phylogeny and historical associations. *Tree, 13*, 356-359.

Page, R. D. M. & Holmes, E. C. (1998). *Molecular evolution: A phylogenetic approach*. Oxford: Blackwell Science.

Palma, R. E., Marquet, P. A. & Boric-Bargetto, D. (2005). Inter- and intraspecific phylogeography of small mammals in the Atacama desert and adjacent areas of northern Chile. *Journal of Biogeography, 32*, 1931-1941.

Parenti, L. R. (1981). Discussion. In G. Nelson, Rosen, D. E., editors. *Vicariance biogeography: A critique*. New York: Columbia University Press; pp. 490-497.

Parenti, L. R. (1991). Ocean basins and the biogeography of freshwater fishes. *Australian Systematic Botany, 4*, 137-149.

Parenti, L. R. (2007). Common cause and historical biogeography. In Ebach, M. C., Tangney, R.S., editors. *Biogeography in a changing world*. Boca Raton: The Systematics Association Special Volume Series 70, CRC Press; pp. 61-82.

Parenti, L. R. & Humphries, C. J. (2004). Historical biogeography, the natural science. *Taxon, 53(4)*, 899-903.

Pielou, E. C. (1992). *Biogeography*. Malabar: Krieger Publishing Company.

Pierrot-Bults, A. C., van der Spoel, S., Zahuranec, B. J. & Johnson, R. K. (1986). Pelagic biogeography. *Unesco Technical Papers in Marine Science, 49,* 1-295.

Pinna, M. C. C. de. (1991). Concepts and tests of homology in the cladistic paradigm. *Cladistics, 7*, 367-394.

Pinna, M. C. C. de. (1996). Comparative biology and systematics: Some controversies in retrospective. *Journal of Comparative Biology, 2*, 3-15.

Platnick, N. I. (1976). Concepts of dispersal in historical biogeography. *Systematic Zoology, 25*, 294-295.

Platnick, N. I. (1988). Systematics, evolution and biogeography: A Dutch treat. *Cladistics, 4*, 308-313.

Platnick, N. I. (1991). On areas of endemism. *Australian Systematic Botany, 4*, xi-xii.

Platnick, N. I. & Nelson, G. (1978). A method of analysis for historical biogeography. *Systematic Zoology, 27*, 1-16.

Platnick, N. I. & Nelson, G. (1988). Spanning-tree biogeography: Shortcut, detour, or dead-end? *Systematic Zoology, 37*, 410-419.

Popper, K. R. (1959). *The logic of scientific discovery.* London: Hutchinson.

Popper, K. R. (1963). *Conjectures and refutations: The growth of scientific knowledge.* London: Routledge.

Porzecanski, A. L. & Cracraft, J. (2005). Cladistic analysis of distributions and endemism (CADE): Using raw distributions of birds to unravel the biogeography of the South American aridlands. *Journal of Biogeography, 32*, 261-275.

Posada, D., Crandall, K. A. & Templeton, A. R. (2000). GeoDis: A program for the cladistic nested analysis of the geographical distribution of genetic haplotypes. *Molecular Ecology, 9*, 487-488.

Posadas, P. & Miranda-Esquivel, D. R. 1999. El PAE (parsimony analysis of endemicity) como una herramienta en la evaluación de la biodiversidad. *Revista Chilena de Historia Natural, 72*, 539-546.

Posadas, P. & Morrone, J. J. (2003). Biogeografía histórica de la familia Curculionidae (Coleoptera) en las subregiones Subantártica y Chilena Central. *Revista de la Sociedad Entomológica Argentina, 62(1-2)*, 75-84.

Rambaut, A. (2001). RHINO version 1.1. Available at http://evolve.zoo.ox.ac.uk/.

Rambaut, A. & Bromham, L. (1998). Estimating divergence dates from molecular sequences. *Molecular Biology and Evolution, 15*, 442-448.

Rambaut, A. & Charleston, M. (2002). *Phylogenetic tree editor and manipulator v1.0 alpha 10.* Oxford: Department of Zoology, University of Oxford.

Rapoport, E. H. (1975). *Areografía: Estrategias geográficas de las especies.* Mexico, D.F.: Fondo de Cultura Económica.

Real, R., Vargas, J. M. & Guerrero, J. C. (1992). Análisis biogeográfico de clasificación de áreas y especies. *Monografías Herpetológicas, 2*, 73-84.

Reig, O. A. (1962). Las interacciones cenogenéticas en el desarrollo de la fauna de vertebrados tetrápodos de América del Sur. *Ameghiniana, 1*, 131-140.

Reig, O. A. (1981). *Teoría del origen y desarrollo de la fauna de mamíferos de América del Sur.* Mar del Plata: Museo Municipal de Ciencias Naturales Lorenzo Scaglia.

Ribichich, A. M. (2005). From null community to non-randomly structured actual plant assemblages: Parsimony analysis of species co-ocurrences. *Ecography, 28*, 88-98.

Riddle, B. R. & Hafner, D. J. (2004). The past and future roles of phylogeography in historical biogeography. In Lomolino, M. V., Heaney, L. R., editors, *Frontiers of biogeography: New directions in the geography of nature.* Sunderland, Massachusetts: Sinauer Associates Inc.; pp. 93-110.

Riddle, B. R. & Hafner, D. J. (2006). A step-wise approach to integrating phylogeographic and phylogenetic biogeographic perspectives on the history of a core North American warm deserts biota. *Journal of Arid Environments, 66,* 435-461.

Riddle, B. R., Hafner, D. J & Alexander, L. F. (2000). Phylogeography and systematics of the *Peromyscus eremicus* species group and the historical biogeography of North American warm regional deserts. *Molecular Phylogenetics and Evolution, 17(2),* 145-160.

Rieppel, O. (1991). Things, taxa and relationships. *Cladistics, 7,* 93-100.

Rieppel, O. (2004). The language of systematics, and the philosophy of 'total evidence'. *Systematics and Biodiversity, 2(1),* 9-19.

Ringuelet, R. A. (1961). Rasgos fundamentales de la zoogeografía de la Argentina. *Physis* (Buenos Aires), *22,* 151-170.

Roig-Juñent, S., Crisci, J. V., Posadas, P. & Lagos, S. (2002). Áreas de distribución y endemismo en zonas continentals. In Costa, C., Vanin, S. A., Lobo, J. M., Meliá, A., editors. *Proyecto de Red Iberoamericana de Biogeografía y Entomología Sistemática PrIBES 2002, Monografías Tercer Milenio, vol. 2.* Zaragoza: Sociedad Entomológica Aragonesa; pp. 247-266.

Rojas Parra, C. A. (2007). Una herramienta automatizada para realizar análisis panbiogeográficos. *Biogeografía, 1,* 31-33.

Ron, S. R. (2000). Biogeographic area relationships of lowland Neotropical rainforest based on raw distributions of vertebrate groups. *Biological Journal of the Linnean Society, 71,* 379-402.

Ronquist, F. (1996). *DIVA, version 1.0: Computer program for MacOS and Win32.* Available at www.systbot.uu.se/personel/f.ronquist.html.

Ronquist, F. (1997a). Dispersal-vicariance analysis: A new approach to the quantification of historical biogeography. *Systematic Biology, 46(1),* 195-203.

Ronquist, F. (1997b). Phylogenetic approaches in coevolution and biogeography. *Zoologica Scripta, 26,* 313-322.

Ronquist, F. (1998). Dispersal-vicariance analysis: A new approach to the quantification of historical biogeography. *Cladistics, 14,* 167-172.

Ronquist, F. (2002). *TreeFitter, version 1.3.* Available at http://morphbank. ebc.uu. se/TreeFitter.

Ronquist, F. & Nylin, S. (1990). Process and pattern in the evolution of species associations. *Systematic Zoology, 39,* 323-344.

Rosen, B. R. (1985). Long-term geographical controls on regional diversity. *Journal of the Open University Geological Society, 6,* 25-30.

Rosen, B. R. (1988a). From fossils to earth history: Applied historical biogeography. In Myers, A. A., Giller, P. S., editors. *Analytical biogeography: An integrated approach to the study of animal and plant distributions.* London and New York: Chapman and Hall; pp. 437-481.

Rosen, B. R. (1988b). Biogeographic patterns: A perceptual overview. In Myers, A. A., Giller, P. S., editors. *Analytical biogeography: An integrated approach to the study of animal and plant distributions.* London and New York: Chapman and Hall; pp. 23-55.

Rosen, B. R. (1988c). Progress, problems and patterns in the biogeography of reef corals and other tropical marine organisms. *Helgolländer Meeresunters, 42,* 269-301.

Rosen, B. R. & Smith, A. B. (1988). Tectonics from fossils?: Analysis of reef-coral and sea-urchin distributions from late Cretaceous to Recent, using a new method. In Audley-Charles, M. G., Hallam, A., editors. *Gondwana and Tethys*. London: Geological Society Special Publication nr. 37; pp. 275-306.

Rosen, D. E. (1978). Vicariant patterns and historical explanation in biogeography. *Systematic Zoology, 27*, 159-188.

Rosen, D. E. (1981). Introduction. In Nelson, G., Rosen, E. E., editors. *Vicariance biogeography: A critique*. New York: Columbia University Press; pp. 1-5.

Rosen, D. E. & Nelson, G. (eds.). (1980). *Vicariance biogeography: A critique.* New York: Columbia University Press.

Ruggiero, A. & Ezcurra, C. (2003). Regiones y transiciones biogeográficas: Complementariedad de los análisis en biogeografía histórica y ecológica. In Morrone, J. J., Llorente, J., editors. *Una perspectiva latinoamericana de la biogeografía*. Mexico, D.F.: Las Prensas de Ciencias, UNAM; pp. 141-154.

Rutschmann, F. (2006). Molecular dating of phylogenetic trees: A brief review of current methods that estimate divergence times. *Diversity and Distributions, 12*, 35-48.

Sanderson, M. J. (1998). Estimating rate and time in molecular phylogenies: Beyond the molecular clock? In Soltis, D. E., Soltis, P. S., Doyle, J. J., editors. *Molecular systematics of plants II: DNA sequencing.* Boston, Dordrecht and London: Kluwer Academic Publishers; pp. 242-264.

Sanderson, M. J. (2002). Estimating absolute rates of molecular evolution and divergence times: A penalized likelihood approach. *Molecular Biology and Evolution, 19(1),* 101-109.

Sanderson, M. J. (2003). R8s: Inferring absolute rates of molecular evolution and divergence times in the absence of a molecular clock. *Bioinformatics, 19*, 301-302.

Salisbury, B. A. (1999). *SECANT: Strongest Evidence Compatibility Analysis Tool. Version 2.2*. New Haven: Department of Ecology and Evolutionary Biology, Yale University.

Sanmartín, I. & Ronquist, F. (2002). New solutions to old problems: Widespread taxa, redundant distributions and missing areas in event-based biogeography. *Animal Biodiversity and Conservation, 25,* 75-93.

Sanmartín, I. & Ronquist, F. (2004). Southern Hemisphere biogeography inferred by event-based models: Plant versus animal patterns. *Systematic Biology, 53(2),* 216-243.

Santos, C. M. D. (2005). Parsimony analysis of endemicity: Time for an epitaph? *Journal of Biogeography, 32,* 1284-1286.

Santos, C. M. D. (2007). On basal clades and ancestral areas. *Journal of Biogeography, 34*, 1470-1471.

Savage, J. M. (1982). The enigma of the Central American herpetofauna: Dispersals or vicariance? *Annals of the Missouri Botanical Garden, 69*, 464-547.

Sclater, P. L. (1858). On the general geographical distribution of the members of the class Aves. *Journal of the Linnean Society, Zoology, 2*, 130-145.

Scotland, R. W. (2000). Taxic homology and the three-taxon statement analysis. *Systematic Biology, 49,* 480-500.

Shields, O. (1996). Plate tectonics or an expanding Earth? *Journal of the Geological Society of India, 47,* 399-408.

Siddall, M. E. (2005). Bracing for another decade of deception: The promise of secondary Brooks parsimony analysis. *Cladistics, 21*, 90-99.

Siddall, M. E. & Perkins, S. L. (2003). Brooks parsimony analysis: A valiant failure. *Cladistics, 19*, 554-564.

Siebert, D. J. & Williams, D. M. (1998). Recycled. *Cladistics, 14*, 339-347.

Simpson, G. G. (1953). *Evolution and geography: An essay on historical biogeography with special reference to mammals.* Eugene: Condon Lecture Series, Oregon State System of Higher Education.

Soest, R. W. M. Van. (1996). Recoding widespread distributions for general area cladogram construction. *Vie et Milieu, 46*, 155-161.

Soest, R. W. M. Van & Hajdu, E. (1997). Marine area relationships from twenty sponge phylogenies: A comparison of methods and coding strategies. *Cladistics, 13(1-2)*, 1-20.

Soltis, D. E., Soltis, P. S. & Milligan, B. G. (1992). Intraspecific chloroplast DNA variation: Systematic and phylogenetic implications. In Soltis, P. S., Soltis, D. E., Doyle, J. J., editors. *Molecular systematics of plants.* New York: Chapman and Hall; pp. 117-150.

Swenson, U., Backlund, A., McLoughlin, S. & Hill, R. S. (2001). *Nothofagus* biogeography revisited with special emphasis on the enigmatic distribution of subgenus *Brassospora* in New Caledonia. *Cladistics, 17(1)*, 28-47.

Swofford, D. L. (2003). *PAUP*: Phylogenetic Analysis Using Parsimony (*and other methods). Version 4.* Sunderland, Massachusetts: Sinauer Associates. Available at http://paup.csit.fsu.edu/.

Szumik, C. A., Casagranda, D. & Roig-Juñent, S. (2006). Manual de NDM/VNDM: Programas para la identificación de areas de endemismo. *Instituto Argentino de Estudios Filogenéticos, 5(3)*, 1-26.

Szumik, C. A., Cuezzo, F., Goloboff, P. A. & Chalup, A. E. (2002). An optimality criterion to determine areas of endemism. *Systematic Biology, 51(5)*, 806-816.

Szumik, C. A. & Goloboff, P. (2004). Areas of endemism: An improved optimality criterion. *Systematic Biology, 53(6)*, 968-977.

Szumik, C. & Roig-Juñent, S. (2005). Criterio de optimación para áreas de endemismo: El caso de América del Sur austral. In Llorente Bousquets, J., Morrone, J. J., editors. *Regionalización biogeográfica en Iberoamérica y tópicos afines: Primeras Jornadas Biogeográficas de la Red Iberoamericana de Biogeografía y Entomología Sistemática (RIBES XII.I-CYTED).* Mexico, D.F.: Las Prensas de Ciencias, UNAM; pp. 495-508.

Taberlet, P., Fumagalli, L., Wust-Saucey, A. G. & Cosson, J. F. (1998). Comparative phylogeography and postglacial colonization routes in Europe. *Molecular Ecology, 7*, 453-464.

Takhtajan, A. (1969). *Flowering plants: Origin and dispersal.* Edimburgh: Oliver and Boyd.

Templeton, A. R. (1998). Nested clade analysis of phylogenetic data: Testing hypotheses about gene flow and population biology. *Molecular Ecology, 7*, 381-397.

Templeton, A. R. (2004). Statistical phylogeography: Methods of evaluating and minimizing inference errors. *Molecular Ecology, 13(4)*, 789-809.

Templeton, A. R., Boerwinkle, E. & Sing, C. F. (1987). A cladistic analysis of phenotypic associations with haplotypes inferred from restriction endenuclease mapping. I. Basic theory and an analysis of alcohol dehydrogenase activity in drosophila. *Genetics, 117*, 343-351.

Templeton, A. R., Crandall, K. A. & Sing, C. F. (1992). A cladistic analysis of phenotypic associations with haplotypes inferred from restriction endenuclease mapping and DNA sequence data. III. Cladogram estimation. *Genetics, 132(2)*, 619-633.

Templeton, A. R., Routman, E. & Phillips, C. A. (1995). Separating population structure from population history: A cladistics analysis of the geographical dfistribution of mitochondrial DNA haplotypes in the tiger salamander, *Ambystoma tigrinum*. *Genetics 140*, 767-782.

Thorne, J. L. & Kishino, H. (2002). Divergence time and evolutionary rate estimation with multilocus data. *Systematic Biology, 51*, 689-702.

Trejo-Torres, J. C. (2003). Biogeografía ecológica de las Antillas: Ejemplos de las orquídeas y las selvas cársticas. In Morrone, J. J., Llorente Bousquets, J., editors, *Una perspectiva latinoamericana de la biogeografía*. Mexico, D.F.: Las Prensas de Ciencias, UNAM; pp. 199-208.

Trejo-Torres, J. C. & Ackerman, J. D. (2001). Biogeography of the Antilles based on a parsimony analysis of orchid distributions. *Journal of Biogeography, 28*, 775-794.

Trejo-Torres, J. C. & Ackerman, J. D. (2002). Composition patterns of Caribbean limestone forests: Are parsimony, classification, and ordination analyses congruent? *Biotropica, 34*, 502-515.

Udvardy, M. D. F. (1969). *Dynamic biogeography*. New York: Van Nostrand.

Upchurch, P. & Hunn, C. A. (2002). "Time": The neglected dimension in cladistic biogeography? In Monegatti, P., Cecca, F., Raffi, S., editors. *International Conference Paleobiogeography and Paleoecology 2001, Piacenza and Castell'Arquato 2001, Geobios* 35 (*mémoire spéciale* 24), pp. 277-286.

Upchurch, P., Hunn, C. A. & Norman, D. B. (2002). An analysis of dinosaurian biogeography: Evidence for the existence of vicariance and dispersal patterns caused by geological events. *Proceedings of the Royal Society of London, Series, B 269*, 613-621.

Veller, M. G. P. Van. (2004). Methods for historical biogeographical analyses: Anything goes? *Journal of Biogeography, 31*, 1552-1553.

Veller, M. G. P. Van & Brooks, D. R. (2001). When simplicity is not parsimonious: *A priori* and *a posteriori* methods in historical biogeography. *Journal of Biogeography, 28*, 1-11.

Veller, M. G. P. Van, Brooks, D. R. & Zandee, M. (2003). Cladistic and phylogenetic biogeography: The art and the science of discovery. *Journal of Biogeography, 30*, 319-329.

Veller, M. G. P. Van, Kornet, D. J. & Zandee, M. (2000). Methods in vicariance biogeography: Assessment of the implementations of assumptions 0, 1, and 2. *Cladistics, 16*, 319-345.

Veller, M. G. P. Van, Zandee, M. & Kornet, D. J. (1999). Two requirements for obtaining valid common patterns under different assumptions in vicariance biogeography. *Cladistics, 15*, 393-406.

Veller, M. G. P. Van, Zandee, M. & Kornet, D. J. (2001). Measures for obtaining inclusive sets of area cladograms under assumptions zero, 1, and 2 with different methods for vicariance biogeography. *Cladistics, 17*, 248-259.

Viloria, Á. 2005. Las mariposas (Lepidoptera: Papilionoidea) y la regionalización biogeográfica de Venezuela. In Llorente Bousquets, J., Morrone, J. J., editors. *Regionalización biogeográfica en Iberoamérica y tópicos afines: Primeras Jornadas Biogeográficas de la Red Iberoamericana de Biogeografía y Entomología Sistemática (RIBES XII.I-CYTED)*. Mexico, D.F.: Las Prensas de Ciencias, UNAM; pp. 441-459.

Wallace, A. R. (1876). *The geographical distribution of animals, with a study of the relations of living and ex tinct faunas as elucidating the past changes of the Earth's surface.* London: Macmillan and Company.

Wegener, A. (1912). Die Entstehung der Kontinente. *Geologische Rundschau, 3,* 276-292.

Wegener, A. (1929). *The origin of continents and oceans.* Dover: Dover Publications.

Weisrock, D. W. & Janzen, F. J. (2000). Comparative molecular phylogeography of North American softshell turtles (*Apalone*): Implications for regional and wide-scale historical evolutionary forces. *Molecular Phylogenetics and Evolution, 14(1),* 152-164.

Welzen, P. C. Van. (1992). Interpretation of historical biogeographic results. *Acta Bot. Neerl., 41,* 75-87.

Welzen, P. C. Van, Turner, H. & Hovenkamp, P. (2003). Historical biogeography of Southeast Asia and the West Pacific, or the generality of unrooted area networks as historical biogeographic hypotheses. *Journal of Biogeography, 30,* 181-192.

Welzen, P. C. Van, Turner, H. & Roos, M. C. (2001). New Guinea: A correlation between accreting areas and dispersing Sapindaceae. *Cladistics, 17,* 242-247.

Westermann, G. E. C. (2000). Biochore classification and nomenclature in paleobiogeography: An attempt at order. *Palaeogeography, Palaeoclimatology, Palaeoecology, 163,* 49-68.

Wiley, E. O. (1980). Phylogenetic systematics and vicariance biogeography. *Systematic Botany, 5,* 194-20.

Wiley, E. O. (1981). *Phylogenetics: The theory and practice of phylogenetic systematics.* New York: Wiley-Interscience.

Wiley, E. O. (1987). Methods in vicariance biogeography. In Hovenkamp, P., Gittenberger, E., Hennipman, E., de Jong, R., Roos, M. C., Sluys, R., Zandee, *M.,* editors. *Systematics and evolution: A matter of diversity.* Utrecht: Institute of Systematic Botany, Utrecht University; pp. 283-306.

Wiley, E. O. (1988a). Parsimony analysis and vicariance biogeography. *Systematic Zoology, 37,* 271-290.

Wiley, E. O. (1988b). Vicariance biogeography. *Annual Review of Ecology and Systematics, 19,* 513-542.

Williams, D. M. (2004). Homologues and homology, phenetics and cladistics: 150 years of progress. In Williams, D. M., Forey, P. L., editors. *Milestones in systematics.* Boca Raton: The Systematics Association Special Volume Series 67, CRC Press; pp. 191-224.

Williams, D. M. & Ebach, M. C. (2004). The reform of palaeontology and the rise of biogeography - 25 years after 'ontogeny, phylogeny, paleontology and the biogenetic law' (Nelson, 1978). *Journal of Biogeography, 31,* 685-712.

Williams, D. M. & Humphries, C. J. (2003). Component coding, three-item coding, and consensus methods. *Systematic Biology, 52(2),* 255-259.

Wojcicki, M. & Brooks, D. R. (2004). Escaping the matrix: A new algorithm for phylogenetic comparative studies of coevolution. *Cladistics, 20,* 341-361.

Wojcicki, M. & Brooks, D. R. (2005). PACT: An efficient and powerful algorithm for generating area cladograms. *Journal of Biogeography, 32,* 755-774.

Yang, Z. (1997). PAML: A program package for phylogenetic analysis by maximum likelihood. *Computer Applications in the Biosciences, 13,* 555-556.

Young, G. C. (1995). Application of cladistics to terrane history - Parsimony analysis of qualitative geological data. *Journal of Southeast Asian Earth Sciences, 11,* 167-176.

Zandee, M. & Roos, M. C. (1987). Component-compatibility in historical biogeography. *Cladistics, 3(4),* 305-332.

Zink, R. M. (1996). Comparative phylogeography in North American birds. *Evolution, 50,* 308-317.

Zink, R. M. (2002). Methods in comparative phylogeography, and their application to studying evolution in the North American aridlands. *Integrative and Comparative Biology, 42,* 953-959.

Zuckerland, E. & Pauling, L. (1965). Molecular disease, evolution and genetic heterogeneity. In Kasha, M., Pullman, B., editors. *Horizons in biochemistry.* London and New York: Academic Press; pp. 189-225.

Zunino, M. & Zullini, A. (2003). *Biogeografía: La dimensión espacial de la evolución.* Mexico, D.F.: Fondo de Cultura Económica.

In: Biogeography
Editors: M. Gailis, S. Kalniš, pp. 63-105

ISBN: 978-1-60741-494-0
© 2010 Nova Science Publishers, Inc.

Chapter 2

CONNECTING BIOGEOGRAPHIC PATTERNS WITH ENERGY AND RESOURCE AVAILABILITY IN MARINE SYSTEMS

Rui Rosa

Laboratório Marítimo da Guia, Centro de Oceanografia, Faculdade de Ciências da
Universidade de Lisboa, Av. Nossa Senhora do Cabo, 939, 2750-374 Cascais, Portugal

ABSTRACT

Three of the most recognized biogeographic paradigms on earth are the latitudinal gradients of species richness (the decreased richness in biological diversity from equatorial to polar regions), Bergmann's rule (the increase of body size in cold climates) and the unimodal pattern of diversity with depth (species richness peaking at intermediate depths). Understanding the causes of such patterns is still one of greatest contemporary challenges for biogeographers and ecologists. Moreover, though the generality and causal predictors of the first two paradigms have been fully debated in the terrestrial biome, their relevance in marine systems is still poorly understood. In this chapter, I review the present knowledge of these broad-scale biogeographic patterns in the marine systems and use cephalopod molluscs as a case study. Latitudinal gradients of species richness are present in coastal cephalopod fauna and both climate and area extent predict much of the diversity variation. Yet, in the open ocean, diversity does not decline monotonically with latitude and is positively correlated to the availability of oceanic resources. Therefore, a much stronger linkage between patterns of cephalopod diversity and bottom-up processes is found in the pelagic ecosystem. Additionally, cephalopod diversity does not show the classical hump-shaped response to depth. It declines sharply from sub-litoral and epipelagic zones to the slope and bathypelagic habitats and then steadily to abyssal depths. This suggests that higher thermal energy availability and productivity in shallow habitats promote diversification rates, and rejects hypotheses such as biome area, environmental stability and mid-domain effect. Climate also seems to play the most important role in structuring the latitudinal distribution of body size in these marine ectotherms. This evidence holds up to the concept of the "temperature-size rule" but does not support hypotheses for Bergmann's rule relating to resource availability, seasonality (or fasting endurance) and competition. Thus, it is evident that species-area-energy theories that have been formulated for the terrestrial biosphere also apply to the marine systems. In fact, although the phyletic composition and life-history characteristics of terrestrial and marine fauna are quite distinct, there is no reason to assume that the causal predictors behind these widespread biogeographic patterns will differ between these two realms.

1. INTRODUCTION

The oldest (from the late eighteenth century) and the most recognized ecological pattern in our planet is the increased richness in biological diversity from polar to equatorial regions (Rohde, 1992; Brown & Lomolino, 1998; Hawkins, 2001), but understanding the causal predictors it is still one of greatest contemporary challenges for ecologists. This paradigm have been scrutinize to a greater extent in terrestrial biosphere over the last decades and several factors have been hypothesized to explain it, such as competition and predation (Pianka, 1966), spatial heterogeneity (MacArthur, 1972), "Rapoport's rule" (Stevens, 1989), environmental stability (Sanders, 1968), ambient energy (Turner et al., 1987), productivity (Wright, 1983), biome area (Terborgh, 1973; Rosenzweig, 1995), evolutionary time (Rohde 1992), energetic-equivalents (Allen et al., 2002), Milankovitch oscillations (Dynesius & Jansson, 2000), geometric constraints (Colwell & Lees, 2000), among other possible causes (for reviews see Pianka, 1966; Rohde, 1992; Rosenzweig, 1995; Whittaker et al., 2003; Willig et al., 2003). Though there have been few efforts to reduce the number of explanations, the best documented and major contenders for a short list of explanations are the climate-related (see Currie et al., 2004) and productivity hypotheses (Gaston, 2000; Hawkins et al., 2003; Mittelbach et al., 2007).

Several climate-based hypotheses have been formulated, including the "physiological tolerance hypothesis", which states that diversity varies according to the species' tolerance to different climatic conditions (Currie et al., 2004) and the "evolutionary rates hypothesis" - high energy-areas accelerate rates of evolution and speciation by shortening generation times and increasing mutation and physiological rates (Rohde, 1992; Allen et al., 2002; Gillooly et al., 2005; Mittelbach et al., 2007), among others. On the other hand, the classic unimodal relationship of maximal diversity at intermediate levels of productivity was, until recently, the most widely documented pattern (Huston, 1979; Tilman, 1982; Rosenzweig, 1995). The available evidence now shows that, this hump shaped relationship is one of many patterns (including U-shaped, positive, and negative) and that none of them predominate (Waide et al., 1999; Mittelbach et al., 2001). Furthermore, patterns are know to change under the influence of spatial scale (Chase & Leibold, 2002), history of community assembly (Fukami & Morin, 2003), disturbance (Kondoh, 2001) and consumers (Worm et al., 2002), among others. The most cited ecological explanation for the positive relationship between productivity and diversity is the "more individuals hypothesis" (Wright, 1983; Srivastava & Lawton, 1998), that suggests that more productive areas can support larger populations, which lower extinction rates and promotes speciation. The way that these mechanisms may explain the broad-scale diversity patterns in the marine biosphere is unclear.

Besides the study of the latitudinal gradients of species richness (LGSR), a considerable effort has been devoted to the study of spatial distributions of body size, especially across large environmental gradients such as those associated with latitude, to understand the organization of ecological communities. Bergmann (1847) was the first to propose an ecogeographic "rule" that stated that smaller endotherms should, in general, abound in warmer areas while larger-bodied species would inhabit in colder climates. Behind his idea was the fact that reduced surface to volume ratio facilitates greater heat conservation and thus allow a lower mass-specific metabolic rate, whereas larger ratios (in smaller individuals) facilitate heat loss.

While Bergmann considered the effect to be interspecific (between closely related species), Rensch (1938) and Mayer (1956) argued that his rule was an entirely intraspecific phenomenon. Since intra- and interspecific gradients in body size may represent different phenomena (Chown & Gaston, 1999), the tendency for geographical size variation within species was formalized as "neo-Bergmannian's rule" by James (1970) and, more recently, as "James's rule" by Blackburn et al. (1999). The latter authors advocated that interspecific approaches (in a monophyletic higher taxon) across a range of latitudes have the best chance of detecting the true pattern. Since the size decrease is expected to result from higher-taxon turnover, phylogenetically restricted data sets can miss Bergmann's pattern. Nonetheless, Bergmann's rule has been tested at various levels, namely within species (e.g. Ashton, 2002a; Yom-Tov & Yom-Tov, 2005; Measey & Van Dongen, 2006), between species within genera (e.g. Taylor & Gotelli, 1994), and between species within a range of higher taxa (e.g. Blackburn & Gaston, 1996; Olalla-Tarraga et al., 2006; Rodríguez et al., 2008). Although the evidence for Bergmann's rule is questionable (McNab, 1971; Mousseau, 1997), the pattern has been demonstrated in taxonomically widespread groups, including mammals (Blackburn & Hawkins, 2004; Rodríguez et al., 2008), birds (Blackburn & Gaston, 1996; Ashton, 2002b; Ramirez et al., 2008), amphibians (Ashton, 2002a; Olalla-Tarraga & Rodriguez, 2007) and reptiles (Ashton & Feldman, 2003; Angilletta et al., 2004; Cruz et al., 2005). Nonetheless, while the pattern holds within the majority of the endotherms, the pattern seems to be more complex in ectotherm species (Atkinson, 1994; Mousseau, 1997; Ashton & Feldman, 2003; Olalla-Tarraga & Rodriguez, 2007).

The mechanisms that have been currently proposed to explain the relationship between latitude and body size have been largely focused on physiological and ecological processes. Besides the climate-based "heat conservation hypothesis" for endotherms and "temperature-size rule" for ectotherms (Atkinson, 1994), that predicts that ambient energy (temperature) is the best environmental explanatory variable, the resource availability (primary productivity) hypothesis assumes that body mass must be maintained by a sufficient food supply and predicts greater body sizes in more productive areas (Rosenzweig, 1968). The seasonality (starvation resistance) hypothesis proposes that large-bodied species are favored in colder, more seasonal environments because they metabolize fat stores at lower weight-specific rates than smaller species (Lindstedt & Boyce, 1985). Although more questionable, a potential biotic factor that may influence Bergmann's rule is competition. Greater body sizes can be attained farther from the equator due to reduced competition (Ashton et al., 2000). Moreover, in marine systems, variation in oxygen availability has also been advocated to explain polar gigantism (Chapelle & Peck, 1999) and size increase in the deep-sea (McClain & Rex, 2001) (but also see Spicer & Gaston, 1999).

Despite the fact that geographic size variations relative to environmental factors have been well studied in the terrestrial biota for at least 150 years, only few large-scale latitudinal surveys on interspecific variation of body size were conducted in marine ecosystems. Other assumed paradigm in the oceans is the parabolic pattern of diversity with depth, i.e., it peaks at intermediate depths (e.g. Sanders, 1968; Rex, 1973; Pineda & Caswell, 1998). Both in pelagic (Angel, 1993) and benthic communities (especially of soft-sediments), this unimodal trend is attributed to an eventual diversifying effect of the more stable nutrient input at intermediate depths, compared to scarcity of food resources in the abyss and the increased seasonality at shallower depths (Rex et al., 2005b). Several other causal factors are thought to shape local deep-sea diversity, such as sediment heterogeneity, oxygen availability, deep-sea currents and catastrophic physical disturbance (see Levin et al., 2001). The mid-domain effect (also known as "boundary constraints model") has also been advocated, i.e., an increased tendency of species ranges to overlap towards the mid-point if there is a random placement of ranges within a spatial domain (Colwell & Lees, 2000). The depth-related patterns are not well understood because the datasets available are too scarce to make generalizations and do not clearly show whether the bathyal depths have higher and lower diversity than coastal systems (Gray, 2001). In this chapter, I review the present knowledge of these broad-scale biogeographic patterns in the marine systems and use cephalopod molluscs as a case study.

2. LATITUDINAL GRADIENTS OF SPECIES RICHNESS IN MARINE SYSTEMS

2.1. Neritic Zone

While several studies with marine groups have described the negative association between diversity and latitude (Valentine, 1966; Macpherson & Duarte, 1994; Roy et al., 1994; Stevens, 1996; Roy et al., 1998; Culver & Buzas, 2000; Rex et al., 2000; Roy et al., 2000; Macpherson, 2002; Astorga et al., 2003; Smith & Gaines, 2003; Rex et al., 2005a), others have failed (Warwick & Ruswahyuni, 1987; Clarke, 1992; Kendall & Aschan, 1993; Dauvin et al., 1994; Lambshead et al., 2000; Ellingsen & Gray, 2002; Mokievsky & Azovsky, 2002). Yet, to understand the contemporary LGSR in the marine systems it is important not only to relate them to contemporary environmental gradients but also to evolutionary processes that have occurred throughout geologic time.

2.1.1. Historical biogeography

Regarding cephalopod biogeography, for instance, the extinction of cuttlefish (sepiids) in western Atlantic (WA) is thought to have occurred in the Caribbean province as a consequence of the sea surface cooling in the WA equatorial region during the Eocene to Oligocene transition (Emiliani, 1966). This hypothesis is corroborated by the exclusive presence of sepiids in post-Eocene strata of the Old World (Khromov, 1998). Another key historical event on cephalopod biogeography occurred around 5 mya, when the North and South American, and Caribbean Plates converged, and the rise of the Central American Isthmus restricted water exchange between the Atlantic and Pacific Oceans (Ibaraki, 1997; Haug & Tiedemann, 1998). The eventual closure of the Central American Seaway, around 1.9 mya, may have led to the formation of geographically isolated octopod populations by vicariance and then to allopatric speciation in the tropical WA (Voight, 1988), where higher diversity is presently found (Figure 1).

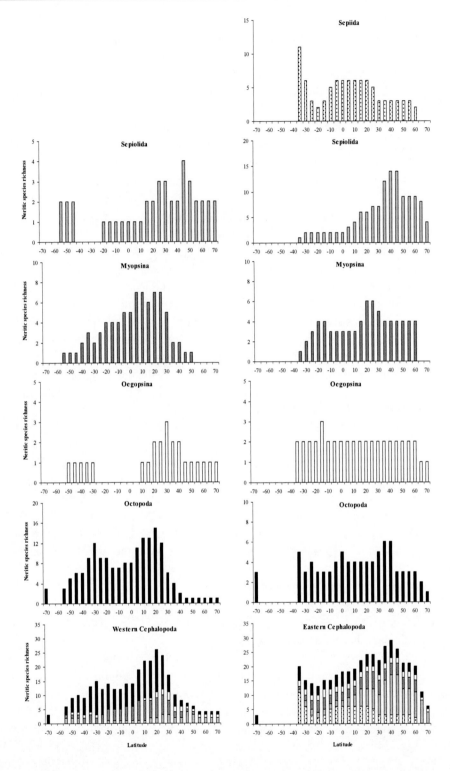

Figure 1. Latitudinal diversity gradient of neritic cephalopods on the Western (left panels) and Eastern (right panels) Atlantic.

In fact, many faunal turnover events (i.e. evolutionary appearance and disappearance of a large proportion of a biota) took place in Late Pliocene in the WA due to a decline in nutrients (and so in marine productivity) associated with Central American Seaway closure and change of ocean circulation (Allmon, 2001). Yet, the emergence of Central American Isthmus also set the stage for glaciation in the Northern Hemisphere at 2.7 million years ago (Haug et al., 1999). In fact, one of the most dominant features of the Earth's past climate are the ice ages in the Quaternary Period, which involve repeated global cooling and expansion of the continental and polar ice sheets and mountain glaciers. These cycles are a consequence of regular orbital eccentricity of the Earth around the sun (Croll-Milankovitch theory), which causes major changes in solar output (Hays et al., 1976). Average temperatures seem to have changed by 7-15 °C over 5-10 years and lasted for periods of hundred years (Hewitt, 1999, 2004) and the sea levels fluctuated dramatically on both global and regional scales, strongly affecting the distributions and diversity of biotas (Brown & Lomolino, 1998). These ice age events are believed to have caused mass species extinctions in the WA as far south as Florida and Bahamas and confined most species in the Caribbean region (Stanley, 1986). During the most recent glacial maximum (LGM) in North America, the ice sheets extended to about 42 degrees north latitude (Riggs et al., 1996). Under this faunal turnover scenario, the shallow-living cephalopod species may also have been subject to localized extinction by Pleistocene glaciation or forced from their typical habitat to refuge beyond the southern-most extent of the glaciers at Long Island Sound. The extinction or geographical range contraction seem coherent with the lower diversity in WA from 35°N towards the pole (Figure 1), and contrasts with the EA where the ice sheets appear to have extended south to 52°N in LGM and caused fewer extinctions and displacements (Vermeij, 1991). The abrupt decrease in western diversity is also noticeably different from the south hemisphere gradient and contradicts Hillebrand's concept of common gradient strength and slope between hemispheres (Hillebrand, 2004a, 2004b). In the EA, the major historical event on the neritic diversity was closure on the Mediterranean Sea during the "Messinian salinity crisis" into an isolated hyperhaline lake about 5.5 mya ("Mare Lago", Krijgsman et al., 1999), with the consequent isolation, extinction of stenohaline species and arise of endemisms (Taviani, 2002; Nesis, 2003). Therefore, it becomes evident that historical geological and climatic events influence the contemporary latitudinal gradients of cephalopod diversity in the Atlantic Ocean.

2.1.2. Species turnovers and geographical boundaries

It is well known that species' range limits are not uniformly distributed in space and cluster at particular shoreline locations (Valentine, 1966). These boundaries are often associated with currents or other hydrographic features and are usually explained by the effect of pronounced physical discontinuities (such as temperature or salininity) on the organism's physiology, and by the hydrodynamic impact on recruitment success (Gaylord & Gaines, 2000). In the north Western Atlantic (WA), the major biogeographic break of cephalopod species occurs along the coast of Florida (Figure 2), where also many other marine organisms show either species-range endpoints or sharp genetic discontinuities (Fischer, 1960; Avise, 1992; Schizas et al., 1999). This concordance is not clear because the wide variety of organisms that show genetic breaks on this area present many different life-histories traits (Avise, 1992). There is no clear evidence of hydrographic factors influencing larval recruitment in southeastern Florida (Collin, 2001), but since a noticeable SST gradient occurs at this latitude, namely between North and South Sargasso Sea, one can advocate that this is primarily a thermal biogeographic boundary.

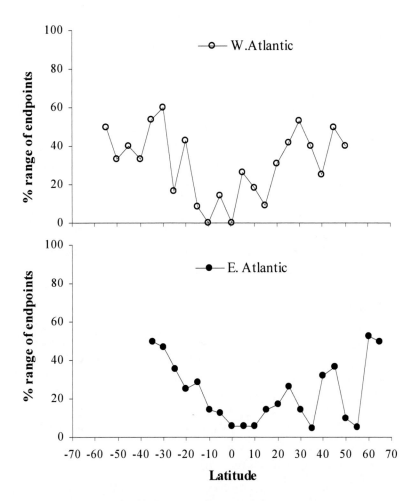

Figure 2. Latitudinal distribution of the range endpoints of neritic cephalopods on the western and eastern Atlantic. Each value represents the ratio (%) between number of species ranges that end at that latitude and the total number of species present at that latitude bin.

Other key endpoint on cephalopod biogeography is located in Cape Hatteras, where the Gulf Stream northwall turns eastwards under the influence of southward flow of the Labrador Current. According to Briggs (1974), this water mass boundary separates the warm-temperate fauna of the south from the cold-temperate fauna of the north. Though this endpoint is a thermal barrier, it is usually explained by the deleterious influence of Gulf Stream oceanographic features on dispersal and recruitment of planktonic larvae (Fischer, 1960; Quattrini et al., 2005).

Many benthic and nektobenthic cephalopods have planktonic paralarvae (term associated with the lack of metamorphosis during post-hatching development, Boletzky, 1974) and, therefore, their distributional ranges and population structures are also highly affected by these current systems. Although many of them are fully capable of living far north physiologically, they seem to be prevented from inhabiting there by these oceanographic constraints. The northern major biogeographic boundary is observed in Nova Scotia shelf (45°N), where the influence of lower arctic SST is well noticed. Similar species–range endpoints and thermal/oceanographic discontinuities are observed throughout the south western Atlantic, namely at latitudes of Cape Frío (analogous to the thermal barrier found in Florida), Patos Lagoon (world's largest chocked lagoon) and Río de la Plata estuary (second largest South American basin). These two locations comprise a large temperate estuarine zone and may constitute a physiologically-induced barrier since cephalopods are unable to osmoregulate and tolerate low salinities, preventing them from invade brackish and estuarine habitats (only with a few loliginid exceptions). Additionally, the coastal circulation on this part of austral South America is dominated by the confluence of Brazil (subtropical) and Malvinas (subantartic) currents, where the two flows turn offshore in a series of large amplitude meanders. A quite distinct scenario is observed in the eastern Atlantic, where the major end-points are situated at the highest latitudes on both hemispheres, in the South Africa coast (35° and 30°S) and north of Norway (>60°N), corresponding to influence of the Benguela current system and the transition between temperate to sub-artic zones, respectively. The absence of major species-turnovers (>40% range of endpoints) along the temperate, sub-tropical and tropical coasts seems to be related with the broader ranges (both latitudinal and bathymetric) of eastern neritic species.

2.1.3. Rapoport effect

The neritic cephalopod diversity is distributed heterogeneously across the Atlantic coasts (Figure1), with the north western tropical and north eastern temperate coastal areas peaking in species richness, while the Artic and Southern Ocean are clearly devoided of biological variation. As already noticed above, large-scale historical events in both Western and Eastern Atlantic have a major influence on the location of the contemporary peaks of diversity. Contrarily to the theory that species' boundaries become more densely packed towards the equator ("Rapoport's latitudinal rule", Stevens, 1989), there was no evidence for a reduction in neritic cephalopod ranges at lower latitudes (Figure 3).

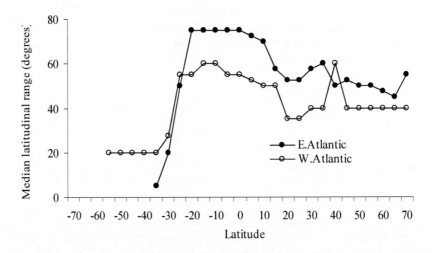

Figure 3. Median latitudinal ranges of Western and Eastern Atlantic neritic cephalopods (based on all species ranges that intersect each particular bin).

Though there is some taxa that exhibit this pattern (e.g., fish, bivalves and mammals, Stevens, 1996; Fortes & Absalao, 2004; Arita et al., 2005), there are also many counterexamples within these same taxa (Rohde et al., 1993; Roy et al., 1994; Rohde, 1996; Roy et al., 1998; Hecnar, 1999; Kerr, 1999; Macpherson, 2003). When Stevens (1989) formulated this rule he argued that the exceptions occurred in species that also did not show LGRS (e.g. migratory birds). Also, he stated that Rapoport phenomenon has a climate-based explanation, with tropical organisms having narrower climatic tolerances in opposition to broad tolerance of organism inhabiting high latitudes where seasonal variation is greater (known as "the climatic hypothesis" or "the seasonal variability hypothesis", Stevens, 1996). However, because tropics are the largest of the world's major bioclimatic zones, it is not surprising to find that cephalopod tropical clades are able to occupy larger areas than lower latitude ones (the largest median latitudinal ranges occur between 20°S and 20°N, Figure 3). Species with planktonic life stages, such as cephalopods, tend to have wider geographical distributions than species lacking it (Vermeij et al., 1990), and these differences in dispersal ability may indeed explain the co-existence of clades with narrow and large latitudinal ranges in the tropics (Rohde, 1998). It has been also argued that the Rapoport's rule (RR) is a local phenomenon that cannot be generalized, because its found with more consistency at high latitudes on the north hemisphere (above approximately 30°-40°N), where the extinction of species adapted to narrow temperature ranges has occurred during the glaciations (Brown, 1995; Rohde, 1996; Gaston et al., 1998; Gaston & Chown, 1999). Interestingly, when we combine the depth component with latitude a strong Rapoport effect was observed in WA. Though the "bathymetric Rapoport's rule" (species increase its median bathymetric range size with increasing latitude) holds true for neritic WA fauna it does not for EA, which point toward the fact that this effect cannot be dismissed but neither be generalized as a rule. The increased depth ranges in north western neritic fauna can allegedly be explained by the conditions of Quaternary glaciations, with the ice sheets preventing the existence of low tolerance, shallow-living stenobathic species and favoring species capable of living in a wide range of depths (eurybathic species). This "differential extinction hypothesis" (see Brown, 1995) is supported by the bathymetric distributions of cephalopod species living in the non-glaciated eastern shelves (where the stenobathic species may have retained the ability to live below the lowered sea-level) but, on the other hand, is refuted by the existence of the exact same Rapoport phenomenon in the south hemisphere, where the effects of glaciation were far less severe. Alternatively, the loss of shallow coral reef habitats to high latitudes has been putted forward to explain the Rapoport effect in shallow-living fish (Smith & Gaines, 2003). The findings corroborate this explanation in two ways. First, the narrow bathymetrical ranges of neritic cephalopods follow closely the latitudinal extent of the shallow Caribean and Brazilian coral faunal regions (isolated due to enormous outflow of Amazon riverine system) with the inflection points coincidental with two major biogeographical boundaries (noticed above). These boundaries definitely have an important restraining impact on coral reef biogeography due to the significant decrease on sea surface temperature (SST, Figure 4) and restricted light penetration associated with water turbidity (specially in the southern hemisphere). Second, the non-occurrence of Rapoport effect in EA, probably associated with the absence of these shallow tropical habitats and the greater influence of upwelling systems. In fact, the dynamics of these coastal ecosystems (e.g. seasonal variations in enrichment and mixing processes; upwelling driven nearshore hypoxia) may explain why the majority of the EA neritic species is not strictly restricted to the continental shelves and overlaps to deeper slope waters (median bathymetric range greater than 400 m along the entire latitudinal gradient, Figure 4).

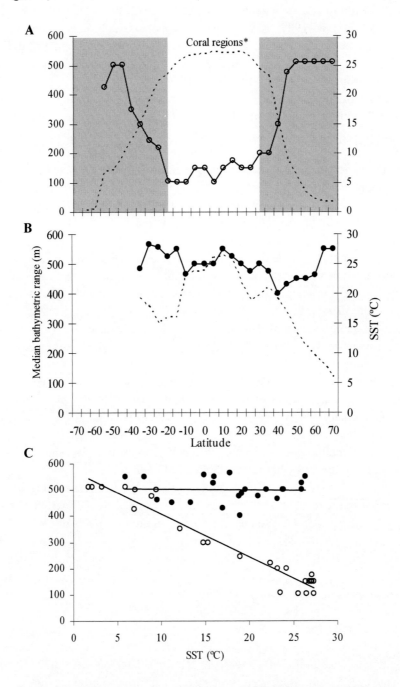

Figure 4. A, B – Latitudinal gradients of sea surface temperature (SST, °C, broken line) and bathymetric ranges (m) of neritic cephalopods in the western (W, open circles) and eastern (E, closed circles) Atlantic Ocean. C – Linear relationships between bathymetric ranges and SST (W: $r^2 = 0.95$; $p<0.0001$; E: $r^2 = 0.00$; $p>0.05$). Since autocorrelation may invalidate correlation estimates, the statistics are only used for comparative purposes. Coral regions* (non-grey zone) - from the Florida to Abrolhos region (Brazil) latitudes.

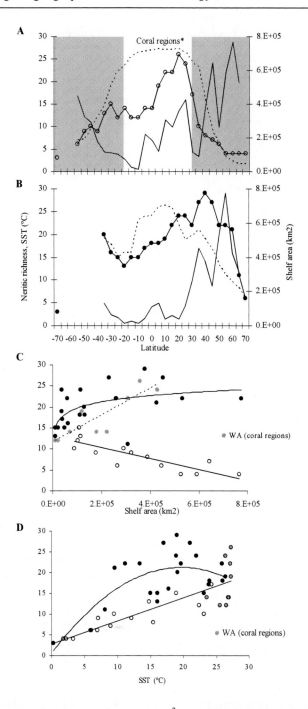

Figure 5. A, B - Latitudinal gradients of shelf area (km^2; solid line), sea surface temperature (SST, ºC, broken line) and neritic species richness in the western (W, open circles) and eastern (E, closed circles) Atlantic Ocean. C – Relationship between diversity and shelf area (W non-coral regions: r^2 = 0.69; p<0.001; W coral regions: r^2 = 0.73; p<0.001; E: r^2 = 0.30; p<0.05). D- Relationship between neritic diversity and SST (all W: r^2 = 0.71; p<0.0001; only W coral regions: r^2 = 0.18; p>0.05; E: r^2 = 0.54; p<0.01). Since autocorrelation may invalidate correlation estimates, the statistics are only used for comparative purposes. Coral regions* (non-grey zone) - from the Florida to Abrolhos region (Brazil) latitudes.

2.1.4. Relationship between coastal diversity, energy availability and shelf area

The increased environmental dynamics and broader ecological tolerance (i.e. non-specialization) of neritic cephalopod eastern fauna seem to be causally linked to the lack of association between bathymetric range and SST (Figure 4C), the weaker and curvilinear species-temperature relationship (Figure 5D), and the lower species turnovers along the EA subtropical and tropical regions (Figure 2). These assumptions are supported by the quite opposite scenario in WA, where SST is the most important environmental predictor for LGSR (Model 1, Table 1), especially in extratropical latitudes (Figure 5D), and for bathymetric ranges (Figure 4C). Western tropical regions contain more species because increased energy availability (and stability) promotes the occurrence of more viable populations of specialized species (e.g. shallow-living reef octopod and loliginid fauna). Though higher levels of solar radiation (and therefore higher temperatures) induce faster evolutionary rates due to faster biochemical processes (Allen et al., 2002; 2006), mutation rates (Gillooly et al., 2005) and shorter generation times ("evolutionary rates hypothesis", Rohde, 1992), eastern tropical coasts do not present higher cephalopod diversities than temperate ones (Figure 1B), which clearly demonstrates that diversity patterns in EA are not solely governed by temperature. In the model 3 (Table 1), shelf area contributed the most to the sum of squares associated with the model. In fact, the EA shelf area increases towards northern temperate latitudes and, consequently, a positively correlation with cephalopod richness is found (Figure 5C). Larger shelf areas should imply higher habitat diversity, higher colonization and speciation rates and lower extinction rates ("geographic area hypothesis", Rosenzweig, 1995). Yet, since the greater Patagonia and Newfoundland-Labrador shelf areas in WA did not favor higher diversification rates compared to the tropical regions, our findings seem at a first glance contradictory in relation to this already controversial concept (see Rohde, 1997; Rosenzweig & Sandlin, 1997; Rohde, 1998). However, by predicting that the complexity and diversity of coral reef regions have a determinant influence on the neritic cephalopod biogeography in WA and by restricting the analysis to their latitudinal extent, we observe that the higher tropical levels of diversity are indeed closely associated to the greater shelf areas in the tropics (grey circles in Figure 5C) and higher coverage of reef habitats. While SST explains the majority of the variance outside the tropics (<25°C), it fails within these regions (grey circles in Figure 5D) where, instead, the shelf area does. By doing this spatial differentiation, area becomes in fact the most important covariate in the modeling of WA diversity (Model 2; Table 1).

Therefore, these findings suggest that cephalopod diversification rates in both Atlantic coasts are greatly influenced by both area and climate. The influence of energy availability on cephalopod diversity is also supported by a positive association with NPP, but the relationship is not as strong as the associations of the other variables. Interestingly, a much stronger linkage between bottom up processes and patterns of cephalopod diversity is found in the open ocean (see below).

2.2. Patterns of Diversity in the Open Ocean

As in the neritic systems, there is no consensus whether the poleward decline in diversity is a pervasive feature in the open-ocean. Again, while some studies demonstrate that tropical regions sustain more pelagic diversity than temperate and polar ones (Bé, 1977; Angel, 1993; McGowan & Walker, 1993; Dodge & Marshall, 1994; Angel, 1997; Macpherson, 2002), others show that species richness is greater at temperate latitudes (McGowan & Walker, 1985; Boltovskoy et al., 1999; Rutherford et al., 1999; Brayard et al., 2005).

Table 1. Models assessing the importance of sea surface temperature (SST), net primary productivity (NPP) and area in predicting cephalopod diversity in Western (Model 1 and 2) and Eastern Atlantic (Model 3 and 4).

	Coefficient	t-ratios	p-value	Total R^2	AIC
Western Atlantic	Model 1 (autocorrelation in model)				
SST	0.5850	6.04*	<0.0001	91.25%	114.89
NPP	0.0111	3.56*	0.0021		
Area	0.000005	2.02	0.0579		
AR(1)	1.0781	6.23*	<0.0001		
AR(2)	-0.6437	-3.73*	0.0014		
	Model 2 (Interaction term in model)				
SST	0.1343	1.28	0.2165	92.68%	106.79
NPP	0.0073	3.31*	0.0035		
Area	-0.000005	-1.39	0.1811		
Interaction Term : Area×Coral	0.000036	6.35*	<0.0001		
Eastern Atlantic	Model 3				
SST	0.40135	3.40*	0.0040	80.33%	102.57
SST^2	-0.05105	-2.61*	0.0196		
NPP	-0.00218	-0.85	0.4087		
NPP^2	0.00001245	2.75*	0.0148		
Log(Area)	2.84189	3.68*	0.0022		
	Model 4				
Latitude	0.10818	3.90*	0.0013	93.59%	96.76
SST	0.51659	4.95*	0.0001		
SST^2	-0.10051	-5.43*	<0.0001		
Log(Area)	1.51164	2.37*	0.0307		

* indicate statistical significance at the 5% level of significance.

2.2.1. Cephalopods as a case of success in the colonization of the open ocean

Pelagic systems are the largest ecosystems on the planet, but when compared with coastal habitats they show much lower biodiversity. This disparity is related to the fact that the environmental instability and the geomorphology of the coastal margins promote more species' isolation and, therefore, speciation and endemisms. Moreover, speciation rates appear to be inversely related to species' distributional ranges, although the generality of this feature requires further testing (Jablonski & Roy, 2003; Lester & Ruttenberg, 2005). It is also known that there are more benthos-associated species than pelagic ones (Grassle & Maciolek, 1992). As an example, in United States, 60% of all marine fish diversity is found in coastal habitats, while 28% in demersal habitats and only 12% in the pelagic systems. This pattern crosses taxonomic boundaries (Angel, 1993). However, almost half (47%) of the Atlantic cephalopod fauna (known to date) are found in the open-ocean, with 55% of that richness restricted to the upper 1000 m of the water column. The great majority of these species are endemic (95%), but some pelagic squids undertake sporadic migrations over the shelf to feed (e.g. *Todarodes sagittatus*). The Atlantic shelves contain around 40% of the species (mainly benthic and nektobenthic), but most of them are not restricted to the shallow waters and are also found in the upper slope, especially in the eastern Atlantic coast (Figure 4B). Only 10% of the benthos diversity is restricted to deep-sea habitats. This disparity is far greater at higher taxonomic levels. Of the 43 families surveyed, only five families (11%) are found along the Atlantic continental shelves (none of them endemic), while 34 (80%) are exclusively present in the open-ocean. By acknowledging the difficulties of sample collection and the expected cryptic diversity, the pelagic contribution for the overall diversity is expected to be even greater. So, the question is why are there so many pelagic cephalopod species? The answer may be found in the distinct evolutionary history of this group when compared, for instance, with other mollusk classes that maintained benthic life strategies throughout time (with some exceptions, e.g. pteropod snails belonging to the orders of Thecosomata and Gymnosomata). Cephalopods are thought to have originated about 500 mya in Cambrian era from a monoplacophoran (gastropod-type) benthic ancestor (Yochelson et al., 1973) and the acquisition of neutral buoyancy and jet propulsion, important evolutionary steps in their early natural history, enable them to be independent from the seafloor, providing an enormous potential for radiation. After the appearance of the coleiod[1] line in the late Devonian, presumably as an evolutionary response to the diversification of early vertebrates and to a previous decrease in cephalopod diversity in the Upper Silurian (Young et al., 1998), Decapodiformes (squids and cuttlefishes) started to radiate earlier in the Paleozoic and had faster evolutionary rates than Octopodiformes (Strugnell et al., 2005). Their earlier diversification during the Paleozoic/Mesozoic transition may have resulted from an invasion of open-ocean habitats where the competition and predation pressures were lower than in shallow waters. The diversification of Octopodiformes is much more recent. The two lineages of the Order Octopoda (Incirrina and Cirrina) shared an earlier pelagic ancestror with the deep-sea vampire squids (Order Vampyromorpha) until the upper Paleozoic, and only diverged in the Mesozoic (Strugnell et al., 2006). The divergence in the two Octopoda lineages led to the adoption of neritic and benthic life-strategies by the majority of incirrates and nektobenthic strategies by deep-sea cirrates. The return to the benthos, with significant morphological and behavioral implications, led to greater speciation rates in coastal areas. Contrary to open-ocean diversity, in which most families have only one genus and many genera with few species, the coastal diversity includes few families that contain almost half of

---

[1] Extant cephalopods are divided in two subclasses: Nautiloidea (*Nautilus* and *Allonautilus*) and Coleiodea (octopuses, squids, cuttlefishes and vampire squids). The latter differs from the former by the reduction and internalization (or even the complete loss) of the shell.

the total species richness. Several causal mechanisms for the neritic diversification are discussed below.

The open-ocean cephalopod fauna can be roughly divided into three distinct ecological groups: 1) highly migratory negatively buoyant species, 2) neutrally buoyant vertical migrators, moving up towards the surface at dusk to feed and going deep at dawn to avoid predation, 3) almost motionless non-muscular and gelatinous deep-sea cephalopods, using bioluminescence and sit and wait strategies to catch prey. Since the large oceanic water masses imply fewer geographical barriers it is not surprising to find that 65% of these pelagic species are amphi-atlantic. They may best be seen as freely interchanging populations united by remarkably strong gene flow (Bucklin et al., 1996; Darling et al., 1999) with their distributional ranges defined more by environmental limits on reproduction success and population growth than by limited dispersal (Norris, 2000).

2.2.2. Relationship between oceanic diversity and resource availability

As in other ecosystems, the maintenance and functioning of the open ocean relies on photosynthetic production, which determines the maximum amount of energy expenditure and sets constrains on the oceanic food-web complexity.

Outside the polar regions (region 1 and 9; Figs. 6, 7), mean NPP explains 44% of the variance in pelagic diversity of cephalopods. The highest large-scale diversity levels are located in two very energetic frontal regions of the Southern and Atlantic Oceans (zone 7), namely the Brazil/Malvinas Confluence and the Agulhas Current and its retroflection along with the upwelling area of the Benguela Current (zone 8). An equivalent subtropical convergence zone is found across the North Atlantic subtropical gyre (at 30-35°N), forming the southern limit to winter-mixing regions of the westerly wind biome. These frontal systems show intense mesoscale activity and exhibit high levels of NPP (Figure 6). Also, diversity is continually higher than 60 species in oceanic regions adjacent to eastern coastal provinces that are strongly controlled by upwelling dominate phytoplankton ecology (Canary, Guinea and Benguela Current Coastal Provinces, see Longhurst, 2007, p.214-273). On the other hand, diversity is markedly lower in the poles, Caribbean Sea (zone 5a), Amazonian Province (zone 5b), and along the western regions of subtropical gyres, namely South Sargasso Sea (zone 3b) and Subtropical South Atlantic (zone 6). In these oligotrophic subtropical zones there is a strong vertical stratification of the water column that limits the supply of nutrients from below the thermocline to the euphotic layer. Carbon fixation by autotrophs is low (NPP is <90 g C m^{-2} y^{-1}; Figure 6), but since subtropical gyres cover globally $>60\%$ of the total ocean surface area, they have a large impact on biogeochemical budgets, accounting for $>30\%$ of the total marine primary production (Longhurst, 1995).

Some studies suggest that the latitudinal increase in species richness from high to low latitudes (see Figure 3 in Angel 1993), followed by a decrease of species dominance, are associated with low productivity and lack of seasonality (Margalef, 1989; Angel, 1997). At oligotrophic regions, primary and second productions are assumed to be closely coupled and their cycles have generally lower seasonal amplitudes, which favors diversification rates (Longhurst & Pauly, 1987). On the other hand, in westerlies biome, the seasonal cycles of production and consumption are uncoupled and more variable, which does not promote specialization and, thus, diversity.

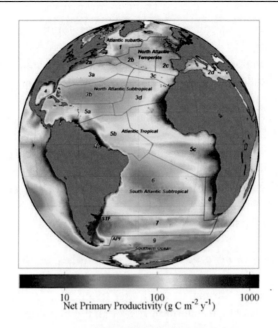

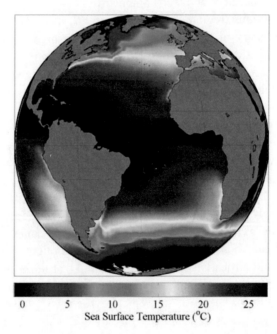

Figure 6. Upper panel - Zoogeographic areas used for the description of oceanic cephalopod fauna (adapted from Backus et al., 1977) and mean net primary productivity averaged over all months (NPP, g C m^{-2} y^{-1}) in the Atlantic Ocean. Lower panel - Mean sea surface temperature (SST, ºC) average over all months. Legend: 1- Atlantic Subartic Province; 2 – North Atlantic Temperate Region: 2a – Slope water; 2b – Northern Gyre, Azores and Britain Province; 2c – Mediterranean Outflow; 2d – Mediterranean Sea; 3 – North Atlantic Subtropical Region: 3a – North Sargasso Sea; 3b - South Sargasso Sea; 3c – N. North African Subtropical Sea; 3d – S. North African Subtropical Sea; 4 – Gulf of Mexico; 5 – Atlantic Tropical Region: 5a – Caribbean Sea and Lesser Antillean Province; 5b – Amazonian Province; 5c – Guinean- Namibian Province. 6 – South Atlantic Subtropical Region: 6 – South Atlantic Subtropical Sea; 7 – Southern Convergence (between sub-tropical front, STF, and Antarctic Polar Front, APF); 8 – Benguela Current Coastal Province; 9 - Southern Ocean.

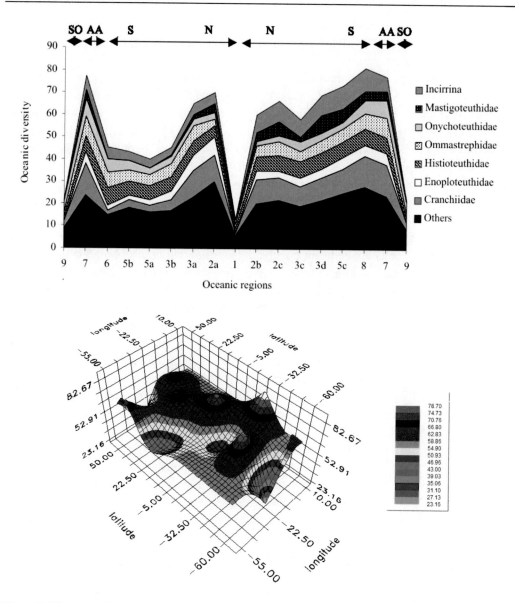

Figure 7. Diversity of oceanic cephalopods in the Atlantic Ocean. Upper panel - The most diverse taxonomic groups are Oegopsida (major families shown) and Octopoda Incirrina, shown by biogeographic areas (adapted from Backus et al. 1977; for abbreviations see Figure 1). SO – Southern Ocean; AA – amphi-atlantic zone; S- south; N- north. Lower panel - The spatial structure in oceanic species richness was best described by a spherical model without nugget effect. Note: the enclosed regions of Gulf of Mexico (zone 4, 43 species) and Mediterranean Sea (zone 2d, 33 species) are not included in either of the panels.

The findings did not support the idea of a latitudinal gradient of cephalopod richness in the open-ocean. Rather, higher diversities at middle (temperate) latitudes and lower diversities in oligotrophic regions of the Trade biome were observed (Figure 7). Similar distributional patterns were also obtained by Cheung et al. (2005, p.19) for commercially important cephalopods. Greater diversification at middle transitional (and more seasonal) latitudes is also observed in many other pelagic groups, such as foraminiferans (Rutherford et al., 1999; Brayard et al., 2005), chaetognaths, euphausiids, pteropods, salpids and ostracods (McGowan & Walker, 1985; Boltovskoy et al., 1999). These latitudinal patterns have been attributed to differences in upper-ocean thermal structure (Figure 6), with middle latitudes exhibiting a weaker permanent thermocline that favors vertical niche partitioning and greater diversification rates. At the poles and equator, the thermocline is either absent or very sharp, respectively (Figure 6), not favoring niche separation and high number of species (Rutherford et al., 1999). The negative effect of high SST on species' eco-physiology has also been put forward to explain the drop in equatorial diversity (illustrated in Figure 8A with the negative trend), but this explanation does not seem plausible since many cephalopod species can reproduce successfully at these temperatures or, on the other hand, can avoid the warmer near-surface temperatures during the daytime by undertaking vertical migrations to deeper waters.

Acknowledging the expected scale dependence in diversity-productivity relationships, the study reveals that there is a significant positive relationship between mean NPP and oceanic richness at regional scales. Other studies have also shown that, at larger scale, species richness increases monotonically with productivity, contrasting with the hump-shaped pattern at smaller scales (Waide et al., 1999; Mittelbach et al., 2001; Chase & Leibold, 2002). The larger amount of energy in highly productive regions may enable a greater number of predatory species to occur (Evans et al., 2005), especially cephalopods. In order to support their high growth rates, short lifespan, semelparity and high metabolic rates, epipelagic squid are highly mobile, voracious and opportunistic predators that are well adapted to the seasonality and spatial patchiness of food resources in the open-ocean. For instance, many are known to make extensive migrations to exploit the latitudinal differences in productivity (Rodhouse & Nigmatullin, 1996).

The richness of other cephalopod groups can also be positively affected by the greater energy availability because it is also associated with environmental heterogeneity (including in productivity itself), which can increase species dissimilarities (Kerr & Packer, 1997; Chase & Leibold, 2002). In fact, there is experimental evidence that seasonal fluctuations in availability of limiting resources can favor biological diversity by the co-existence of different ecotypes via frequency-dependent competition (Spencer et al., 2007). Environments in which resources occur in pulses facilitate the co-existence of subdominant species with competitively dominant ones and, therefore, tend towards higher species richness (Sommer, 1984; Sommer, 1985; Grover, 1988). Outside the poles, the significant positive relationship between the range in NPP and cephalopod oceanic richness (Figure 8D) supports this idea. Additionally, increased productivity also results in different resource types that can sustain a larger number of more specialist species (DeAngelis, 1994; Abrams, 1995).

The more productive and seasonally variable regions in the Atlantic Ocean support the largest cephalopod populations (northern and southern Atlantic regions represent 84% of total fishery catches while central ones represent only 16%, Fishstat 2004). These figures may be indicative of higher diversity because higher resource availability enables larger populations, which buffer extinction and promotes species richness ("more individual hypothesis" Wright, 1983; Srivastava & Lawton, 1998). Underlying this hypothesis is the concept that abundance and energy availability are closely linked (Hurlbert, 2004; Evans et al., 2005; Pautasso & Gaston, 2005) and that the number of species arising per unit time is positively related to the number of individuals in the metacommunity (see "neutral theory of biodiversity", Hubbell, 2001).

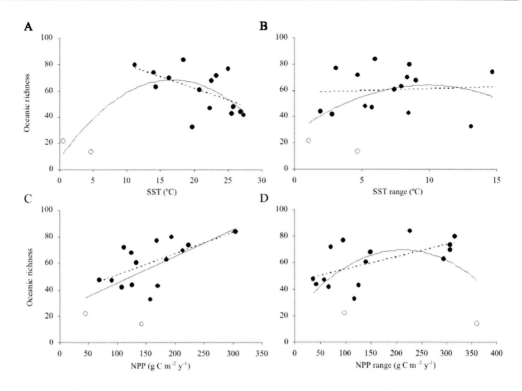

Figure 8. Relationships between oceanic diversity of cephalopods (number of species) within the Atlantic zoogeographic regions and a) sea surface temperature (SST, °C); b) SST range; c) net primary productivity (NPP, g C m^{-2} y^{-1}); d) range of NPP range. Solid lines represent the best fit for all regions and dotted line represent the best fit excluding the Artic and Southern Ocean (Regions 1 and 9 represented with open circles). All regressions are adjusted for area of each region.

3. LATITUDINAL VARIATION OF BODY SIZE IN THE MARINE BIOTA

Despite the fact that geographic size variations relative to environmental factors have been well studied in the terrestrial biota for at least 150 years, only few large-scale latitudinal surveys on interspecific variation of body size were conducted in marine ecosystems, namely in fish (Macpherson & Duarte, 1994), amphipods (Poulin & Hamilton, 1995; Chapelle & Peck, 1999), gastropod (Frank, 1975; Olabarria & Thurston, 2003) and bivalve mollusks (Roy & Martien, 2001). Besides the ambiguous findings in size clines, only two of these studies have covered both hemispheres (Poulin & Hamilton, 1995, Chapelle & Peck, 1999). In fact, while bathymetric variations in body size has been fairly well documented (Rex & Etter, 1998; Rex et al., 1999; McClain & Rex, 2001; McClain, 2004; McClain et al., 2006; Rex et al., 2006), very little is known about broad-scale latitudinal trends in body size in marine fauna, and much less what factors drive them.

Latitude has a significant effect on the mean body size of coastal cephalopods (Figure 9), but that the increase toward the poles (i.e. Bergmann's rule) was more robust in some groups (e.g. squids) than in others (e.g. sepiolids), and changed between hemispheres (e.g. sepiids) and Atlantic margins (e.g. octopods). These differences illustrate the greater complexity of patterns shown by ectotherms, as already seen in the terrestrial systems (Hawkins & Lawton,

1995; Ashton & Feldman, 2003; Olalla-Tarraga & Rodriguez, 2007). Nonetheless, Bergmann's rule was supported at higher taxonomic (class) level.

One important determinant on the latitudinal-size distributions of coastal marine fauna are biogeographic barriers (Roy & Martien, 2001). In fact, species' range limits are not uniformly distributed in space and cluster at particular shoreline locations (Valentine, 1966). These boundaries are often associated with pronounced physical discontinuities on the organism's physiology (Gaylord & Gaines, 2000) and dispersal abilities (Frank, 1975). Yet, the size-latitude relationships of the different cephalopods groups are not significantly influenced by provincial boundaries (Figure 10). Neritic cephalopods are known to have planktonic paralarval stages (term associated with the lack of metamorphosis during post-hatching development) and, therefore, they may be less susceptible to size variations among different provinces than species lacking such larval dispersal mechanisms.

Depth range had a significant association with body size of neritic Cephalopoda, which reveals the importance to account the effect of depth on size even in coastal-restricted surveys. Yet, quite opposite significant associations in the WA (positive) and EA (negative) were obtained (Table 2 and 3).

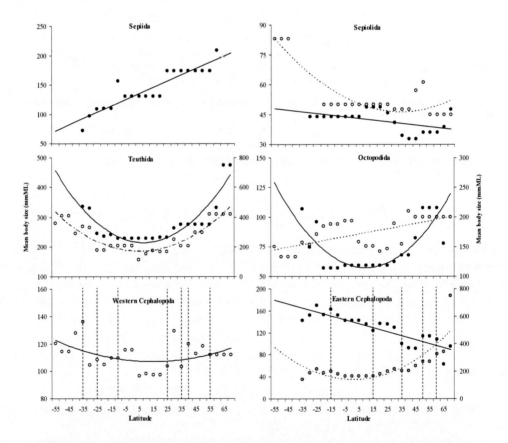

Figure 9. Latitudinal variation in mean body size (maximum mantle length, MML) of coastal cephalopods in the Western (WA, open circles, left Y-axis in panels C, D) and Eastern Atlantic (EA, closed circles, right Y-axis in panels C, D). Grey circles in panel F represent the latitudinal-size relationship of the Class Cephalopoda in EA after excluding the Order Sepiolida. The faunal provinces used here are defined in Valentine (1973) and Watling & Gerk (2000).

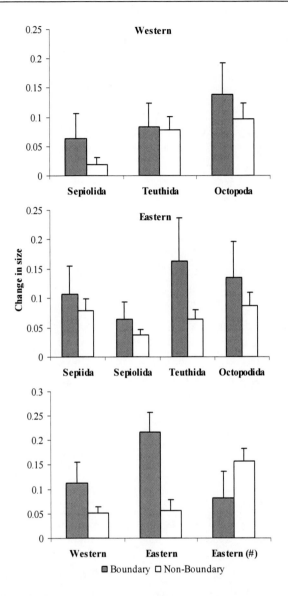

Figure 10. Body size change between adjacent 5° latitudinal bins that cross provincial boundaries and between non-boundary adjacent bins in the Western and Eastern Atlantic. Difference letters represent significant differences (*t*-test, p<0.05). # - represent the size change of the Class Cephalopoda after excluding the Order Sepiolida.

Curiously, besides increasing their size, western neritic cephalopod fauna is known to increase its depth range with increasing latitude (Figure 4), thus following Rapoport's bathymetric rule (Rosa et al., 2008). The smaller bathymetric ranges around the WA tropical regions probably reflect the influence of the environmental stability of coral regions on the ecology of shallow-living cephalopods. This hypothesis was supported by the absence of Rapoport's bathymetric rule in the EA margins, where coral regions are absence. In contrast, the greater environmental (upwelling-driven) dynamics along EA shelves may induce a greater ecological plasticity in cephalopod fauna, which seems to be reflected on the larger bathymetric ranges (Figure 4).

The opposite depth-size associations among Atlantic margins do not support or reject the assumingly most prominent adaptive feature of the marine benthos (Gage & Tyler, 1991), which is the small size of most deep-sea species compared to their shallow-living counterparts ("size-structure hypothesis" in Thiel, 1975, 1979). Yet, some argue against the generality of such phenomenon, since body size has been reported to decrease, increase or show no association with depth (see reviews in Gage & Tyler, 1991, Rex & Etter, 1998). Though this study only comprised coastal habitats, it is worth noting that the majority of the neritic cephalopods are not strictly limited to the continental shelves and some undertake seasonal migration towards the shelf break or to the upper slope (up to 500-700 m) after breeding in more favorable onshore waters (e.g. coastal myopsid squids, incirrate octopuses *Octopus* and *Eledone*) (Rosa & Sousa Reis, 2004). Additionally, the nature of the stronger and positive depth-size relationships (and also Bergmann's rule) in squids in both margins also derive, in part, from the contribution of a larger-sized group (Suborder Oegopsina) that is periodically abundant in coastal habitats (e.g. genera *Illex* and *Todaropsis*) or make major migratory incursions to the shelves for feeding (e.g. genus *Todarodes*) when oceanographic conditions are suitable (Boyle & Rodhouse, 2005). These squids are highly mobile predators (with large bathymetric ranges) that are well adapted to the seasonality of food resources and particularly abundant in highly productive regions at middle (temperate) latitudes (Rosa et al., in press). Therefore, besides environmental forcing and physiological constraints (discussed below), the size-depth-latitude relationships also reflect fundamental aspects of the cephalopod feeding ecologies and life histories.

3.1. Connecting Thermal Energy, Resource and Habitat Availability with Body Size

Although the climate-based heat conservation hypothesis (see Introduction) may be plausible for endotherms (thermoregulators), it does not explain Bergmann size clines in marine ectotherms (thermoconformers), such as cephalopods. Alternatively, the prevalence of negative associations between thermal energy availability (SST) and body size in cephalopods (Table 2 and 3) supports the concept of the "temperature-size rule" (Atkinson, 1994). At lower temperatures (higher latitudes), ectotherms exhibit lower growth rates, delayed maturation but also grow to a larger body size, while at higher temperatures they grow faster, mature at smaller sizes but their adult asymptotic body size is reduced. There is evidence that cephalopod fauna show this life-history tradeoff. For instance, the giant octopus (*Enteroctopus dofleini*) lives at average temperatures of 10°C in the NE Pacific Ocean, achieves maturity with 10-15 kg (~1020 days of age), grow over 5 m length (more than 50 kg) and its life span varies between 4 and 5 years (Hartwick, 1983). In contrast, the pigmy octopus (*Octopus joubini*) from WA tropical shallow waters (average lifetime temperatures of 25 °C), matures in 182 days (with 30 grams) and attains a maximum total length of 15 cm (Hanlon, 1983). Outside the neritic province, the two most striking examples of cold-associated gigantism are the two largest invertebrates of the oceans, namely the colossal (*Mesonychoteuthis hamiltoni*) and giant squids (*Architeuthis* sp.). Knowledge on the biology and ecology of those organisms is scarce. Nonetheless, the first is known to be a reclusive inhabitant of the circumpolar Antarctic region that can attain more than 495 kg (O'Shea & Bolstad, 2008), while the second is a widespread large oceanic predator (up to 16m of total length) that, assumingly, lives at mesopelagic depths (around 10-13°C) and has a life span of several years (Landman et al., 2004). Though many small-sized counterexamples in polar/deep-sea habitats can be pointed out (e.g. small octopus *Bathypolypus arcticus* in North WA), the greater disparity of sizes in these cold environments is unequivocal. This evidence is also supported by the general increase in size disparity (variance) towards the poles shown in the present survey (Figure 2). Other inter-specific studies also point out that larger size at maturity in cephalopods is a result of longer life spans (Van Heukelem, 1976; Forsythe, 1984; Wood & O'Dor, 2000) rather than faster growth rates (Calow, 1987).

Table 2. Models assessing the importance of depth range, sea surface temperature (SST), net primary productivity (NPP), SST and NPP range and shelf area in predicting cephalopod body size variation in Western Atlantic margins. For each variable the t-ratios are included and Akaike Information Criteria (*AIC*) is also given. Restricted maximum likelihood was used to estimate the parameters of each of the models.

Western	Model	Depth range	SST	NPP	SST range	NPP range	Shelf Area	AR order	Total R^2	AIC
Sepiolida	1	5.98**						1	0.92	125.17
	2		-1.45					1	0.83	142.60
	3			-0.06				1	0.81	145.30
	4				-0.06	1.27		1	0.83	144.77
	5						-0.39	1	0.80	139.89
Theutida	1	2.01[a]						1	0.69	206.91
	2		-2.35*					1	0.70	206.17
	3			-0.03				1	0.62	211.41
	4				-0.36	0.76		1	0.63	212.70
	5						0.65	1	0.63	210.91
Octopodida	1	2.72*						1	0.53	205.95
	2		-1.14					1	0.41	212.06
	3			-0.50				1	0.38	213.22
	4				1.49	-0.92		1	0.42	213.26
	5						2.19*	1	0.48	201.47
All	1	2.05[a]						-	0.15	191.76
	2		-2.20*					-	0.17	191.17
	3			0.58				-	0.01	195.60
	4				1.70	2.67*		-	0.41	184.05
	5						-0.03	-	0.00	189.55

Models:

M1 - evaluates size-depth clines;

M2 - evaluates the "temperature-size rule";

M3 - evaluates the "resource availability" hypothesis;

M4 - evaluates the "seasonality hypothesis";

M5 - evaluates habitat availability as a surrogate for competition.

* indicate statistical significance at the 5% level of significance.

** indicate statistical significance at the 1% level of significance.

[a] – marginally significant (p=0.05).

Table 3. Models assessing the importance of depth range, sea surface temperature (SST), net primary productivity (NPP), SST and NPP range and shelf area in predicting cephalopod body size variation in Eastern Atlantic margins. For each variable the t-ratios are included and Akaike Information Criteria (AIC) is also given. Restricted maximum likelihood was used to estimate the parameters of each of the models.

Eastern	Model	Depth range	SST	NPP	SST range	NPP range	Shelf Area	AR order	Total R^2	AIC
Sepiida	1	-2.66[a]						1	0.69	148.26
	2		0.01					1	0.69	181.99
	3			-0.29				1	0.11	199.04
	4				-0.32	0.30		1	0.69	183.80
	5						-0.84	1	0.71	181.19
Sepiolida	1	-0.13						2	0.80	104.86
	2		-1.16					2	0.82	103.20
	3			0.24				2	0.80	104.80
	4				-5.48 **	3.06 **		-	0.72	107.39
	5						-0.74	2	0.87	91.37
Theutida	1	7.69 **						2	0.94	228.80
	2		-5.97 **					2	0.83	251.02
	3			-1.44				1	0.73	258.90
	4				-0.41	-0.48		1	0.71	262.47
	5						-0.33	2	0.65	247.81
Octopodida	1	-0.45						1	0.45	206.98
	2		-2.07[a]					1	0.50	204.41
	3			-0.54				1	0.45	206.95
	4				0.48	-0.46		1	0.45	208.89
	5						2.03 [a]	1	0.52	203.77
All	1	-0.76						1	0.63	177.28
	2		-0.74					1	0.63	192.20
	3			1.35				1	0.64	190.87
	4				-1.15	0.66		1	0.64	193.33
	5						-7.34 **	2	0.71	181.19
All (#)	1	-5.37**						1	0.74	251.66
	2		-4.94**					-	0.55	260.88
	3			-0.17				1	0.46	268.83
	4				-0.49	-0.13		1	0.47	270.47
	5						-0.97	1	0.82	199.30

For the purposes of Models 1-5 see footnotes in Table 2. (#) - excluding Sepiolida.
* indicate statistical significance at the 5% level of significance.
** indicate statistical significance at the 1% level of significance.
[a] – marginally significant (p=0.05).

Although the temperature-size rule applies to the majority of ectotherms (>80%, Atkinson, 1994), the underlined mechanisms are still poorly understood (Atkinson & Sibly, 1997; Angilletta & Dunham, 2003). Although genetic divergence has been associated with latitudinal clines in body size (Partridge & Coyne, 1997; Gockel et al., 2001; de Jong &

Bochdanovits, 2003), phenotypic plasticity seems to be a major contributor. In fact, non-adaptive plasticity may be associated with thermal effect on growth and differentiation, namely on the size of cells (Partridge et al., 1994; Van Voorhies, 1996), number of cells (James et al., 1997; Noach et al., 1997), both (Zwaan et al., 2000) and at supra-cellular levels (e.g. organs, Nijhout, 2003). These temperature-induced size changes can be interpreted as an integrated adaptive suite of acclimatory responses at all levels of organization to "maintain aerobic scope and regulate oxygen supply" ("MASROS hypothesis", Atkinson et al., 2006). This concept followed the "oxygen-limited thermal tolerance" model (Pörtner, 2002), that links thermal tolerance windows directly to oxygen supply and energy demand in parallel to adjustments at the molecular and membrane level (Hochachka & Somero, 2002). Thus, warm-induced cephalopod size decrease at lower (tropical) latitudes may be a strategy to mitigate oxygen limitation (i.e. lower maintenance costs due to the positive relationship between metabolic rates and temperature), which is also exacerbated by the reduction in oxygen solubility with increasing temperature (Chapelle & Peck, 1999; Woods, 1999). Although growth is initially faster at higher temperatures it may slow down at a smaller size due to insufficient resource (oxygen) acquisition (Atkinson & Sibly, 1997). Reduced adult size at increased temperature and lower oxygen levels has also been observed in terrestrial ectotherms (Frazier et al., 2001).

Besides temperature, increased primary productivity has also been shown to have a positive effect on body size (Aava, 2001) and considered as a potential explanation of Bergmann' rule in terrestrial habitats (Rosenzweig, 1968). At a given temperature, growth rates and size at maturity increases as food or resource availability increases (Atkinson & Sibly, 1997). Yet, the differences in resource availability (as NPP) did not explain much of variation of mean body size (Table 2 and 3), but since it was the first time that this predictor was tested in marine systems, the present data is definitely insufficient to reject this hypothesis. Since feeding, behavior and reproduction of neritic cuttlefish, octopuses and squids are closely associated with the seabed characteristics, the larger continental shelves (habitat availability) could also explain body size variation by reducing competition. Species would adopt smaller body sizes in more equatorial areas (with smaller shelves) because of the increased inter- and intra-specific competition for resources (Ashton et al., 2000). This hypothesis seems to find some support among the order Octopodida, which curiously is the group more closely associated with the sea bed. In fact, contrarily to the nektobenthic cuttlefishes and squids, octopods are known to have a strictly benthic life strategy. However, besides shelf area being significantly intercorrelated with temperature, there is no direct evidence that competition (contrary to predation, see below) is a major driver in the growth and population dynamics of cephalopods.

Seasonality (or fasting endurance) has also been advocated to explain Bergmann's rule in both endo- and ecthoherms, with large-bodied species being favored in colder and more variable environments because they can store more energy reserves (namely fat) and use them to enhance survival during seasonal shortage of resources (Boyce, 1979). Small-bodied species may have low starvation resistance, because fat reserves increase more rapidly with body size than does metabolic rate (Calder, 1984; Lindstedt & Boyce, 1985). We did not find any evidence to support such hypothesis since environmental seasonality (here as SST and NPP range) did not explain much of the variance of cephalopod size. Moreover, though lipids in the digestive gland have been hypothesized to be possible metabolic substrates and a site for energy storage in cephalopods (Seibel et al., 2000; Rosa et al., 2004a; Rosa et al., 2005), their ability to accumulate them and use them remains highly controversial (Moltschaniwskyj & Johnston, 2006).

Cephalopods are voracious carnivorous with many different feeding strategies (including cannibalism) that enable them to feed opportunistically on a wide range of prey (Rosa et al., 2004b), and as already pointed out, many of them also evolved migratory behaviors to exploit the seasonality of food resources. Thus, the growth of cephalopods (in the wild) seems to be primarily limited by predation (other potential driver of Bergmann's rule) rather than food shortage (Wood & O'Dor, 2000). Predation is more likely to limit growth of cephalopods since the consumption by marine mammals, sea birds and fish is widespread, with some feeding exclusively on them (e.g. some elasmobranchs, Boyle & Rodhouse, 2005). For instance, the abundance of octopuses has been shown to be inversely related with the number of predatory fish present (Aronson, 1986). Yet, in order for the predation hypothesis to explain the present Bergmman's rule, the predation intensity for most cephalopod species must be positively correlated with latitude, which is not possible to be tested since relevant data are not available.

4. DEPTH-RELATED PATTERNS OF DIVERSITY

There is no terrestrial equivalent to the extensive, cold and dark deep-sea habitats. Its distinctiveness has many ecological implications (e.g. bioluminescence is the primary source of light and communication) and may hold the key to understanding global diversity patterns. Diversity typically shows a parabolic pattern with depth, i.e., it peaks at intermediate depths (Sanders, 1968; Rex, 1973, 1981; Pineda & Caswell, 1998). Yet, cephalopod diversity is highest above the first 200 meters, namely in sub-litoral and epipelagic regions (Figure 11) and declines with depth. The vast majority (~60%) of benthic and nektobenthic diversity (e.g. cuttlefishes and loliginids) is restricted to the upper 500 m of depth and to the upper 1000 m for oceanic fauna. This greater spread in the pelagic biota is associated with diel vertical migrations of the most abundant oegopsid squids, between epi- and mesopelagic zones (e.g. Hunt & Seibel, 2000; Watanabe et al., 2006). A continuous decrease in diversity with depth has also been observed in pelagic fishes (Smith & Brown, 2002). Although it is expected that the distributional range of the lesser known bathyal and abyssal fauna is underestimated, such limitations are not expected to change the observed decrease of biodiversity with depth.

Though its difficult to compare the almost two-dimensional terrestrial area to the three-dimensional aquatic domain, we don't believe there is an effect of water volume on diversity ("biome area hypothesis", Rosenzweig, 1995), because the majority of cephalopod species avoids the immense deep-sea biome and is restricted to the smaller volume of shallower productive waters. For example, the continental shelf only represents 3% of the global ocean area and even less of the ocean's volume. While the greater environmental stability of the deep-sea has not permitted extensive niche diversification ("environmental stability hypothesis" Sanders 1968), the higher energy availability at coastal and shallow waters may enhance speciation ("more-individual hypothesis", see above). Additionally, since feeding, behavior and reproduction of the neritic cephalopods are closely associated with the seabed characteristics, the greater spatial heterogeneity along the continental shelves may explain the higher faunal diversification rates ("habitat heterogeneity hypothesis", Grassle & Sanders, 1973).

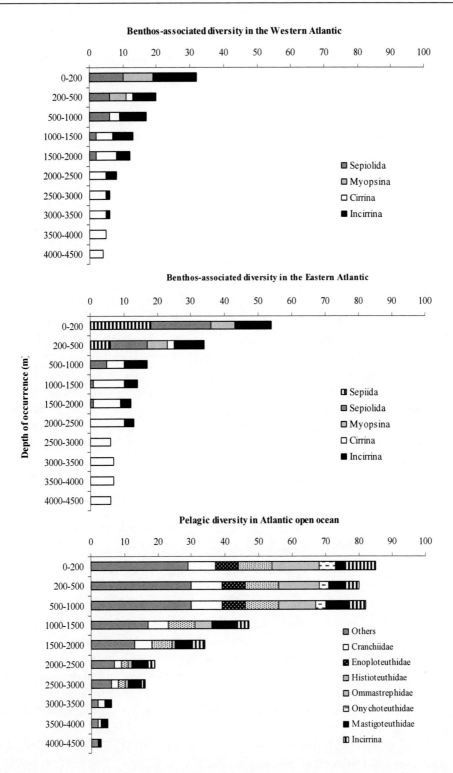

Figure 11. Bathymetric gradient of: benthos associated species diversity in the western (A) and eastern (B) Atlantic coasts, and pelagic species diversity (C) in Atlantic open ocean. Diversity is estimated as the number of coexisting species ranges in pre-defined depth increments.

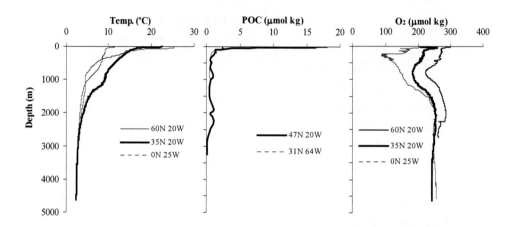

Figure 12. Patterns of temperature gradient (°C), particulate organic carbon (POC, μmol kg) and oxygen concentration (O$_2$, μmol kg) in the North Atlantic Ocean (from Comprehensive Ocean Atmosphere Data Set - COADS). POC is used as a food availability indicator.

Cephalopods do not show smaller bathymetric range sizes toward the "spatial boundaries" and larger ranges at the intermediate depth (mid-domain effect). Instead, they show a linear (pelagic species) or a polynomial (benthos-related species, with range uniformity below 2500 m, Rosa et al. in press) increase in range size with depth. The different patterns between these two groups may reflect the distinct locomotory abilities of the two different life strategies. Mid-domain effects also do not seem to account for diversity-depth patterns in other invertebrate groups (McClain & Etter, 2005) and fish (Kendall & Haedrich, 2006). It is also known that depth-related trends may vary between regions (Kendall & Haedrich, 2006), but the present knowledge of cephalopod deep-sea biogeography does not allow us to make a smaller-scale approach. Although it is expected that the distributional range of the lesser known bathyal and abyssal fauna is underestimated, such limitations are not expected to change the distinct decrease of biodiversity with depth.

4.1. Relationship between Depth-Related Changes in Diversity and Energy Availability

The inverse relationship between depth and diversity in both benthos-related and pelagic cephalopods strongly indicates the influence of the same causal mechanisms. In fact, the vertical gradients in the physical environment are pronounced (Figure 12).

In the pelagic realm, these gradients dictate dictate pervasive behavioral (e.g. bioluminescence and sit-and-wait predation strategies), locomotory (e.g. fin or medusoid swimming instead of inefficient high jet speed propulsion) and morphological adaptations (e.g. reduced high-density musculature; coelomic chambers and ammonium chloride retention as buoyancy mechanisms), leading to a strong depth-related decline in metabolism, above and beyond the effect of temperature. The general decrease in pelagic cephalopod metabolic rates with depth has been attributed to a reduced need for locomotory capabilities associated with visually cued predator/prey interactions in the light-limited deep-sea ("the visual interaction hypothesis", Childress, 1995; Seibel et al., 1997; Seibel, 2007), while the possible influences of hydrostatic pressure, reduced food availability and oxygen levels have been ruled out. This pattern is not evident in benthic organisms, including cephalopods (Seibel & Childress, 2000), where lifestyles and temperature-normalized metabolic rates are similar across the depth range. Nevertheless, temperature has been hypothesized to influence the abundance and diversity of organisms via its effects on biochemical processes (Allen et al. 2002, 2006), mutation rates (Gillooly et al. 2005), and lifestyles ("metabolic niche hypothesis", Clarke & Gaston, 2006). Contrary to the patterns observed in soft-sediment benthic diversity, the cephalopod richness decreases along the permanent thermocline and becomes more stable at lower bathyal and abyssal regions. In fact, of the several possible environmental factors influencing the depth distributions of benthos-related and pelagic cephalopods (some illustrated in Figure 12), food availability (measured as particulate organic carbon) and temperature (independent of metabolism) seem to be the best causal predictors for depth-related patterns. However, given the differential responses of pelagic and benthic cephalopod lifestyles and metabolism to depth, these environmental influences must mediate diversity via some mechanism other than biochemical rate processes or lifestyles. Although not quantified, temperature-species relationships should be stronger with depth than with latitude, emphasizing the stronger influence of the sharp vertical thermal gradient than the smoother and more seasonal horizontal (latitudinal) one.

5. CONCLUSION

The latitudinal gradients of species richness in cephalopod mollusks are present in both Atlantic coasts, but quite distinct from each other. Besides the evolutionary history, I discuss that the contemporary environmental gradients, shelf area and coral habitat extent can predict much of the diversity variation. However, even in the WA, where there is a stronger species-energy relationship, the north hemisphere diversity gradient seems primarily set by glaciation periods and contemporary oceanographic boundaries, rather than organism's physiological responses to climate. Since environmental gradients cannot increase or decrease species richness by themselves (Wiens & Donoghue, 2004) but rather promote or emphasize differences in diversification rates (Stevens, 2006), a combined approach between historical and contemporary processes is essential to fully understand large-scale patterns of cephalopod diversity.

Moreover, I also showed that contrary to the coastal environment, no classical latitudinal gradient in species richness is found in the open-ocean, since higher diversity levels are found at middle latitudes. While coastal patterns of cephalopod diversity can be robustly predicted by climate (sea surface temperature) and non-climate (spatial area) variables, pelagic richness follows the spatial changes in surface ocean net primary productivity. The significant linear increase in cephalopod diversity with mean productivity and the occurrence of higher population densities (fisheries data) in more productive regions of the Atlantic Ocean are consistent with the "more individuals" hypothesis. At the same time, the positive response of species richness with NPP range outside the Polar Regions also supports the idea that diversity can be favored by seasonal fluctuations in availability of limiting resources. Both benthos-related and pelagic cephalopod diversity does not increase at bathyal and mesopelagic depths, which has been described in several macrofaunal groups, including bivalves and gastropod mollusks (Rex et al., 2005a). Our findings clearly show that diversity peaks along the continental shelves and in the epipelagic zone (< 200m), and decreases with depth. It suggests that higher environmental energy availability and productivity in shallow and surface waters promotes diversification rates. Finally, Bergmman's rule seems to be a general trend for Cephalopoda, but the latitudinal-body size patterns were more complex at lower taxa (order levels). Additionally, thermal energy availability (SST) seemed to play the most important role in structuring the distribution of body size along the continental shelves of the Atlantic Ocean. Our findings did not support predictions of hypotheses for Bergmman's rule relating to resource availability, seasonality and competition.

Thus, it is evident that species-area-energy theories that have been formulated for the terrestrial biosphere also apply to the marine systems. Although the phyletic composition and life-history characteristics of terrestrial and marine fauna are quite distinct, there is no reason to assume that the causal predictors behind these widespread biogeographic patterns will differ between these two realms.

REFERENCES

Aava, B. (2001). Primary productivity can affect mammalian body size frequency distributions. *Oikos, 93*, 205-213.

Abrams, P. A. (1995). onotonic or unimodal diversity-productivity gradients: what does competition theory predict? *Ecology, 76*, 2019-2027.

Allen, A. P., Brown, J. H. & Gillooly, J. F. (2002).Global biodiversity, biochemical kinetics, and the energetic-equivalence rule. *Science, 297*, 1545-1548.

Allen, A. P., Gillooly, J. F., Savages, V. M. & Brown, J. H. (2006). Kinetic effects of temperature on rates of genetic divergence and speciation. *Proceedings of the National Academy of Sciences of the United States of America, 103*, 9130-9135.

Allmon, W. D. (2001). Nutrients, temperature, disturbance, and evolution: a model for the late Cenozoic marine record of the western Atlantic. *Palaeogeography, Palaeoclimatology, Palaeoecology, 166*, 9-26.

Angel, M. V. (1993). Biodiversity of the pelagic ocean. *Conservation Biology, 7*, 760-772.

Angel, M. V. (1997). Pelagic biodiversity. In: Ormond, R. F. G., Gage J. D. & Angel M. V., (Eds.), *Marine biodiversity* (pp. 35-68). Cambridge: Cambridge University Press.

Angilletta, M. J. & Dunham, A. E. (2003). The temperature-size rule in ectotherms: simple evolutionary explanations may not be general. *American Naturalist, 162*, 332-342.

Angilletta, M. J., Niewiarowski, P. H., Dunham, A. E., Leache, A. D. & Porter, W. P. (2004). Bergmann's clines in ectotherms: Illustrating a life-history perspective with sceloporine lizards. *American Naturalist, 164*, E168-E183.

Arita, H. T., Rodriguez, P. & Vazquez-Dominguez, E. (2005). Continental and regional ranges of North American mammals: Rapoport's rule in real and null worlds. *Journal of Biogeography, 32*, 961-971.

Aronson, R. B. (1986). Life history and den ecology of *Octopus briareus* Robson in a marine lake. *Journal of Experimental Biology and Ecology, 95*, 37-56.

Ashton, K. G. (2002a). Do amphibians follow Bergmann's rule? *Canadian Journal of Zoology, 80*, 708-716.

Ashton, K. G. (2002b). Patterns of within-species body size variation of birds: strong evidence for Bergmann's rule. *Global Ecology and Biogeography, 11*, 505-523.

Ashton, K. G. & Feldman, C. R. (2003). Bergmann's rule in nonavian reptiles: Turtles follow it, lizards and snakes reverse it. *Evolution, 57*, 1151-1163.

Ashton, K. G., Tracy, M. C. & de Queiroz, A. (2000). Is Bergmann's rule valid for mammals? *American Naturalist, 156*, 390-415.

Astorga, A., Fernandez, M., Boschi, E. E. & Lagos, N. (2003). Two oceans, two taxa and one mode of development: latitudinal diversity patterns of South American crabs and test for possible causal processes. *Ecology Letters, 6*, 420-427.

Atkinson, D. (1994). Temperature and organism size – a biological law for ectotherms? *Advances in Ecological Research, 25*, 1-58.

Atkinson, D., Morley, S. A. & Hughes, R. N. (2006) From cells to colonies: at what levels of body organization does the 'temperature-size rule' apply? *Evolution & Development, 8*, 202-214.

Atkinson, D. & Sibly, R. M. (1997). Why are organisms usually bigger in colder environments? Making sense of a life history puzzle. *Trends in Ecology & Evolution, 12*, 235-239.

Avise, J. C. (1992). Molecular population structure and the biogeographic history of a regional fauna: a case history with lessons for conservation biology. *Oikos, 63*, 62-76.

Backus, R. H., Craddock, J. E., Haedrich, R. L. & Robison, B. H. (1977). Atlantic mesopelagic zoogeography. *Memoirs of the Sears Foundation for Marine Research, 1*, 266-287.

Bé, A. W. H. (1977). An ecological, zoogeographic and taxonomic review of recent planktonic foraminifera. In: Ramsey, A. T. S., (Ed.), *Oceanic micropalaeontology, Vol 1.* (pp. 1-100). London: Academic Press.

Bergmann, C. (1847). Ueber die Verhältnisse der Wärmeökonomie der Thiere zu ihrer Grösse (Concerning the relationship of heat conservation of animals to their size). *Gottinger Studien, 3*, 595-708.

Blackburn, T. M. & Gaston, K. J. (1996). Spatial patterns in the body sizes of bird species in the New World. *Oikos, 77*, 436-446.

Blackburn, T. M., Gaston, K. J. & Loder, N. (1999). Geographic gradients in body size: A clarification of Bergmann's rule. *Diversity and Distributions, 5*, 165-174.

Blackburn, T. M. & Hawkins, B. A. (2004). Bergmann's rule and the mammal fauna of northern North America. *Ecography, 27*, 715-724.

Boletzky, S. V. (1974). The "larvae" of Cephalopoda. A review. *Thalassia, 10*, 45-76.

Boltovskoy, D., Gibbons, M. J., Hutchings, L. & Binet, D. (1999). General biological features of the South Atlantic. In: D. Boltovskoy (Ed.), *South Atlantic zooplankton* (pp. 1-42). Leiden: Backhuys.

Boyce, M. S. (1979). Seasonality and patterns of natural selection for life histories. *American Naturalist, 1979*, 569-583.

Boyle, P. & Rodhouse, P. G. (2005). *Cephalopods. Ecology and Fisheries*. Oxford: Blackwell Publishing, .

Brayard, A., Escarguel, G. & Bucher, H. (2005). Latitudinal gradient of taxonomic richness: combined outcome of temperature and geographic mid-domains effects? *Journal of Zoological Systematics and Evolutionary Research, 43*, 178-188.

Briggs, J. C. (1974). *Marine zoogeography*. New York: McGraw-Hill.

Brown, J. H. (1995). *Macroecology*. Chicago: Univ. Chicago Press.

Brown, J. H. & Lomolino, M. V. (1998). *Biogeography* 2nd ed. Sunderland, MA: Sinauer.

Bucklin, A., LaJeunesse, T. C., Curry, E., Wallinga, J. & Garrison, K. (1996). Molecular genetic diversity of the copepoda, Nannocalanus minor: genetic evidence of species and population structure in the N. Atlantic Ocean. . *Journal of Marine Research, 54*, 285-310.

Calder, W. A. (1984). *Size, function and life history*. Cambridge: Harvard University Press.

Calow, P. (1987). Fact and theory: an overview. In: Boyle, P. R., (Ed.), *Cephalopod life cycles: comparative reviews. Vol. 2* (pp. 351-366). London: Academic Press.

Chapelle, G. & Peck, L. S. (1999). Polar gigantism dictated by oxygen availability. *Nature, 399*, 114-115.

Chase, J. M. & Leibold, M. A. (2002). Spatial scale dictates the productivity-biodiversity relationship. *Nature, 416*, 427-430.

Childress, J. J. (1995). Are there physiological and biochemical adaptations of metabolism in deep-sea animals? *Trends in Ecology and Evolution, 10*, 30-36.

Chown, S. L. & Gaston, K. J. (1999). Exploring links between physiology and ecology at macro-scales: the role of respiratory metabolism in insects. *Biological Reviews, 74*, 87-120.

Clarke, A. (1992). Is there a latitudinal diversity cline in the sea? . *Trends in Ecology & Evolution, 7*, 286-287.

Clarke, A. & Gaston, K. J. (2006). Climate, energy and diversity. *Proceedings of the Royal Society B-Biological Sciences, 273*, 2257-2266.

Collin, R. (2001). The effects of mode of development on phylogeography and population structure of North Atlantic *Crepidula* (Gastropoda: Calyptraeidae). *Molecular Ecology, 10*, 2249-2262.

Colwell, R. K. & Lees, D. C. (2000). The mid-domain effect: geometric constraints on the geography of species richness. *Trends in Ecology & Evolution, 15*, 70-76.

Cruz, F. B., Fitzgerald, L. A., Espinoza, R. E. & Schulte, J. A. (2005). The importance of phylogenetic scale in tests of Bergmann's and Rapoport's rules: lessons from a clade of South American lizards. *Journal of Evolutionary Biology, 18*, 1559-1574.

Culver, S. J. & Buzas, M. A. (2000). Global latitudinal species diversity gradient in deep-sea benthic foraminifera. *Deep-Sea Research Part I-Oceanographic Research Papers, 47*, 259-275.

Currie, D. J., Mittelbach, G. G., Cornell, H. V., Field, R., Guegan, J. F., Hawkins, B. A., Kaufman, D. M., et al. (2004). Predictions and tests of climate-based hypotheses of broad-scale variation in taxonomic richness. *Ecology Letters, 7*, 1121-1134.

Darling, K. F., Wade, C. M., Kroon, D., Leigh Brown, A. J. & Bijma, J. (1999). The diversity and distribution of modern planktonic foraminiferal small subunit ribosomal RNA genotypes and their potential as tracers of present and past ocean circulations. *Paleoceanography, 14*, 3-12.

Dauvin, J. C., Kendall, M., Paterson, G., Gentil, F., Jirkov, I. & Sheader, M. (1994). An initial assessment of polychaete diversity in the northeastern Atlantic Ocean. *Biodiversity Letters, 2*, 171-181.

de Jong, G. & Bochdanovits, Z. (2003). Latitudinal clines in *Drosophila melanogaster*: body size, allozyme frequencies, inversion frequencies, and the insulin-signalling pathway. *Journal of Genetics, 82*, 207-223.

DeAngelis, D. L. (1994). Relationships between the energetics of species and large-scale species richness. In: C. G. Jones & J. H. Lawton (Eds.), *Linking species and ecosystems* (pp. 263-272). New York: Chapman and Hall.

Dodge, J. D. & Marshall, H. G. (1994). Biogeographic analysis of the armoured planktonic dinoflagellate *Ceratium* in the North Atlantic and adjacent seas. *Journal of Phycology, 30*, 905-922.

Dynesius, M. & Jansson, R. (2000). Evolutionary consequences of changes in species' geographical distributions driven by Milankovitch climate oscillations. *Proceedings of the National Academy of Sciences of the United States of America, 97*, 9115-9120.

Ellingsen, K. E. & Gray, J. S. (2002). Spatial patterns of benthic diversity: is there a latitudinal gradient along the Norwegian continental shelf? *Journal of Animal Ecology, 71*, 373-389.

Emiliani, C. (1966). Isotropic paleotemperatures. *Science, 154*, 851-857.

Evans, K. L., Greenwood, J. J. D. & Gaston, K. J. (2005). Dissecting the species-energy relationship. *Proceedings of the Royal Society B-Biological Sciences, 272*, 2155-2163.

Fischer, A. G. (1960). Latitudinal variations in organic diversity. *Evolution 14*, 64-81.

Forsythe, J. W. (1984). *Octopus joubini* (Mollusca: Cephalopoda): a detailed study of growth through the full life cycle in a closed seawater system. *Journal of Zoology, 202*, 393-417.

Fortes, R. R. & Absalao, R. S. (2004). The applicability of Rapoport's rule to the marine molluscs of the Americas. *Journal of Biogeography, 31*, 1909-1916.

Frank, P. W. (1975). Latitudinal variation in the life history features of the black turban snail *Tegula funebralis* (Prosobranchia: Trochidae) *Marine Biology, 31*, 181-192.

Frazier, M. R., Woods, H. A. & Harrison, J. F. (2001). Interactive effects of rearing temperature and oxygen on the development of *Drosophila melanogaster*. *Physiological and Biochemical Zoology, 74*, 641-650.

Fukami, T. & Morin, P. J. (2003). Productivity-biodiversity relationships depend on the history of community assembly. *Nature, 424*, 423-426.

Gage, J. D. & Tyler, P. A. (1991). *Deep-sea biology: a natural history of organisms at the deep-sea floor*. Cambridge: Cambridge University Press.

Gaston, K. J. (2000). Global patterns in biodiversity. *Nature, 405*, 220-227.

Gaston, K. J., Blackburn, T. M. & Spicer, J. I. (1998). Rapoport's rule: time for an epitaph? *Trends in Ecology & Evolution, 13*, 70-74.

Gaston, K. J. & Chown, S. L. (1999). Why Rapoport's Rule Does Not Generalise. *Oikos, 84*, 309-312.

Gaylord, B. & Gaines, S. D. (2000). Temperature or transport? Range limits in marine species mediated solely by flow. *American Naturalist, 155*, 769-789.

Gillooly, J. F., Allen, A. P., West, G. B. & Brown, J. H. (2005). The rate of DNA evolution: Effects of body size and temperature on the molecular clock. *Proceedings of the National Academy of Sciences of the United States of America, 102*, 140-145.

Gockel, J., Kennington, W. J., Hoffmann, A., Goldstein, D. B. & Partridge, L. (2001). Nonclinality of molecular variation implicates selection in maintaining a morphological cline of Drosophila melanogaster. *Genetics, 158*, 319-323.

Grassle, J. F. & Maciolek, N. J. (1992). Deep-sea species richness: regional and local diversity from quantitative bottom samples. *American Naturalist, 139*, 313-341.

Grassle, J. F. & Sanders, H. L. (1973). Life histories and the role of disturbance. *Deep Sea Research, 20*, 643-659.

Gray, J. S. (2001). Marine diversity: the paradigms in patterns of species richness examined. *Scientia Marina, 65,* 41-56.

Grover, J. P. (1988). Dynamics of competition in a variable environment – experiments with two diatom species. *Ecology, 69,* 408-417.

Hanlon, R. T. (1983). Octopus joubini. In: P. R. Boyle (Ed.), *Cephalopod life cycles: species accounts. Vol 1* (pp. 293-310). London: Academic Press.

Hartwick, B. (1983). *Octopus dofleini.* In: P. R. Boyle (Ed.), *Cephalopod life cycles: species accounts. Vol 1* (pp. 277-292). London: Academic Press.

Haug, G. H., Sigman, D. M., Tiedemann, R., Pedersen, T. F. & M., S. (1999). Onset of permanent stratification in the subarctic Pacific Ocean. *Nature, 401,* 779-782.

Haug, G. H. & Tiedemann, R. (1998). Effect of the formation of the Isthmus of Panama on Atlantic Ocean thermohaline circulation. *Nature, 393,* 673-676.

Hawkins, B. A. (2001). Ecology's oldest pattern? *Trends in Ecology & Evolution, 16,* 470-470.

Hawkins, B. A., Field, R., Cornell, H. V., Currie, D. J., Guegan, J. F., Kaufman, D. M., Kerr, J. T., et al. (2003). Energy, water, and broad-scale geographic patterns of species richness. *Ecology, 84,* 3105-3117.

Hawkins, B. A. & Lawton, J. H. (1995). Latitudinal gradients in butterfly body sizes: is there a general pattern? *Oecologia, 102,* 31-36.

Hays, J. D., Imbrie, J. & Shackleton, N. J. (1976). Variations in the Earth's orbit: pacemaker of the ice ages. *Science, 194,* 1121-1132.

Hecnar, S. J. (1999). Patterns of turtle species' geographic range size and a test of Rapoport's rule. *Ecography, 22,* 436-446.

Hewitt, G. M. (1999). Post-glacial re-colonization of European biota. *Biological Journal of the Linnean Society, 68,* 87-112.

Hewitt, G. M. (2004). Genetic consequences of climatic oscillations in the Quaternary. *Philosophical Transactions of the Royal Society of London Series B-Biological Sciences, 359,* 183-195.

Hillebrand, H. (2004a). On the generality of the latitudinal diversity gradient. *American Naturalist, 163,* 192-211.

Hillebrand, H. (2004b). Strength, slope and variability of marine latitudinal gradients. *Marine Ecology-Progress Series, 273,* 251-267.

Hochachka, P. W. & Somero, G. N. (2002). *Biochemical adaptation: mechanisms and process in physiological evolution.* Oxford: Oxford University Press.

Hubbell, S. P. (2001). *The unified neutral theory of biodiversity and biogeography.* Princeton, N.J.: Princeton University Press.

Hunt, J. C. & Seibel, B. A. (2000). Life history of *Gonatus onyx* (Cephalopoda: Teuthoidea): ontogenetic changes in habitat, behavior and physiology. *Marine Biology, 136,* 543-552.

Hurlbert, A. H. (2004). Species–energy relationships and habitat complexity in bird communities. *Ecology Letters, 7,* 714-720.

Huston, M. A. (1979). A general hypothesis of species diversity. *American Naturalist, 113,* 81-101.

Ibaraki, M. (1997). Closing of the Central American Seaway and Neogene coastal upwelling along the Pacific coast of South America. *Tectonophysics, 281,* 99-104.

Jablonski, D. & Roy, K. (2003). Geographical range and speciation in fossil and living molluscs. *Proceedings of the Royal Society B-Biological Sciences, 270,* 401-406.

James, A. C., Azevedo, R. B. R. & Partridge, L. (1997). Genetic and environmental responses to temperature of Drosophila melanogaster from a latitudinal cline. *Genetics, 146*, 881-890.

James, F. C. (1970). Geographic size variation in birds and its relationship to climate. *Ecology, 51*, 365-390.

Kendall, M. A. & Aschan, M. (1993). Latitudinal gradients in the structure of macrobenthic communities: a comparison of Arctic, temperate and tropical sites. *Journal of Experimental Marine Biology and Ecology, 172*, 157–169.

Kendall, V. J. & Haedrich, R. L. (2006). Species richness in Atlantic deep-sea fishes assessed in terms of the mid-domain effect and Rapoport's rule. *Deep-Sea Research Part I-Oceanographic Research Papers, 53*, 506-515.

Kerr, J. T. (1999). Weak links: 'Rapoport's rule' and large-scale species richness patterns. *Global Ecology and Biogeography, 8*, 47-54.

Kerr, J. T. & Packer, L. (1997). Habitat heterogeneity as a determinant of mammal species richness in high energy regions. *Nature, 385*, 252–254.

Khromov, D. N. (1998). Distribution patterns of Sepiidae. *Smithsonian Contributions to Zoology, 586*, 191-206.

Kondoh, M. (2001). Unifying the relationships of species richness to productivity and disturbance. *Proceedings of the Royal Society B: Biological Sciences, 268*, 269-271.

Krijgsman, W., Hilgen, F. J., Raffi, I., Sierro, F. J. & Wilson, D. S. (1999). Chronology, causes and progression of Messinian salinity crisis. *Nature, 400*, 652-655.

Lambshead, P. J. D., Tietjen, J., Ferrero, T. & Jensen, P. (2000). Latitudinal diversity gradients in the deep sea with special reference to North Atlantic nematodes. *Marine Ecology-Progress Series, 194*, 159-167.

Landman, N. H., Coshran, J. K., Cerrato, R., Mak, J., Roper, C. F. E. & Lu, C. C. (2004). Habitat and age of the giant squid (*Architeuthis sanctipauli*) inferred from isotopic analyses. *Marine Biology, 144*, 685-691.

Lester, S. E. & Ruttenberg, B. I. (2005). The relationship between pelagic larval duration and range size in tropical reef fishes: a synthetic analysis. *Proceedings of the Royal Society B: Biological Sciences, 272*, 585-591.

Levin, L. A., Etter, R. J., Rex, M. A., Gooday, A. J., Smith, C. R., Pineda, J., Stuart, C. T., et al. (2001) Environmental influences on regional deep-sea species diversity. *Annual Review of Ecology and Systematics, 32*, 51-93.

Lindstedt, S. L. & Boyce, M. S. (1985). Seasonality, fasting endurance, and body size in mammals. *American Naturalist, 125*, 873-878.

Longhurst, A. (1995). Seasonal cycles of pelagic production and consumption. *Progress in Oceanography, 36*, 77-168.

Longhurst, A. (2007). *Ecological geography of the sea. 2nd ed.* Amsterdam: Academic Press.

Longhurst, A. R. & Pauly, D. (1987). *The ecology of the tropical oceans.* Orlando, Florida: Academic Press.

MacArthur, R. H. (1972). *Geographical ecology: patterns in the distribution of species.* Princeton: Princeton Univ. Press.

Macpherson, E. (2002). Large-scale species-richness gradients in the Atlantic Ocean. *Proceedings of the Royal Society of London Series B-Biological Sciences, 269*, 1715-1720.

Macpherson, E. (2003). Species range size distributions for some marine taxa in the Atlantic Ocean. Effect of latitude and depth. *Biological Journal of the Linnean Society, 80*, 437-455.

Macpherson, E. & Duarte, C. M. (1994). Patterns in species richness, size, and latitudinal range of East Atlantic fishes. *Ecography, 17* 242-248.

Margalef, R. (1989). *Ecologia*. Barcelona: Omega.

Mayr, E. (1956). Geographical character gradients and climatic adaptation. *Evolution, 10*, 105-108.

McClain, C. R. (2004). Connecting species richness, abundance and body size in deep-sea gastropods. *Global Ecology and Biogeography, 13*, 327-334.

McClain, C. R., Boyer, A. G. & Rosenberg, G. (2006). The island rule and the evolution of body size in the deep sea. *Global Ecology and Biogeography, 33*, 1578-1584.

McClain, C. R. & Etter, R. J. (2005). Mid-domain models as predictors of species diversity patterns: bathymetric diversity gradients in the deep sea. *Oikos, 109*, 555-566.

McClain, C. R. & Rex, M. A. (2001). The relationship between dissolved oxygen concentration and maximum size in deep-sea turrid gastropods: an application of quantile regression. *Marine Biology, 139*, 681-685.

McGowan, J. A. & Walker, P. W. (1985). Dominance and diversity in an oceanic system. *Ecological Monographs, 55*, 103-118.

McGowan, J. A. & Walker, P. W. (1993). Pelagic diversity patterns. In: R. E. Ricklefs & D. Schluter (Eds.), *Species diversity in ecological communities: historical and geographical perspectives* (pp. 203-214). Chicago: Chicago University Press.

McNab, T. A. (1971). On the ecological significance of Bergmann's rule. *Ecology, 52*, 845-854.

Measey, G. J. & Van Dongen, S. (2006). Bergmann's rule and the terrestrial caecilian *Schistometopum thomense* (Amphibia: Gymnophiona: Caeciliidae). *Evolutionary Ecology Research, 8*, 1049–1059.

Mittelbach, G. G., Schemske, D. W., Cornell, H. V., Allen, A. P., Brown, J. M., Bush, M. B., Harrison, S. P., et al. (2007). Evolution and the latitudinal diversity gradient: speciation, extinction and biogeography. *Ecology Letters, 10*, 315-331.

Mittelbach, G. G., Steiner, C. F., Scheiner, S. M., Gross, K. L., Reynolds, H. L., Waide, R. B., Willig, M. R., et al. (2001) What is the observed relationship between species richness and productivity? *Ecology, 82*, 2381-2396.

Mokievsky, V. & Azovsky, A. (2002). Re-evaluation of species diversity patterns of free-living marine nematodes. *Marine Ecology-Progress Series, 238*, 101-108.

Moltschaniwskyj, N. A. & Johnston, D. (2006). Evidence that lipid can be digested by the dumpling squid *Euprymna tasmanica*, but is not stored in the digestive gland. *Marine Biology, 149*, 565-572.

Mousseau, T. A. (1997). Ectotherms follow the converse to Bergmann's rule. *Evolution, 51*, 630-632.

Nesis, K. N. (2003). Distribution of recent Cephalopoda and implications for plio-pleistocene events. *Berliner Paläobiologische Abhandlungen, 3*, 199-224.

Nijhout, H. F. (2003). The control of body size in insects. *Developmental Biology, 161*, 1-9.

Noach, E. J., de Jong, G. & Scharloo, W. (1997). Phenotypic plasticity of wings in selection lines of *Drosophila melanogaster*. *Heredity, 79*, 1-9.

Norris, R. D. (2000). Pelagic species diversity, biogeography, and evolution. *Paleobiology, 26,* 236-258.

O'Shea, S. & Bolstad, K. (2008). *Giant Squid and Colossal Squid Fact Sheet. http://www.tonmo.com/science/public/giantsquidfacts.php (accessed 31 May 2008)*

Olabarria, C. & Thurston, M. H. (2003). Latitudinal and bathymetric trends in body size of the deep-sea gastropod *Troschelia berniciensis* (King). *Marine Biology, 143,* 723-730.

Olalla-Tarraga, M. A. & Rodriguez, M. A. (2007). Energy and interspecific body size patterns of amphibian faunas in Europe and North America: anurans follow Bergmann's rule, urodeles its converse. *Global Ecology and Biogeography, 16,* 606-617.

Olalla-Tarraga, M. A., Rodriguez, M. A. & Hawkins, B. A. (2006). Broad-scale patterns of body size in squamate reptiles of Europe and North America. *Journal of Biogeography, 33,* 781-793.

Partridge, L., Barrie, B., Fowler, K. & French, V. (1994). Evolution and development of body size and cell size in *Drosophila melanogaster* in response to temperature. *Evolution, 48,* 1269-1276.

Partridge, L. & Coyne, J. A. (1997). Bergmann's rule in ectotherms: is it adaptive? *Evolution, 51,* 632-635.

Pautasso, M. & Gaston, K. J. (2005). Resources and global avian assemblage structure in forests. *Ecology Letters, 8,* 282-289.

Pianka, E. R. (1966). Latitudinal gradients in species diversity: a review of concepts. *American Naturalist, 100,* 33-46.

Pineda, J. & Caswell, H. (1998). Bathymetric species-diversity patterns and boundary constraints on vertical range distributions. *Deep-Sea Research Part Ii-Topical Studies in Oceanography, 45,* 83-101.

Pörtner, H. O. (2002). Climate variation and the physiological basis of temperature dependent biogeography: systemic to molecular hierarchy of thermal tolerance in animals. *Comparative Biochemistry and Physiology A, 132,* 739-761.

Poulin, R. & Hamilton, W. J. (1995). Ecological determinants of body size and clutch size in amphipods: a comparative approach. *Functional Ecology, 9,* 364-370.

Quattrini, A. M., Lindquist, D. G., Bingham, F. M., Lankford, T. E. & Govoni, J. J. (2005). Distribution of larval fishes among water masses in Onslow Bay, North Carolina: implications for cross-shelf exchange. *Fisheries Oceanography, 14,* 413-431.

Ramirez, L., Dinis-Filho, J. A. F. & Hawkins, B. A. (2008). Partitioning phylogenetic and adaptive components of the geographical body-size pattern of New World birds. *Global Ecology and Biogeography, 17,* 100-110.

Rensch, B. (1938). Some problems of geographical variation and species formation. *Proceedings of the Linnean Society of London, 150,* 275-285.

Rex, M. A. (1973). Deep-sea species diversity: decreased gastropod diversity at abyssal depths. *Science, 181,* 1051-1053.

Rex, M. A. (1981). Community structure in the deep-sea benthos. *Annual Review of Ecology and Systematics, 12* 331-353.

Rex, M. A., Crame, J. A., Stuart, C. T. & Clarke, A. (2005a). Large-scale biogeographic patterns in marine mollusks: A confluence of history and productivity? *Ecology, 86,* 2288-2297.

Rex, M. A. & Etter, R. J. (1998). Bathymetric patterns of body size: implications for deep-sea biodiversity. *Deep Sea Research Part II: Topical Studies in Oceanography, 11,* 103-127.

Rex, M. A., Etter, R. J., Clain, A. J. & Hill, M. S. (1999). Bathymetric patterns of body size in deep-sea gastropods. *Evolution, 53*, 1298-1301.

Rex, M. A., Etter, R. J., Morris, J. S., Crouse, J., McClain, C. R., Johnson, N. A., Stuart, C. T., et al. (2006). Global bathymetric patterns of standing stock and body size in the deep-sea benthos. *Marine Ecology-Progress Series, 317*, 1-8.

Rex, M. A., McClain, C. R., Johnson, N. A., Etter, R. J., Allen, J. A., Bouchet, P. & Waren, A. (2005b) A source-sink hypothesis for abyssal biodiversity. *American Naturalist, 165*, 163-178.

Rex, M. A., Stuart, C. T. & Coyne, G. (2000). Latitudinal gradients of species richness in the deep-sea benthos of the North Atlantic. *Proceedings of the National Academy of Sciences of the United States of America, 97*, 4082-4085.

Riggs, S. R., Snyder, S. W., Hine, A. C. & Mearns, D. L. (1996). Hard bottom morphology and relationship to the geologic framework: mid-Atlantic continental shelf. *Journal of Sediment Research, 66*, 830-846.

Rodhouse, P. G. & Nigmatullin, C. M. (1996). Role as consumers. *Philosophical Transactions of the Royal Society of London: Biological Sciences, 351*, 1003-1022.

Rodríguez, M. A., Olalla-Tárraga, M. A. & Hawkins, B. A. (2008). Bergmann's rule and the geography of mammal body size in the Western Hemisphere. *Global Ecology and Biogeography, 17*, 274-283.

Rohde, K. (1992). Latitudinal gradients in species diversity: the search for the primary cause. *Oikos, 65*, 514–527.

Rohde, K. (1996). Rapoport's rule is a local phenomenon and cannot explain latitudinal gradients in species diversity. *Biodiversity Letters, 3*, 10-13.

Rohde, K. (1997). The larger area of the tropics does not explain latitudinal gradients in species diversity. *Oikos, 79*, 169-172.

Rohde, K. (1998). Latitudinal gradients in species diversity. Area matters, but how much? *Oikos, 82* 184-190.

Rohde, K., Heap, M. & Heap, P. (1993). Rapoport's rule does not apply to marine teleosts and cannot explain latitudinal gradients in species richness. *American Naturalist, 142*, 1-16.

Rosa, R., Costa, P. R., Bandarra, N. & Nunes, M. L. (2005). Changes in tissue biochemical composition and energy reserves associated with sexual maturation of *Illex coindetii* and *Todaropsis eblanae*. *Biological Bulletin, 208*, 100-113.

Rosa, R., Costa, P. R. & Nunes, M. L. (2004a). Effect of sexual maturation on the tissue biochemical composition of *Octopus vulgaris* and *O. defilippi* (Mollusca: Cephalopoda). *Marine Biology, 145*, 563-574.

Rosa, R., Dierssen, H. M., Gonzalez, L. & Seibel, B. A. (2008). Ecological biogeography of cephalopod mollusks in the Atlantic Ocean: historical and contemporary causes of coastal diversity patterns. *Global Ecology and Biogeography, 17*, 600-610.

Rosa, R., Dierssen, H. M., Gonzalez, L. & Seibel, B. A. (in press) Large-scale diversity patterns of cephalopods in the Atlantic open ocean and deep-sea. *Ecology*.

Rosa, R., Marques, A. M., Nunes, M. L., Bandarra, N. & Reis, C. S. (2004b). Spatial-temporal changes in dimethyl acetal (octadecanal) levels of *Octopus vulgaris* (Mollusca, Cephalopoda): relation to feeding ecology. *Scientia Marina, 68*, 227-236.

Rosa, R. & Sousa Reis, C. (2004). *Cephalopods of the Portuguese coast*. Cascais [in Portuguese]: Prémio do Mar- Rei D. Carlos. Câmara Municipal de Cascais.

Rosenzweig, M. L. (1968). The strategy of body size in mammalian carnivores. *American Midland Naturalist, 80*, 299-315.

Rosenzweig, M. L. (1995). *Species diversity in space and time*. Cambridge, MA: Cambridge University Press.

Rosenzweig, M. L. & Sandlin, E. A. (1997). Species diversity and latitudes: listening to area's signal. *Oikos, 80*, 172-176.

Roy, K., Jablonski, D. & Valentine, J. W. (1994). Eastern Pacific Molluscan Provinces and Latitudinal Diversity Gradient: No Evidence for "Rapoport's Rule". *Proceedings of the National Academy of Sciences of the United States of America, 91*, 8871-8874.

Roy, K., Jablonski, D. & Valentine, J. W. (2000). Dissecting latitudinal diversity gradients: functional groups and clades of marine bivalves. *Proceedings of the Royal Society of London Series B-Biological Sciences, 267*, 293-299.

Roy, K., Jablonski, D., Valentine, J. W. & Rosenberg, G. (1998). Marine latitudinal diversity gradients: Tests of causal hypotheses. *Proceedings of the National Academy of Sciences of the United States of America, 95*, 3699-3702.

Roy, K. & Martien, K. K. (2001). Latitudinal distribution of body size in north-eastern Pacific marine bivalves. *Journal of Biogeography, 28*, 485-493.

Rutherford, S., D'Hondt, S. & Prell, W. (1999). Environmental controls on the geographic distribution of zooplankton diversity. *Nature, 400*, 749-753.

Sanders, H. L. (1968). Marine benthic diversity: a comparative study. *American Naturalist, 102*, 243-282.

Schizas, N. V., Street, G. T., Coull, B. C., Chandler, G. T. & Quattro, J. M. (1999). Molecular population structure of the marine benthic copepod Microarthridion littorale along the southeastern and Gulf coasts of the USA. *Marine Biology, 135*, 399-405.

Seibel, B. A. (2007). On the depth and scale of metabolic rate variation: scaling of oxygen consumption rates and enzymatic activity in the Class Cephalopoda (Mollusca). *Journal of Experimental Biology, 210*, 1-11.

Seibel, B. A. & Childress, J. J. (2000). Metabolism of benthic octopods (Cephalopoda) as a function of habitat depth and oxygen concentration. *Deep-Sea Research Part I, 47*, 1247-1260.

Seibel, B. A., Hochberg, F. G. & Carlini, D. B. (2000). Life history of *Gonatus onyx* (Cephalopoda: Teuthoidea): deep-sea spawning and post-spawning egg care. *Marine Biology, 137*, 519-526.

Seibel, B. A., Thuesen, E. V., Childress, J. J. & Gorodezky, L. A. (1997). Decline in pelagic cephalopod metabolism with habitat depth reflects differences in locomotory efficiency. *Biological Bulletin, 192*, 262-278.

Smith, K. F. & Brown, J. H. (2002). Patterns of diversity, depth range and body size among pelagic fishes along a gradient of depth. *Global Ecology and Biogeography, 11*, 313-322.

Smith, K. F. & Gaines, S. D. (2003). Rapoport's bathymetric rule and the latitudinal species diversity gradient for Northeast Pacific fishes and Northwest Atlantic gastropods: evidence against a causal link. *Journal of Biogeography, 30*, 1153-1159.

Sommer, U. (1984). The paradox of the plankton: fluctuations of phosphorus availability maintain diversity of phytoplankton in flow-through cultures. *Limnology and Oceanography, 29*, 633-636.

Sommer, U. (1985). Comparison between steady-state and non-steady-state competition: experiments with natural phytoplankton. *Limnology and Oceanography, 30*, 335-346.

Spencer, C. C., Saxer, G., Travisano, M. & Doebeli, M. (2007). Seasonal resource oscillations maintain diversity in bacterial microcosms. *Evolutionary Ecology Research, 9*, 775-787.

Spicer, J. I., & Gaston, K. J. (1999) Amphipod gigantism dictated by oxygen availability? *Ecology Letters, 2*, 397-401.

Srivastava, D. S. & Lawton, J. H. (1998). Why more productive sites have more species: An experimental test of theory using tree-hole communities. *American Naturalist, 152*, 510-529.

Stanley, S. M. (1986). Anatomy of a regional mass extinction: Plio-Pleiostocene decimation of the western Atlantic bivalve fauna. *Palios, 1*, 17-36.

Stevens, G. C. (1989). The latitudinal gradient in geographical range: how so many species coexist in the tropics. *American Naturalist, 133*, 240-256.

Stevens, G. C. (1996). Extending Rapoport's Rule to Pacific Marine Fishes. *Journal of Biogeography, 23*, 149-154.

Stevens, R. D. (2006). Historical processes enhance patterns of diversity along latitudinal gradients. *Proceedings of the Royal Society B-Biological Sciences, 273*, 2283-2289.

Strugnell, J., Jackson, J., Drummond, A. J. & Cooper, A. (2006). Divergence time estimates for major cephalopod groups: evidence from multiple genes. *Cladistics, 22*, 89-96.

Strugnell, J., Norman, M., Jackson, J., Drummond, A. J. & Cooper, A. (2005). Molecular phylogeny of coleoid cephalopods (Mollusca:Cephalopoda) using a multigene approach; the effect of data partitioning on resolving phylogenies in a Bayesian framework. *Molecular Phylogenetics and Evolution, 35*, 426-441.

Taviani, M. (2002). The Mediterranean benthos from late Miocene up to present: ten million years of dramatic climatic and geologic vicissitudes. . *Biologia Marina Mediterranea, 9*, 445-463.

Taylor, C. M. & Gotelli, N. J. (1994). The macroecology of *Cyprinella*: correlates of phylogeny, body size and geographical range. *American Naturalist, 144*, 549-569.

Terborgh, J. (1973). On the notion of favourableness in plant ecology. *American Naturalist, 107*, 481–501.

Thiel, H. (1975). The size structure of the deep-sea benthos. *Internationale Revue der Gesamten Hydrobiologie, 60*, 576–606.

Thiel, H. (1979). Structural aspects of the deep-sea benthos. *Ambio Special Report, 6*, 25-31.

Tilman, D. (1982) *Resource competition and community structure.* . Princeton, New Jersey: Princeton University Press.

Turner, J. R. G., Gatehouse, C. M. & Corey, C. A. (1987). Does solar energy control organic diversity? Butterflies, moths and the British climate. *Oikos, 48*, 195-205.

Valentine, J. W. (1966). Numerical analysis of marine molluscan ranges on the extratropical northeastern Pacific shelf. *Limnology and Oceanography, 11*, 198-211.

Van Heukelem, W. F. (1976). *Growth, bioenergetics and life-span of Octopus cyanea and Octopus maya. PhD. Thesis.* Unpublished PhD. Thesis, University of Hawaii, Honolulu.

Van Voorhies, W. A. (1996). Bergmann size clines: a simple explanation for their occurrence in ectotherms. *Evolution, 50*, 1259-1264.

Vermeij, G. (1991). Anatomy of an invasion: the trans-Arctic interchange. *Paleobiology, 17*, 281-307.

Vermeij, G. J., Palmer, A. R. & Lindberg, D. R. (1990). Range limits and dispersal of mollusks in the aleutian islands, Alaska. *Veliger, 33*, 346-354.

Voight, J. R. (1988). Trans-Panamanian geminate octopods (Mollusca: Octopoda). *Malacologia, 29*, 289-294.

Waide, R. B., Willig, M. R., Steiner, C. F., Mittelbach, G., Gough, L., Dodson, S. I., Juday, G. P., et al. (1999). The relationship between productivity and species richness. *Annual Review of Ecology and Systematics, 30*, 257-300.

Warwick, R. M. & Ruswahyuni, R. (1987). Comparative study of the structure of some tropical and temperate marine -soft-bottom macrobenthic communities. *Marine Biology, 95*, 641-649.

Watanabe, H., Kubodera, T., Moku, M. & Kawaguchi, K. (2006). Diel vertical migration of squid in the warm core ring and cold water masses in the transition region of the western North Pacific. *Marine Ecology Progress Series, 315*, 187-197.

Whittaker, R. J., Willis, K. J. & Field, F. (2003). Climatic-energetic explanations of diversity: a macroscopic perspective. In: T. M. Blackburn & K. J. Gaston (Eds.), *Macroecology: concepts and consequences* (pp. 107-129). Oxford Blackwell Science.

Wiens, J. J. & Donoghue, M. J. (2004). Historical biogeography, ecology and species richness. *Trends in Ecology & Evolution, 19*, 639-644.

Willig, M. R., Kaufman, D. M. & Stevens, R. D. (2003). Latitudinal gradients of biodiversity: Pattern, process, scale, and synthesis. *Annual Review of Ecology Evolution and Systematics, 34*, 273-309.

Wood, J. B. & O'Dor, R. K. (2000). Do larger cephalopods live longer? Effects of temperature and physiology on interspecific comparisons of age and size maturity. *Marine Biology, 136*, 91-99.

Woods, H. A. (1999). Egg-mass size and cell size: Effects of temperature on oxygen distribution. *American Zoologist, 39*, 244–252.

Worm, B., Lotze, H. K., Hillebrand, H. & Sommer, U. (2002). Consumer versus resource control of species diversity and ecosystem functioning. *Nature, 417*, 848-851.

Wright, D. H. (1983). Species-energy theory: an extension of species-area theory. *Oikos, 41*, 496–506.

Yochelson, E. L., Flower, R. H. & Webers, G. F. (1973). The bearing of the new Late Cambrian monoplacophoran genus *Knightoconus* upon the origin of the Cephalopoda. *Lethaia, 6*, 275-310.

Yom-Tov, Y. & Yom-Tov, J. (2005). Global warming, Bergmann's rule and body size in the masked shrew *Sorex cinereus* Kerr in Alaska. *Journal of Animal Ecology, 74*, 803-808.

Young, R. E., Vecchione, M. & Donovan, D. T. (1998). The evolution of coleiod cephalopods and their present biodiversity and ecology. *South African Journal of Marine Science, 20*, 393-420.

Zwaan, B. J., Azevedo, R. B. R., James, A. C., Van 't Lande, J. & Partridge, L. (2000). Cellular basis of wing size variation in *Drosophila melanogaster*: A comparison of latitudinal clines on two continents. *Heredity, 84*, 338-347.

In: Biogeography
Editors: M. Gailis, S. Kalninš, pp. 105-137

ISBN: 978-1-60741-494-0
© 2010 Nova Science Publishers, Inc.

Chapter 3

PHYLOGENETIC BIOGEOGRAPHY
OF AFROTROPICAL DROSOPHILIDAE

Amir Yassin[1,2] and Jean R. David[1]

[1]Centre National de la Recherche Scientifique (CNRS); Laboratoire Evolution, Génomes et Spéciation (LEGS); av. de la Terrasse, 91198 Gif-sur-Yvette, France.

ABSTRACT

The Drosophilidae is one of the most evolutionary successful muscomorphan families, with nearly 4,000 species showing astonishing morphological and ecological diversity. It is also unique in biological sciences thanks to the wealth of over 100 years of genetic and ecophysiological research, culminated in the sequencing of the complete genomes of 12 of its species. The aim of this Chapter was to trace the evolution of the Drosophilidae in the Afrotropical region in light of the accumulated faunistic and ecological observations and recent advances in the family phylogenetics and African paleoenvironmental studies. First, a reanalysis of the geographical distribution of 527 species belonging to 31 genera and two subfamilies revealed three main centers of endemism: West Africa (WA), East and South Africa (ESA) and the Insular Indian Ocean (IIO). Species richness shows a longitudinal cline in WA and a latitudinal cline in ESA, with Upper Guinea and South Africa being the most speciose, respectively. However, there are only four endemic genera in Africa, showing its old affinities with other continents. We have thus reviewed the origin and evolution of major drosophilid genera and species groups by reanalyzing recently published sequences of the nuclear gene *Amyrel* in the Schizophora, the Drosophilidae and the Afrotropical subgenus *Zaprionus*. Relative rate test showed no departure from the assumption of strict molecular clock, and thus *Amyrel* evolutionary rate was calibrated using the fossil record. The results show the family to have originated in the Early Eocene (50 MYA) and diversified in the Middle to Late Eocene. We also suggest that the Drosophilidae may be of Euramerican origin. Although Africa was an isolated continent from the Cretaceous to the Early Miocene, it may have been inhabited by early drosophiline radiations (*Chymomyza* and the *latifasciaeformis* group of *Scaptodrosophila*) in the Late Eocene. These taxa are

[2]Corresponding author: e-mail: yassin@legs.cnrs-gif.fr; current address: American Museum of Natural History (AMNH); Division of Invertebrate Zoology; 79th St. Central Park West, New York, NY 10024-5192, USA. Tel: +1.212.769.56.94; fax: +1.212.769.52.77; e-mail: ayassin@amnh.org

associated with palm trees, a dominant feature of the African Eocene environment. Molecular dating of the basal genus *Lissocephala* whose species breed early in premature fig fruits (syconia) show the genus to start diversification in the Oligocene. More 'derived' lineages (*Zaprionus* and Old World *Sophophora*) breed later on mature and ripening syconia. This ecological succession may recapitulate a phylogenetic trend towards an increase in alcohol tolerance and the utilization of sweet resources. Neogene climatic and tectonic changes have proposed many speciation models of Afrotropical drosophilids: the ecotonal and refugial hypotheses. A reanalysis of *Amyrel* divergence and geographical distribution in three clades (*Drosophila melanogaster* subgroup, *D. montium* group and the genus *Zaprionus*) showed that no single mode of speciation has prevailed. Although in an analysis at both the family and continental scales only general patterns are relevant, we hope that our review will serve as a lightening rod for future integrative research aiming to understand the biogeography of the Drosophilidae in Africa and in other regions.

Keywords: Africa, Cenozoic evolution, Pleistocene refugia, ecotonal speciation, breeding sites, molecular clock; *Amyrel* gene.

I. INTRODUCTION

The family Drosophilidae (Diptera) comprises nearly 4,000 species that show the most morphological and ecological diversity among muscomorphan flies. The geographic distribution of the family is cosmopolitan, found in almost every possible habitat on the planet except in the poles, but most species are found in tropical rainforests. Many of its species can easily be bred in the laboratory, providing an unique opportunity for evolutionary biology studies from ethological and ecophysiological experiments to comparative genomics (Markow & O'Grady 2007). The early discovery of genetic variation in geographical populations in *Drosophila* coupled with the identification of many sibling species with partial reproductive isolation or range overlap, were one of the bases that led to the formulation of the modern evolutionary synthesis (Mallet 2006).

Almost one eighth of the family diversity is found in Africa, among which 80% are endemic to the tropical fauna. The habitat diversity of this fauna is astonishing (Lachaise & Tsacas 1983), including association with specific fruits, flowers, fungi, leaves and stems, decaying wood, cercopid spittle masses and bee nests. Some species are also entomophagous, predating homopterans mealy bugs and aquatic dragonfly eggs and larvae of *Simulium* and chironomids. Early Drosophila systematists of the Afrotropical fauna were mostly exploring museum materials (e.g., Coquillett 1901; Adams 1905; Séguy 1938; Collart 1937; Duda 1939, 1940), but the great species richness and ecological diversity of this fauna has not been discovered before the field investigations of Burla (1954) in Côte d'Ivoire. In the 1970s, our laboratory (then called Laboratoire de Biologie et Génétique Evolutive) conducted numerous investigations in the Afrotropical region including Republic of Congo, Central African Republic, Côte d'Ivoire, Cameroon and Gabon, as well as Mauritius and the islands of the South Atlantic Ocean. The geographical distribution and ecological observations were summarized in two comprehensive papers (Tsacas et al. 1981; Lachaise & Tsacas

1983). At that time, only a single phylogenetic hypothesis of the Drosophilidae has been proposed on the bases of a few anatomical characters (Throckmorton 1962, 1975). One of the most important findings of this period was the confirmation of the Afrotropical origin of the *Drosophila melanogaster* species subgroup by the discovery of seven endemic species close to the two cosmopolitans (*D. melanogaster* and *D. simulans*) (Lachaise et al. 1988; David et al. 2007). Five out of the twelve drosophilid with complete sequenced genomes belong to this subgroup, with six geographical representatives of one of its species, *D. simulans* (Drosophila 12 Genomes Consortium 2007).

Since the early 1980s, many other Afrotropical countries have been prospected by our laboratory (mainly by the late Dr. Daniel Lachaise), including Kenya, Tanzania, Malawi, Madagascar, Seychelles, South Africa and Sao Tome and Principe. This resulted in almost doubling the number of described African species. Many new phylogenetic hypotheses, first morphological (Okada 1989; Grimaldi 1990) and then molecular (DeSalle 1992; Pélandakis & Solignac 1993; Russo et al. 1995; Tatarenkov et al. 2001; Remsen & O'Grady 2002; Robe et al. 2005; Da Lage et al. 2007), have been proposed for the Drosophilidae; as well as for a number of non-*Drosophila* genera (Grimaldi 1987a; Katoh et al. 2000; Hu & Toda 2001, 2002; Sidorenko 2002; Sultana et al. 2006; O'Grady & DeSalle 2008; Yassin et al. 2008a).

Phylogenetic biogeography is the search of an explanation of the geographical context of taxa formation (Van Veller et al. 2003). It stems from the evolutionary (Darwinian) assumption that each taxon has a center of origin in which it remained and/or from which it dispersed. By mapping the geographical distribution of current taxa on a phylogeny one can trace back the origin of each internal node (i.e. each speciation event). Dating is thus crucial for the analysis, as it can shed light on which geological event has played a role in a particular speciation (e.g., vicariance or dispersal). Following the same rationale, mapping the current ecological niche on the phylogeny can also be relevant in explaining past speciations when sympatry or the expansion of a certain host are invoked. Paleogeography and paleoecology are of equal impact on shaping the modern distribution of taxa as the evolution of niches and habitats are interrelated. Regarding the uniqueness of the Drosophilidae in comparative and ecological studies, such a knowledge is very significant.

The aim of this chapter is to put previous and new knowledge on the Afrotropical species within the context of their dated phylogenetic relationships as they are currently understood. We start by analyzing the taxonomic, regional and phylogenetic diversity of the African fauna. Because most African species belong to cosmopolitan or subcosmopolitan species groups or genera, their historical biogeography can not be explained without reviewing higher relationships between these ultraspecific taxa. We discuss thus the phylogenetic placement of the family Drosophilidae in light of recent molecular and comparative studies in the Acalyptrata. Then we propose a new hypothetical 'center of origin' for the Drosophilidae and discuss the worldwide evolution of the family until the formation of extant ultraspecific taxa. Next, we investigate the possible origin of some particular breeding niche in Africa in light of the recent paleobotanical discoveries. Finally, we discuss the relative role of Neogene tectonic and climatic changes in shaping the modern diversity, as can be deduced from recent molecular phylogenies of three Afrotropical clades.

II. DIVERSITY OF AFROTROPICAL DROSOPHILIDS

Data on contemporary Afrotropical drosophilids and their classification can be found in TAXODROS (Bächli 2008). We have used these data to reconstruct a species-country presence/absence matrix, after minor revisions for synonymy and accurate distributions from the literature (cited in Bächli 2008). In sum, the data show that Africa contains currently 527 drosophilid species classified under 31 genera and two subfamilies, and distributed in 51 countries. Countries were considered here geographical units, as only general continent-level patterns were investigated. A detailed microspatial analysis of species richness and centers of endimicity using latitudinal and longitudinal grids will be given elsewhere. The number of species per country is given in Figure 1. Côte d'Ivoire is the richest fauna (188 spp.), followed by South Africa (111), Democratic Republic of Congo (94), Cameroon (89), Uganda (85), Kenya (83), Tanzania (73), Malawi (63), Republic of the Congo (57), Zimbabwe (54) and Gabon (47). Among the adjacent islands, Madagascar bears the largest fauna (64 spp.), followed by Sao Tome and Principe (37), Seychelles (35), Mauritius (31), Reunion (28) and the islands of the South Atlantic Ocean (Saint Helena and Tristan da Cunha, 12 spp. each). Four species-rich areas can thus be recognized: Upper Guinea, Lower Guinea, East Africa including Madagascar and South Africa.

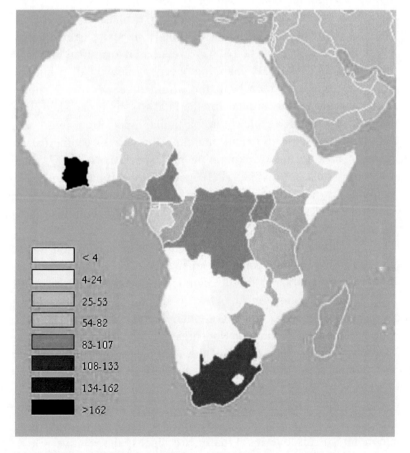

Figure 1. Diversity of Afrotropical drosophilids shown as the number of species per countries (species richness).

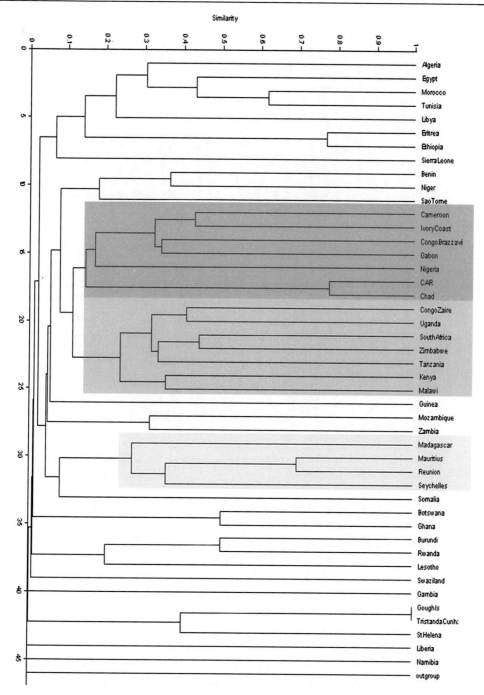

Figure 2. UPGMA areagram of Afrotropical drosophilids highlighting three main biogeographical regions: West Africa (dark gray), East and South Africa (gray), and the insular Indian Ocean (light gray).

When only species that are unique to certain country were retained to correct for endemism, equatorial Africa (Lower Guinea and East-Africa) appeared to be more homogeneous. As with species richness, Côte d'Ivoire was shown to bear the highest number of endemic species (53 spp.), followed by South Africa (34), Democratic Republic of Congo and Tanzania (15 each), Cameroon and Zimbabwe (13 each), Kenya and Malawi (11 each), and Nigeria and Uganda (7 each). Among the adjacent islands, Madagascar bore the highest endemism (25 spp.), followed by Seychelles (12), Tristan da Cunha (8), and Mauritius and Sao Tome and Principe (5 each).

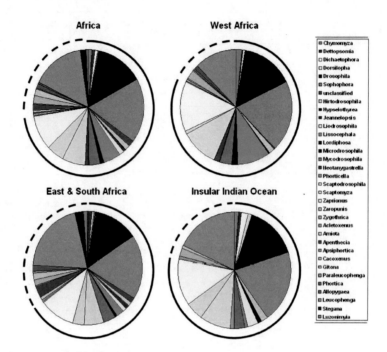

Figure 3. Generic and subgeneric diversity of Afrotropical drosophilids according to biogeographical regions. Solid line = subfamily Drosophilinae; dashed line = subfamily Steganinae.

In order to investigate the relationship between these areas, cosmopolitan species (those occurring in more than 9 countries) and singletone species (those with single occurrence) were eliminated. An UPGMA areagram was then reconstructed from the corrected matrix (Figure 2). From this, three main Afrotropical endemism clusters can be recognized: West Africa (Cameroon, Côte d'Ivoire, Gabon and Republic of Congo), East and South Africa (Democratic Republic of Congo, Kenya, Malawi, South Africa, Tanzania, Uganda, Zimbabwe) and finally the insular Indian Ocean (Madagascar, Mauritius, Reunion and Seychelles). Combining the two previous diversity analyses, an East-West (longitudinal) cline of endemism is found in West Africa, with Upper Guinea bearing many unique taxa, whereas in East and South Africa the endemism cline is mostly North-South (latitudinal). Within this final cluster, the basal position of Malawi is consistent with previous observations of the great heterogeneity of its fauna (Chassagnard et al. 1997).

Phylogenetic diversity is usually measured from the branch lengths of a molecular tree (Mooers 2007). This, however, is difficult to estimate in African drosophilids as molecular sequences are available for only 14% of species. A less accurate but still efficient alternative is to use ultraspecific taxonomic ranks (i.e. genera and subgenera). It must thus be kept in mind that taxonomic ranks, in contrast to taxonomic categories, lack any comparative criterion (Dubois 2007).

Four drosophilid genera are endemic to the Afrotropical region: *Allopygaea, Jeannelopsis* and *Zaropunis*. *Allopygaea* (3 spp.) belongs to the subfamily Steganinae and is endemic to West Africa. It is close to the cosmopolitan genus *Paraleucophenga*, represented in the African fauna by a single species *P. semiplumata*, which is widespread in East and South Africa (Tsacas 2000). *Jeannelopsis* (4 spp.) is close to *Dichaetophora*. One of its species is endemic in West Africa, whereas the remaining species are endemic in East and South Africa (Tsacas 1990). *Zaropunis* is a monotypic genus, i.e. containing a single species, *Z. melanope*, widespread throughout the mainland. It is close to the genus *Zaprionus* (Tsacas 1990).

The genera *Hypselothyrea* and *Neotanygastrella* are only found in West Africa, both belonging to the subfamily Drosophilinae. The genera *Apsiphortica, Paraleucophenga, Stegana* and *Luzonomyia* are only found in East and South Africa, all belonging to the subfamily Steganinae. In the insular Indian Ocean, the genera *Phorticella* (Drosophilinae) and *Acletoxenus* (Steganinae) are found. All the abovementioned genera are Paleotropical to cosmopolitan. In sum, East and South Africa appears to be the most phylogenetically diversified cluster. It is the most speciose as well (297 spp. *vs.* 247 in West Africa). This is in concordance with the fact it is the most environmentally diversified and can also be attributed to general latitudinal cline of species diversity.

III. THE ORIGIN AND EVOLUTION OF THE DROSOPHILIDAE

3.1. Origin of the Drosophilidae

The family Drosophilidae is one of the nearly 120 families of the suborder Brachycera, (flies with reduced antennae). It belongs to the muscomorphan section Schizophora of flies that pupate within the penultimate larval puparia and whose heads are provided with an inflatable membraneous sac (the ptilinium) that help in the emergence from the puparium. Schizophoran flies are classified under two subsections: Acalyptrata, flies whose the posterior margin of the forewings lack posterior lobes (calypters); and Calyptrata, flies with calypters. The monophyly of acalyptrate flies is questionable, and the subsection is widely thought to be paraphyletic in respect to Calyptrata (Grimaldi & Engel 2004; Yeates et al. 2007; but see McAlpine 1989). Within Acalyptrata, the Drosophilidae belongs to the superfamily Ephydroidea, where it forms a monophyletic clade with the family Curtonotidae (Grimaldi, 1990).

The fossil record supports an origin of Schizophoran about 65 MYA at the boundary between the Cretaceous and the Tertiary (Grimaldi & Engel 2004). However, there has been a much disparity in the strict molecular clock estimates of the divergence age between the two schizophoran subsections, going from 99 MYA using larval haemolymph proteins (Beverley & Wilson 1984) to 81 MYA using 28S rDNA sequences (Wiegmann et al. 2003). Relaxed molecular clock estimates of the 28S sequences resulted in a divergence age around 50 MYA (Wiegmann et al. 2003), which is lesser than the fossil estimate.

Molecular phylogenetic relationships within the Drosophilidae have been extensively investigated. Da Lage et al. (2007) have recently conducted a phylogenetic revision of the family using the *Amyrel* gene, a remote paralog of the amylase gene family. This study, which represents the most comprehensive revision in terms of number of species (164 spp.) with

1,515 characters (base pairs, bp), has been followed by the addition of 16 species mainly of the genus *Zaprionus* (Yassin et al. 2008a). Moreover, Maczkowiak & Da Lage (2006) sequenced *Amyrel* in a number of brachyceran flies and showed that it has originated by duplication in Schizophora, before the split between Acalyptrata and Calyptrata. Instead of using this unique gene to investigate phylogenetic relationships within the Schizophora, Maczkowiak & Da Lage (2006) used McAlpine's (1989) monophyletic hypotheses for both Acalyptrata and Calyptrata to trace and understand the molecular evolution of *Amyrel*.

We have examined the molecular clock hypotheses of the *Amyrel* sequences in the Drosophilidae using Tajima's (1989) relative rate test, and the results (not shown) did not show a signification departure from neutrality. *Amyrel* sequences seem thus adequate not only to reconstruct phylogenies within the Schizophora and the Drosophilidae, but also to estimate divergence times using strict clock. Figure 4 shows the phylogenetic relationships between schizophoran superfamilies using Maczkowiak & Da Lage's (2006) *Amyrel* sequences, with the tree root calibrated at 65 MYA as suggested by the fossil record. The results clearly reconfirm the paraphyly of Acalyptrata, and suggest the divergence between Calyptrata and their closest acalyptrates around 55 MYA. The sister relationship between the Drosophilidae and the Curtonotidae suggested by Grimaldi (1990) is also reconfirmed, and the Drosophilidae appears to have originated during the Early Eocene around 50 MYA. Throckmorton (1975) was the first to attempt to estimate the origin date of the Drosophilidae, and he noted: "the family Drosophilidae may be 50 million years old, or older."

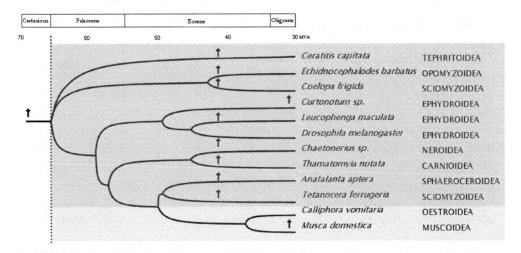

Figure 4. *Amyrel* chronogram of schizopheran superfamilies. *Amyrel* sequences are reanalyzed from Maczkowiak & Da Lage (2006). Fossil species (†) are from Evenhuis (1994), and are plotted according to their estimated age. Note the paraphyly of Acalyptrata (dark gray) in respect to Calyptrata (light gray).

3.2. Phylogeny of the Drosophilidae

Traditional Drosophila systematics has usually classified the drosophilids under two subfamilies: the Drosophilinae and the Steganinae. Earlier phylogenetic investigations regarded the Steganinae as a primitive paraphyletic group (Throckmorton 1962, 1975).

Evidence for the monophyly of the Steganinae came mainly from morphological cladistic studies (Grimaldi 1990; Sidorenko 2002) and from studies of the nuclear gene 28S rDNA (Remsen & O'Grady 2002). The monophyly of the Drosophilinae is far more supported by morphological and molecular data. Throckmorton (1962) divided the subfamily into two radiations: a 'primitive' paraphyletic radiation comprising the subgenus *Sophophora* of *Drosophila* and its allied genera, and a 'derived' monophyletic radiation comprising the subgenus *Drosophila s.s.* and its allied genera. We are going to refer to these two radiations as Lower and Higher Drosophilinae, respectively. Later molecular analyses mostly agreed with Throckmorton's (1962) general division and many of the subdivisions he proposed, confirming the value of his heuristic deductions.

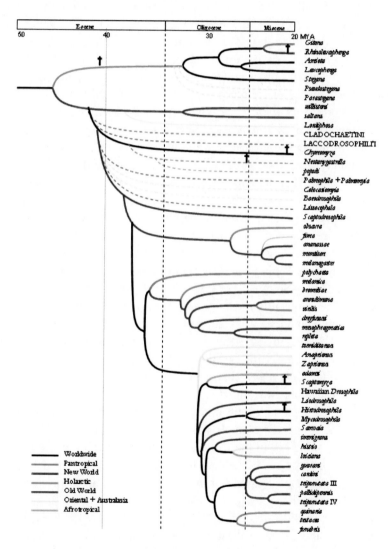

Figure 5. *Amyrel* chronogram of genera and major species groups of the Drosophilidae. *Amyrel* sequences reanalyzed from Da Lage et al. (2007). Fossil species (†) are from Evenhuis (1994). Thick lines = *Amyrel* branches; thin lines = non-*Amyrel* molecular branches; dashed lines = morphological branches (from Throckmorton 1975; Grimaldi 1990). Geographical distributions are mapped on the chronogram.

Figure 5 shows the phylogeny of the Drosophilidae as it can be understood today in light of recent morphological and molecular studies. The main chronological backbone of this phylogeny, however, is our estimates from the *Amyrel* sequences of Da Lage et al. (2007) after calibration with the fossil record. When these sequences were not available for a particular species group of relevant interest, they have been replaced by other nuclear sequences. This was the case for the Steganinae (28S rDNA from Remsen & O'Grady 2002) and the genera *Chymomyza* (*Gpdh* from Kwiatowski et al. 1997), *Scaptodrosophila* (*amd* from Tatarenkov et al. 2001) and *Lordiphosa* (*Adh* from Katoh et al. 2000). Mitochondrial sequences were discarded as at this deep phylogenetic level they may be misleading, except in the case of *Lissocephala*, where 16S rDNA sequence divergence estimates were used from Harry et al. (1996; no GenBank accession numbers). However, none of the molecular studies have included genera as many as the morphological revision of Grimaldi (1990). Some interesting branches were thus grafted on the molecular tree besides their closest relatives from Grimaldi's (1990) cladogram. Future molecular phylogenetic investigations adding more taxa will certainly add accuracy and precisions on the general patterns given here.

Currently, six radiations can be recognized in the drosophilid phylogeny: the subfamily Steganinae, New World *Sophophora*, the genus *Chymomyza*, the genus *Scaptodrosophila*, Old World *Sophophora* and *Drosophila s.s.*, each with a number of allied genera. The last five radiations are firmly Drosophilinae, but the branching order between the second to the fifth radiation is highly debatable. Here, we follow the *Amyrel* divergence estimates but it should be kept in mind that phylogenetic hypotheses of Lower Drosophilinae can be overturned in the future. Two out of the six radiations belong to the subgenus *Sophophora*, now containing 10 species groups: seven in the Old World (*ananassae*, *dentissima*, *dispar*, *fima*, *melanogaster*, *montium* and *obscura*), two in the New World (*saltans* and *willistoni*), and one (*populi*) in Alaska, Russia and Scandinavia. Throckmorton (1962) was the first to question the monophyly of *Sophophora*. Later molecular phylogenies starting from single nuclear genes (Pélandakis & Solignac 1993; Katoh et al. 2000) up to whole nuclear genomes (Drosophila 12 Genomes Consortium 2007) failed to strongly support its monophyly too (but see for nuclear and mitochondrial weak support DeSalle 1992; Kwiatowski & Ayala 1998; Tatarenkov et al. 2001; O'Grady & Remsen 2002; O'Grady & Kidwell 2002; O'Grady & DeSalle 2008). This can be an artifact of the known particular codon usage bias in nuclear genomes of New World *Sophophora*, but we do not agree with this explanation for many reasons. First, codon usage bias was found in the two species groups of New World radiations meaning that it has a strong phylogenetic signal that has prevailed for at least 34 million years (Powell 2003). Second, there is no single molecular study to our knowledge investigating coding bias in the Steganinae or in the Curtonotidae to examine its ancestry. Third, there is substantial evidence from the modifications of abdominal sternites in male drosophilids, a character of strong phylogenetic signal (Grimaldi 1990), that can firmly distinguish the six radiations of the Drosophilinae (Wheeler 1960, Throckmorton 1962; and see a recent investigation in the Curtonotidae Pollock 2002). Fourth, Wheeler (1947, 1952) mentioned an anteromedian pouch in the vaginal wall in the *willistoni* group of New World *Sophophora*, and in an Alaskan species of the genus *Chymomyza*. This pouch has not been observed in any species of the other main radiations (Throckmorton 1962), although being found in the three curtonotid genera, where in the highly derived genus *Cyrtona* it has been modified for ovovivipary (Meier et al. 1997; Pollock 2002).

The two subfamilies have split around 46 MYA in concordance with the earliest drosophiline fossil, *Electrophortica succina*, from the Middle Eocene Baltic amber (48-40 MYA), that Hennig (1965) hesitantly classified under the Steganinae. 28S rDNA-dependent estimates show that the Steganinae, although may be present as an *Electrophortica*-like lineage through the Eocene, have been diversified in the Early Oligocene. Early steganine branches, the genera *Parastegana* and *Pseudostegana*, are endemic in Southeast Asia (Toda & Peng 1992; Sidorenko 2002). The genera *Stegana*, *Leucophenga*, *Amiota* and *Phortica* are worldwide, but they have many allied genera in the Old World. It appears thus that the subfamily has diversified in Asia through the Oligocene and has subsequently colonized America during the Early Miocene (22 MYA) by the subtribes Gitonina (including the genera *Gitona* and *Cacoxenus*) and Acletoxinina (including the genera *Acletoxenus*, *Mayagueza*, *Pseudiastata*, *Rhinoleucophenga* and *Hyalistata*). This is in concordance with the discovery of a fossil *Hyalistata* in the Early Miocene Dominican amber (Grimaldi 1993). These subtribes represent the most ecologicall derived steganines, as with the exception of *Gitona* whose larvae are leaf miners, all other genera are predators of hemipterans. Gitoninan genera are Pantropical, whereas all acletoxininans are Neotropical, with the exception of the Paleotorpical genera *Acletoxinina* and *Luzonimyia* (Grimaldi 1988).

3.3. Lower Drosophilinae

Unlike the Steganinae, the Drosophilinae shows a much older diversity, with a distinct biogeographical trends. Earlier lineages diverging from 46 to 42 MYA (New World *Sophopohora* and *Lordiphosa*, the tribe Cladochaetini *sensu* Grimaldi 1990, the subtribe Colocasiomyina *sensu* Grimaldi 1990 and Grimaldi et al. 2003, and the *Chymomyza* genus group *sensu* Grimaldi 1990) are exclusively American (the Cladochaetini), showing a disjunctive American-Asian distribution (New World *Sophophora* and *Lordiphosa*, and the Colocasiomyina) or cosmopolitan (the *Chymomyza* genus group). However, the divergence between New World sophophorans and the Holarctic *Lordiphosa* has taken place in the Late Eocene (38 MYA), and we think that this radiation is basal or as old as the tribe Cladochaetini. Colocasiomyina is a paraphyletic subtribe with the American genera (*Palmomyia* and *Palmophila*) being basal to the Asian genera (*Baeodrosophila* and *Colocasiomyia*) (Grimaldi et al. 2003; Sultana et al. 2006). All these genera have evolved strict relation with Araceae and Pandanacae. The genus *Chymomyza* is cosmopolitan and it has been divided into five species groups. One of these groups, *aldrichii*, is entirely American (Grimaldi 1986), with a single South African exception (Tsacas 1990). The *aldrichii* group can be considered basal to the other groups, judging from the vaginal pouch in one of its Alaskan species (Wheeler 1952). The oldest drosophiline fossil belongs to the allied genus *Neotanygastrella* and has been found in the Late Oligocene Mexican amber, and there is also an allied fossil genus *Protochymomyza* that has been found in the Early Miocene Dominican amber (Grimaldi 1987b). *Chymomyza* and *Neotanygastrella* breed in wood trunks. The sophophoran species group *populi* belongs also to this *Chymomyza* genus group and it is found in Alaska, Russia and Scandinavia (Throckmorton 1975). This witnesses the old diversity of this genus group in the New World and the old relationships between America and Asia.

From 41 to 37 MYA radiations took another pattern, mainly in the Old World (the genera *Lissocephala* and *Scaptodrosophila*), but relationships with the New World has not been fully interrupted as the infratribe Laccodrosophiliti (including the genera *Laccodrosophila* and *Zapriothrica*, Grimaldi 1990) can be considered a Neotropical vicariant of *Scaptodrosophila* (Throckmorton 1975). African *Lissocephala* breed exclusively in figs (Moraceae), but the ecological niche of their Asian relatives (including the genus *Mulgravea*) remains obscure, save *L. powelli* which breeds in the nephric grooves of land crabs (Harry et al. 1996). Old World *Sophophora* appears basal to Higher Drosophilinae. There is also a great increase in these lineages towards fruit breeding. Although present as a lineage since the Late Eocene, diversification of Old World sophophorans started in the Late Oligocene by the budding of the mainly Holarctic *obscura* group (27 MYA). Further diversification has taken place through the Late Oligocene to Early Miocene.

3.4. Higher Drosophilinae

There are two main radiations of Higher Drosophilinae: the *virilis-repleta* and the *immigrans-Hirtodrosophila* (Throckmorton 1962, 1975). Diversification in the *virilis-repleta* radiation appears older, with the most basal species group (*polychaeta*) being Pantropical. The *bromeliae* and *melanica* groups are Oriental or Holarctic respectively, whereas most of the higher groups being Neotropical (*annulimana, dreyfussi, mesophragmatica* and *repleta*) or Holarctic (*virilis*). The *immigrans-Hirtodrosophila* is certainly the most diverse at the generic level. It has originated in Southeast Asia where a burst of generic diversification has taken place. It may have reached the Americas during the Late Oligocene. This period has probably also been characterized by successful transpacific migrations, witness the diversity of Hawaiian drosophilids. There is now a great evidence that this microcosm of morphological and phyletic diversification in the Hawaiian islands is the legacy of a single colonization event by a *Scaptomyza*-like ancestor (O'Grady & DeSalle 2008). O'Grady & DeSalle (2008) also think that the cosmopolitan genus *Scaptomyza* has originated in Hawaii. Representative of many Higher drosophiline genera were found in the Early Miocene Dominican amber (Grimaldi 1987b), a period during which most of the *Drosophila* species groups known today have been formed.

3.5. Historical Biogeography

Throckmorton (1975) suggested from current distribution of extant species the family Drosophilidae to have originated in the tropics. However, the exact location of the ancestral drosophilid is hard to determine. The family has originated in the middle Eocene, a period of homogeneous warm climate, with temperate forests extending right to the poles and tropical rainforests extending as far north as 45or 60° (Tiffney & Manchester 2001). As mentioned above, the oldest drosophilid fossil *Electrophortica* was found in the Middle Eocene Baltic amber (48-40 MYA). Europe has been separated from Asia by the Turgai strait, a large body of water from the Caspian Sea to the paleo-Arctic region, from the Middle Jurassic to the Late Eocene (160-33 MYA). This has acted as a geographical barrier for all transeurasian migrations (Tiffney & Manchester 2001). In the meantime, Europe was a tropical archipelago bathed by the warm waters of the circumequatorial Tethys Sea. It was connected to the west to North America via the North Atlantic Land Bridge which sank in the Middle Eocene. An Euramerican origin of the Drosophilidae would thus appear most plausible. Although European ancestral groups might become extinct during the Oligocene "grande coupure", American early groups (New World *Sophophora* and Cladochaetini) prevailed until today.

The worldwide spread of the family has taken place at the boundary between Middle and Late Eocene, most probably via the Bering Strait which was open to terrestrial migrations until its closure in the Late Miocene (7-5 MYA). In the Middle Eocene, both Alaska and Siberia were dominated by a subtropical climate, but their tropical flora started to retreat southward throughout the Late Eocene, and to be replaced by deciduous woodlands. This is in concordance with the gradual global cooling starting during the Middle Eocene. It is at this period that ancestors of *Lordiphosa* and the *Chymomyza* genus group started to expand throughout Asia up to Africa via trans-Tethyan filter routes, along with other *in situ* originating genera (*Lissoceophala* and *Scaptodrosophila*) and Old World *Sophophora*. The sophophoran group *populi* may be a biogeographical relict of the Bering route.

An intriguing aspect of the present hypothesis is the high ecological specialization of extant species of the early drosophiline lineages. Most Drosophilidae are generalist saprophagous or micromycophagous, and thus a generalist breeding strategy was hypothesized to be ancestral to the whole family. Lower Drosophilinae however show great specialization in particular flower and cut tree trunk breeding. Flower-breeding genera include *Diathoneura* and some *Cladochaeta* (tribe Cladochaetini), Laccodrosophiliti, and many *Scaptodrosophila*. Genera of the infratribe Colocasiomyina breed exclusively in the inflorescence of Araceae plants (Grimaldi et al. 2003; Sultana et al. 2006). Tree trunk breeders include the *Chymomyza* genus group and the *latifasciaeformis* species group of the genus *Scaptodrosophila* (Wheeler 1952; Burla 1955; Lachaise 1975). Tree trunk breeding is of a particular biogeographical interest as trunks (especially Araceae) might have helped as floating arks in transoceanic dispersal during Middle Eocene (Morley 2003; Cuenca et al. 2008). African *Lissocephala* are a classical example of highly specialized fig breeders (Lachaise & Tsacas 1983; Harry et al. 1996), whereas a single Oriental species was reported to breed exclusively in the nephritic grooves of crabs (Okada 1985). Information on the breeding sites of New World *Sophophora* and *Lordiphosa* are scarce, but most of their species as other early drosophilines are not readily attracted to standard fermenting fruit traps in contrast to many Old World *Sophophora* and Higher Drosophilinae (Pipkin 1965; Lastovka & Máca 1978). Information on the ecology of the Curtonotidae are quasi-negligible, although they have been hypothesized to breed in animal dung (Tsacas 1977; Meier et al. 1997; but see Kirk-Spriggs 2008).

As shown from the lineage-through-time plot in Figure 6, major drosophilid radiations took place during the Oligocene, without significant deviation from Yule's process of constant rate of cladogenesis (Nee 2006). Most of these radiations, either in the Steganinae or in the Drosophilinae, occurred in Southeast Asia. This can be explained by two paleontological observations. First, the Oligocene climate was a period of global cooling, resulting in the evolution of temperate forests and more open ecosystems that intergraded with the tropical rainforests in Asia. The evolution of the Steganinae can firmly reflects the role of natural selection across an environmental gradient in diversification (i.e. ecotonal speciation), as early lineages such as *Stegana* breed in humid close environments, whereas more derived lineages such as *Amiota* and *Leucophenga* breeding in drier and more open environments. The second important Oligocene event was the elevation of the Tibetan plateau due to India-Eurasian collision around 35 MYA (Ali & Aitchison 2008) resulting in the strengthening of monsoonal seasonal climate that continued throughout the Miocene (Zhisheng et al. 2001; Harris 2006).

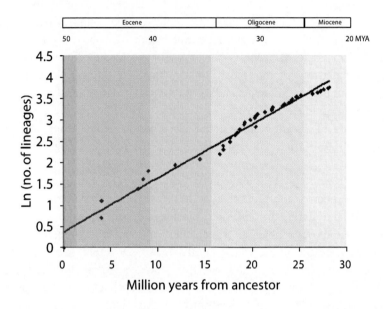

Figure 6. Lineage-through-time plot of *Amyrel* diversification of genera and species groups of the Drosophilidae.

IV. AFRICA: THE SHAPING OF THE MODERN FAUNA

4.1. Evolution in the 'Palmae Province'

When the Drosophilidae first appeared in the Middle Eocene, Africa had already been an isolated continent since its separation from its last remnant connection with a Gondwanan continent, South America, in the Middle Cretaceous (110 MYA) (Sanmartín & Ronquist 2004). Such geographic isolation continued until the Early Miocene, when finally Africa became connected to Eurasia through the Middle East (23-16 MYA). This has resulted in the evolution of a high degree of endemicity in African fauna and flora for lineages descending from Gondwanan vicariant ancestors (e.g., reptiles) or sporadic Laurasian colonists (e.g., mammals). Gheerbrant & Rage (2006) recognized seven trans-Tethysian dispersal phases from Laurasia to Africa during its isolation using mammalian fossils. The three older phases (65, 58 and 55 MYA) are characterized by Euramerican faunas and they all precede the divergence of the Drosophilidae and the Curtonotidae (50 MYA). The next four phases (48, 40, 37 and 34 MYA) are of southeast European and Asiatic faunas. It is during the last three phases that ancestral Lower Drosophilinae might have reached the continent (*Chymomyza*, *Neotanygastrella*, *Lissocephala*, *Scaptodrosophila* and Old World *Sophophora*). These taxa would have been able to spread throughout the continent where a tropical palm-dominated rainforest ecosystem prevailed by that time (Jacobs 2004). Palm diversity increased dramatically during the Late Eocene in concordance with the dependence of the early drosophiline arrivals (*Chymomyza*, *Neotanygastrella* and *Scaptodrosophila*) on them, but it started to decline in the Early Oligocene, with the establishment of a more 'modern' mesic rainforest floras in West Africa as the continent continued to drift northward. Moreover, ancestral lineages might be able to spread to Madagascar, as the island was connected to the mainland via the Mozambique Channel Bridge from 46 to 26 MYA (McCall 1997). However, extant representatives of *Chymomyza*, *Neotanygastrella*, *Lissocephala* and *latifasciaeformis* are only or mostly found in West Africa, especially Upper Guinea.

4.2. The Rise of Lowland Forests and a Phylogenetic Perspective on the Evolution of Fig-Breeding Strategies

The second wave of drosophilid colonization of Africa originated in the Oligocene and resulted in the endemic evolution of the subgenus *Zaprionus s.s.* (30 MYA), *Drosophila adamsi* (30 MYA) and the *D. loiciana* species complex (26 MYA). However, nothing would say that the ancestors of these taxa attained the continent by that time. Indeed, the dating reflects the time these taxa split from their Asian closest relatives. Colonization of Africa by these taxa might thus have taken place in the Late Oligocene or the Early Miocene, when Africa became completely connected with Eurasia.

Old World *Sophophora* are the dominant taxonomic group in the Afrotropical fauna (18% of species richness), with two species groups out of seven being exclusively endemic in Africa (the *fima* and *dentissima* groups). With the exception of *dentissima* and the Australasian *dispar* groups, there is a wealth of molecular studies investigating the other groups. There are, however, many species (mostly Oriental or Australasian) that are not assigned to any of these groups. All studies agree that the cosmopolitan *obscura* group is the most basal (27 MYA). This group is firmly of Old World origin, but the exact origin is opaque to intuition, although one of its five subgroups (*microlabis*) is endemic in East African mountains and is usually placed basal to the remaining subgroups along with the Palearctic subgroup *subobscura* (Cariou et al. 1988; Gao et al. 2007). A rapid radiation took place at the Oligocene-Miocene boundary (24 MYA) for the higher species groups (*ananassae*, *fima*, *melanogaster* and *montium*). Most and basal species of the threes groups, except *fima*, are Oriental and Australasian, and their African representatives have subsequently colonized the continent in the Middle (*melanogaster* group) to Late Miocene (*montium* and *ananassae*).

The adaptive success of all of the previous groups in the African tropics can be attributed to their fruit breeding niche. In West Africa, the Oligocene was a period of floral turnover, as palm forests were gradually replaced by lowland mesic forests (Jacobs 2004). A major component of tropical rainforest are figs (Moraceae). Due to their high diversity and strict dependence on agaonid wasps in pollination, figs constitute a permanent food source in the tropics for pollinating and non-pollinating organisms (Cook & Rasplus 2003). Forty seven percent of fruit-breeding drosophilid diversity in the Afrotorpical region were reared from Moraceae, mainly the genus *Ficus* (Lachaise & Tsacas 1983). Daniel Lachaise has thoroughly studied the succession of drosophilid species breeding on the fig fruit (the syconium) (Lachaise 1977; Lachaise et al. 1982; Lachaise & Tsacas 1983). He recognized two obligate breeders: the genus *Lissocephala* and the *fima* species group of Old World *Sophophora*.

Lissocephala species breed in the early reproductive stages of the fig (floral immaturity), whereas the *fima* species breed in the fallen and decaying syconium. Between the two phases, waves of facultative breeders, mainly of the genus *Zaprionus* and other Old World *Sophophora* (species groups *ananassae*, *melanogaster* and *montium*), breed in the syconium during floral maturity and early post-sexual period. The origin and evolution of such a stable ecological system is a very interesting query.

Fruit-breeding strategy requires mainly the metabolic evolution of tolerance to alcoholic stress in fermenting fruits. Merçot et al. (1994) showed that the activity of alcohol dehydrogenase (ADH) is highly correlated with alcohol tolerance in drosophilid species breeding in artificial (e.g., breweries) or natural (e.g., fruits) high sweet resources, whereas no relation was observed for species using non-sweet substrates (e.g., flowers, fungi and decaying plant material). So one can expect primitive drosophiline niches (anthophily and xylophily) to not require high ADH activity. This may be the case for the ancestral *Lissocephala* when it first shifted to immature syconia. As fruit ripening proceeds, alcohol concentration in the substrate increases, requiring higher tolerance in more derived drosophilids. *Lissocephala* species represent the oldest fig-breeding lineage in Africa. Calibrating 16S rDNA pairwise sequence divergences from Harry et al. (1998) shows major *Lissocephala* diversification to take place in the Oligocene to Middle Miocene in synchrony with Afrotropical *Ficus* (Rønsted et al. 2005, 2007) (Figure 7). *Lissocephala* specialization on fig has required the convergent evolution of many morphological and behavioral strategies that facilitated the penetration of the syconium via ostioles or wasp-bore tunnels (Lachaise & McEvey 1990; Harry et al. 1998). This has resulted in the formation of species-specific ontogenic niches in different *Ficus* species, although adults are usually generalist. In contrast, facultative and obligate fig-breeding in *Zaprionus* and Old World *Sophophora* was the result of biochemical evolution in alcohol tolerance via intensification of function (culminating in the *fima* group) rather than morphological changes. Indeed, in these later taxa that have mainly diversified in the Miocene (see below), neither a *Ficus* species-specific relation nor an ontogenic niche separation has evolved (Figure 7). From a phylogenetic perspective, the drosophilid breeding succession on the syconium recapitulates the age of the diversification of their lineages (Figure 8).

4.3. Tectonic and Climatic Changes During the Neogene

In Africa, Miocene and Pliocene tectonic and climatic changes have been deeply investigated as they are closely related to hominid evolution (Maslin & Christenen 2007). Two main interrelated environmental aspects with deep impact on extant Afrotropical drosophilids arose in the Miocene. First was aridity as the climate continued to cool throughout the Miocene, culminating in the glaciations episodes in the Quaternary. This has favored the rapid expansion of grasses and savannahs, a major biome in Africa, that started in the Middle Miocene (16-12 MYA) and predominated in the Late Miocene (9-7 MYA). However, climatic changes were not gradual. They rather fluctuated permitting episodic floral turnovers at the forest/savannah boundaries. Second was orogenesis as the collision between the Afro-Arabian and Eurasian plates has induced high tectonic and volcanic activity throughout the Miocene. In the Early Miocene, the African biota was homogeneous with Asia, but the formation of the Red Sea and the Great Ethiopian Rift (19-12 MYA) in Eastern

Africa placed geographical barriers for further biotic exchanges. The Late Miocene witnessed major rifting in East Africa (Tanganyika 12 MYA, Malawi 7 MYA). Major uplifting took place during the Pliocene and Early Pleistocene (5-2 MYA) in East and South Africa. In West Africa, the Cameroon Volcanic Line started to form the islands of the Golf of Guinea (Principe 31 MYA; Sao Tome 13 MYA; Annobon 5 MYA), and continued on the inland including Mt Cameroon (since Late Miocene) (Burke 2001). Tectonic uplifting is also known to intensify aridity during the Late Miocene, Pliocene and the Quaternary (Sepulchre et al. 2006). Large African deserts originated also in the Miocene: Namib (16 MYA) and Sahara (10 MYA). However, no desert-adapted drosophilids are known from Africa.

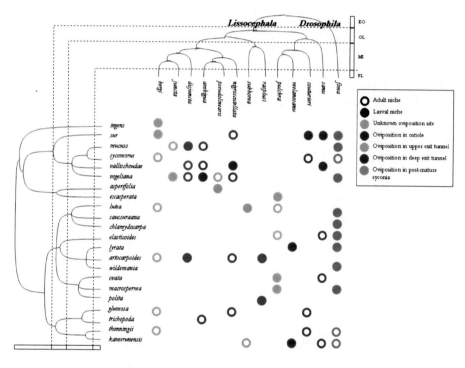

Figure 7. Comparison of phylogenetic relationships between *Ficus*-breeding Afrotropical drosophilids (genus *Lissocephala* and the *Drosophila fima* species group) and between their fig (*Ficus*) host. *Lissocephala* chronogram was inferred from 16S rDNA sequences (from Harry et al. 1996), whereas that of *Ficus* was compiled from Rønsted et al. (2005, 2007). *Lissocephala* species, breeding in premature syconia, show higher degree of host-specificity than *D. fima*, which is breeding in post-mature syconia. Note the independence of the evolution of oviposition site in *Lissocephala* (mapped) from both the *Lissocephala* and *Ficus* phylogenies.

Many hypotheses have been proposed to explain the high species diversity in tropical rainforests, among which the refugia and the ecotonal hypotheses prevail (Moritz et al. 2000). The refugia hypothesis states that at time of Pleistocene glacial episodes (ice ages), tropical rainforests retracted to fragmented refugia (usually in mountains or riversides). This has resulted in the isolation of the populations of many organisms, which allowed the independent accumulation of different genetic variation in the isolated populations, leading to the evolution of reproductive incompatibility at times of secondary contact during the interglacial episodes. Glaciations were thus thought to play as speciation pump. The glacial refuge hypothesis was first proposed to explain the diversity in the Neotropics (Haffer 1969). Shortly after, the hypothesis was brought to African biodiversity (Livingstone 1975, 1982), and was strongly advocated to explain speciation in Afrotropical drosophilids (Tsacas et al. 1981; Lachaise et al. 1988; Chassagnard et al. 1997; Lachaise & Chassagnard 2001). According to Linder (2001), the centers of floral endemism he has determined in his study reflects the African refugias (Upper Guinea, Lower Guinea, Kivu, East African coast, Zambezi-Congo watershed, Huilla, and the Cape). Although much of these presumed refugias correspond to the centers of endemisim of Afrotropical drosophilids, a critical test would be dating for speciation events (during the Pleistocene) that have resulted in the presence of sister taxa endemic to the different refugias. Dating test for the ecotonal hypothesis, *i.e.* diversification by natural selection across an environmental gradient, does not imply diversification to have taken place in the Pleistocene, but rather to coincide with episodes of climatic turnovers. A knowledge of the current ecology and physiology of the current taxa is thus crucial for the ecotonal hypothesis, rather than distribution across presumed refugias. We have tested these hypotheses thus using recent *Amyrel* phylogenies of African *Sophophora* (Da Lage et al. 2007) and *Zaprionus* (Yassin et al. 2008a).

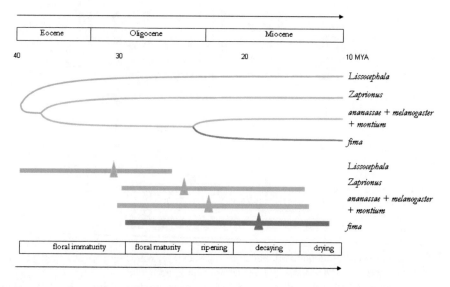

Figure 8. Principle of ecological recapitulation where the chronological succession of drosophilid species breeding in *Ficus* (below) reflects a phylogenetic trend (above) in utilization of sweet resources (i.e. alcohol tolerance). Triangles signal maximum egg production (data from Lachaise & Tsacas 1983).

V. TEMPO AND MODE OF SPECIATION
IN AFROTROPICAL DROSOPHILIDS

5.1. The *Melanogaster* Species Group

The *melanogaster* group contains nearly 65 species classified under 10 species subgroups, of which only two are present in Africa (*flavohirta* and *melanogaster*). The Oriental and Australasian subgroups form together a paraphyletic clade in respect to the

African subgroups (Da Lage et al. 2007). The *flavihorta* subgroup contains a single species *D. flavohirta* which breeds in the flower of *Eucalyptus* (Tsacas & Lachaise 1983; McEvey et al. 1989; Nicolson 1994). It is found mainly in Australia and in high latitude Africa (South Africa and Madagascar). The *melanogaster* subgroup contains nine species, endemic in Africa, with the exception of two cosmopolitan species (*D. melanogaster* and *D. simulans*) (Lachaise et al. 1988, 2004; David et al. 2007). The dated *Amyrel* phylogeny of the African *melanogaster* subgroups is given in Figure 9 (Da Lage et al. 2007). The split of the two subgroups occurred in the Early Miocene (21 MYA), whereas diversification of the *melanogaster* subgroup took place in the Late Miocene (9 MYA). As has been previously presumed, the *melanogaster* subgroup might have colonized Africa in the Middle Miocene (Lachaise et al. 1988; Pool & Aquadro 2006).

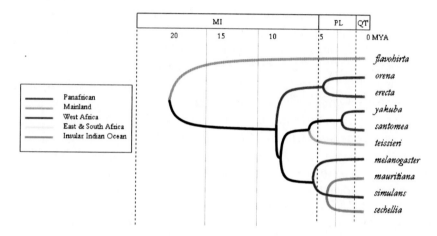

Figure 9. *Amyrel* chronogram of Afrotropical species of the *Drosophila melanogaster* species group. *Amyrel* sequences were reanalyzed from Da Lage et al. (2007). Geographical distributions are mapped on the chronongram.

The phylogeny of the *melanogaster* subgroup (Figure 9) shows three main rapid radiations (9-8 MYA): the *erecta* species complex, the *yakuba* complex and the *melanogaster* supercomplex comprising both the *melanogaster* and the *simulans* complexes). The most basal clade is the *erecta* species complex (9 MYA) containing two species (*D. erecta* and *D. orena*) that have diverged in Early Pliocene (4.6 MYA) and are endemic in West Africa. *D. erecta* breeds seasonally in *Pandanus* fruits, but shows a generalist behavior in the absence of *Pandanus*. The ecology of *D. orena* is the less known among all species of the subgroup, since the species is only known from a single isofemale strain collected from Mt Lefo in Cameroon (Tsacas & David 1978). Many attempts have been carried to recollect the species from other places in West Africa or even in its type locality since its discovery, but all these attempts were unsuccessful (M. Veuille, *pers. comm.*). However, the type locality was described as close to a river surrounded by autochthonous tropical flora and 19th century *Eucalyptus* plantations (Tsacas & David 1978) and another myrtaceous plant (Lachaise et al. 1988). Merçot et al. (1994) showed that ADH activity and alcohol tolerance of *D. orena* was the least among the nine fruit-breeding species of the subgroup, approaching that of those breeding in non-sweet substrates. This also has been suggested by the difficulty by which this species is reared in the laboratory.

The *yakuba* complex contains three species: *D. teissieri*, *D. yakuba* and *D. santomea*. *D. teissieri* forms the earliest branch (6 MYA) and it is confined to the continental rainforests, being very abundant in West Africa and exceedingly rare in East Africa (Lachaise et al. 1988). *D. yakuba* is found throughout the mainland, the insular Indian Ocean and Sao Tome. In the later island, it is found in lowland, whereas the third species of the complex, *D. santomea*, is endemic in the highlands (Lachaise et al. 2000). *D. yakuba* is more abundant in the savannahs and open environments, whereas *D. santomea* is a montane species. This means that both species are arid adapted in concordance with their estimated divergence date of 3.4 MYA (the divergence was estimated using non-calibrated molecular clock to be ca. 0.5 MYA by Cariou et al. 2001). It is known that at 3.48 MYA there is evidence for a change to a drier climate (Plana 2004).

The *melanogaster* supercomplex comprises two species complexes: *melanogaster* and *simulans*. The former is monotypic, whereas the latter consists of three species, two of which are endemic to the insular Indian Ocean (*D. simulans* and *D. mauritiana*). *D. melanogaster* and *D. simulans* are cosmopolitan and human commensals (Lachaise & Silvain 2004). In Africa, *D. melanogaster* is widespread throughout the continent, whereas *D. simulans* is more abundant in East Africa. Lachaise et al. (1988) and Lachaise & Silvain (2004) suggested *D. melanogaster* to have originated in West Africa and *D. simulans* in East Africa. However, there is strong evidence now from many molecular population genetics analyses that *D. melanogaster* and *D. simulans* have originated in East Africa and Madagascar, respectively (Veuille et al. 2004; Pool & Aquadro 2006). The split between the two complexes has taken place in the Late Miocene (6 MYA), according to our estimation. The *simulans* complex has undergone an insular radiation about 4 MYA with *D. mauritiana* in Mauritius and *D. sechellia* in Seychelles. The latter species is known to breed exclusively in *Morinda citrifolia*.

Lachaise et al. (2000) and Cariou et al. (2001) argued that the *melanogaster* subgroup has originated in the Cameroon Volcanic Line, because basal branches (the *erecta* and *yakuba* complexes) are more abundant in West Africa, and because two species (*D. orena* and *D. santomea*) are endemic to its mountains. We disagree with their hypothesis for two reasons. First, the predominance of basal branches in West African rainforests can reflect ancestral habitat rather than geographical origin. This is supported by the fact that divergence estimates of the split of the basal branches is coincident with the rise of aridity episode (9-8 MYA) in Africa during the Late Miocene, meaning the simultaneous retraction of the close-environment adapted early branches with West African rainforests and the evolution to more open-environment adaptation in the ancestral lineage of the *melanogaster* supercomplex in East Africa. Second, both montane endemic species are of recent origin (4.6 and 3.4 MYA, respectively) and their phylogenetic position does not suggest their suitability for hypothesizing the origin of the subgroup as a whole. *D. melanogaster* shows many physiological traits suggesting its adaptation to more arid environments than *D. simulans* (David et al. 2004), and this can be explained in light of the timing and the origin of the two species (Pool & Aquadro 2006). In sum, speciation in the *melanogaster* subgroup reflects the ecotonal model rather than the refugial model. However, the significant structuring of Afrotropical populations of some of its species (*D. melanogaster*: David & Bocquet 1975; David & Capy 1988; Glinka et al. 2003; Baudry et al. 2004; Haddrill et al. 2005; Pool & Aquadro 2006; Stephan & Li 2007; *D. simulans*: Capy et al. 1993; Baudry et al. 2006; Schofl & Schlötterer 2006; *D. teissieri*: Lachaise et al. 1981; Cobb et al. 2000; *D. yakuba*: Cariou et al. 2001) suggests that Pleistocene glaciations might have an impact at the intra-specific level.

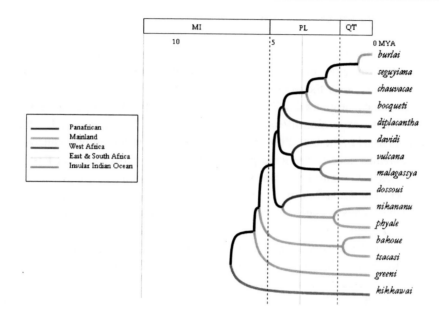

Figure 10. *Amyrel* chronogram of Afrotropical species of the *Drosophila montium* species group. *Amyrel* sequences were reanalyzed from Da Lage et al. (2007). Geographical distributions are mapped on the chronongram.

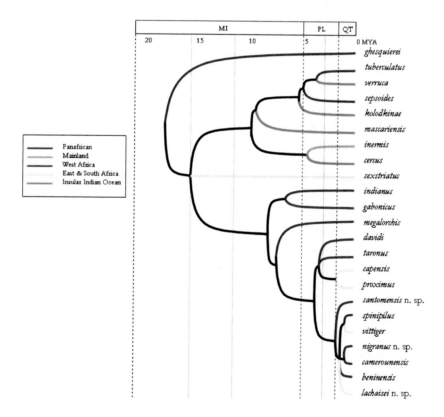

Figure 11. *Amyrel* chronogram of Afrotropical species of the genus *Zaprionus*. *Amyrel* sequences were reanalyzed from Yassin et al. (2008a). Geographical distributions are mapped on the chronongram.

VI. CONCLUSION

The Drosophilidae fauna of Africa is highly diversified and represents one eighth of the family total diversity, with a high level of endemism (80%). Analyses of species richness and endemism of this fauna recognize five main centers: Upper Guinea, Lower Guinea, East Africa, South Africa and the insular Indian Ocean. These centers can be classified into three distinct regions: West Africa (with a longitudinal cline of species richness), East and South Africa (with a latitudinal cline) and the insular Indian Ocean. Afrotropical species are classified under 31 genera and two subfamilies, of which only 4 genera are endemic. This reflects the relative young age of this fauna in comparison to Neotropical or Oriental ones.

Dated molecular estimates from recently published *Amyrel* sequences show that the Drosophilidae has split from its closed acalyptrate family (the Curtonotidae) in the Early Eocene (50 MYA). During the Middle and Late Eocene, seven radiations, each with a number of allied genera, took place: the subfamily Steganinae, New World *Sophophora*, *Chymomyza*, *Scaptodrosophila*, Old World *Sophophora* and *Drosophila s.s.* but the branching order of these radiations remains questionable. The *Amyrel* phylogeny suggests New World *Sophophora* to be the earliest drosophiline radiation, and there may be a morphological evidence for this. If this turns to be true, the family can be hypothesized to be of an Euramerican origin. The subgenus *Drosophila s.s.* and its allied genera forming the most derived radiation.

During the early evolution of the Drosophilidae, Africa was an isolated continent until its final connection with Eurasia in the Early Miocene. However, there are some signals that the drosophilid fauna might be of Late Eocene origin. The ecological diversity in Africa can recapitulate the phylogeny of its species, as it has been presumed that alcohol tolerance and breeding in sweet substrates (e.g., fermenting fruits) are a derived habitat. Many early drosophilids (*Chymomyza*, *Neotanygastrella* and the *latifasciaeformis* species group of the genus *Scaptodrosophila*) breed in palm tree trunks in concordance with the palm-dominant rainforest in Africa that prevailed until the Oligocene (the 'Palmae province'). Fig-breeding Afrotropical drosophilids also reflect this principle as the oldest lineage (the genus *Lissocephala*) breeds in premature syconia and shows a more *Ficus* species-specific niche, whereas the derived lineages (*Zaprionus* and Old World *Sophophora* especially the *fima* species group) breed on fallen and rotten fruits. Finally, the Neogene was a period of great climatic and tectonic changes that had a deep impact on the shaping of the present diversity.

Two speciation models have been previously proposed to explain the origin and evolution of great diversity in tropical rainforests: the Pleistocene refugial and the ecotonal hypotheses. These were tested using recent *Amyrel* phylogenies in three Afrotorpical clades (the *melanogaster* and *montium* species groups of Old World *Sophophora*, and the genus *Zaprionus*). The results clearly indicate that no general pattern dominates in the three clades. The *melanogaster* subgroup shows an ecotonal pattern originated in the Late Miocene, with Pleistocene glaciations impact only observed at the intra-specific level. The *montium* group has diversified during the Pliocene and the Quaternary, but the geographical distribution of its current taxa is not restricted to presumed refugias. The *inermis s.s.* group of *Zaprionus* shows a mainly dispersal (insular) model. The *vittiger* subgroup of *Zaprionus* shows a typical Pleistocene refugial model. However, taxonomic groups that originated in the Pliocene (*montium* and *vittiger*) are more diversified than those of earlier origin (*melanogaster* and *inermis s.s.*).

We are aware that in a family-scale synthesis such as the Drosophilidae, our analyses have many caveats. First, divergence times were roughly estimated from using direct distance methods. Unlike other probabilistic methods (maximum likelihood and Bayesian inference), distances do not take into account different possible tree topologies and do not give confidence intervals. However, we were more concerned about the general timing and place of origin of different clades rather than about the exact branching order. Second, our hypotheses are mainly based on a single gene (*Amyrel*) that, albeit representing the longest nuclear sequence shared by the highest number of drosophilid and schizopheran species, is only available for 5% of the family. This seems like composing an epic poem using only two letters. Molecular phylogenies and ecology of many unexplored groups and genera (e.g., subfamily Steganinae and Lower Drosophilinae) are highly needed, and these may deeply overturn our hypotheses. Third, at the species-level, *Amyrel* phylogeny represents a gene tree rather than a species tree. A multi-locus population genetics approach would thus be extremely useful in disentangling the Quaternary impacts on the modern drosophilid fauna, and to define boundaries between nominal species of uncertain status. Nevertheless, our findings are mostly consistent with many geological, ecophysiological and morphological data. We hope thus that our review will serve as a lightening rod for future integrative research aiming to understand the biogeography of the Drosophilidae in Africa and in other regions.

REFERENCES

Adams, C. F. (1905). Diptera africana, I. *The Kansas University Science Bulletin, 3,* 149-208.

Ali, J. R. & Aitchison, J. C. (2008). Gondwana to Asia: Plate tectonics, paleogeography and the biological connectivity of the Indian sub-continent from the Middle Jurassic through latest Eocene (166-35 Ma). *Earth-Science Reviews, 88,* 145-166.

Bächli, G. (2008). TAXODROS, the database on Taxonomy of Drosophilidae. http://taxodros.unizh.ch/, as of 2008, August 1st.

Baudry, E., Viginier, B. & Veuille, M. (2004). Non African populations of *Drosophila melanogaster* have a unique origin. *Molecular Biology and Evolution, 21,* 1482-1491.

Baudry, E., Derome, N., Huet, M. & Veuille, M. (2006). Contrasted polymorphism patterns in a large sample of populations from the evolutionary genetics model *Drosophila simulans. Genetics, 173,* 759-767.

Beverley, S. M. & Wilson, A. C. (1984). Molecular evolution in *Drosophila* and the Higher Diptera. II. A time scale for fly evolution. *Journal of Molecular Evolution, 21,* 1-13.

Burke, K. (2001). Origin of the Cameroon Line of volcano-capped swells. *The Journal of Geology, 109,* 349-362.

Burla, H. (1954). Zur kenntnis der Drosophiliden der Elfenbeinküste (Französisch West-Afrika). *Revue Suisse de Zoologie, 61(suppl),* 1-218.

Burla, H. (1955). The order of attraction of *Drosophila* species to cut palm trees. *Ecology, 36,* 153-155.

Capy, P., Pla, E. & David, J. R. (1993). Phenotypic and genetic variability of morphometrical traits in natural populations of *Drosophila melanogaster* and *D. simulans.* I. Geographic variations. *Genetics, Selection, Evolution, 25,* 517-536.

Cariou, M. L., Lachaise, D., Tsacas, L., Sourdis, J., Krimbas, C. & Ashburner, M. (1988). New African species in the *Drosophila obscura* species group: genetic variation, differentiation and evolution. *Heredity, 61*, 73-84.

Cariou, M. L., Silvain, J. F., Daubin, V., Da Lage, J. L. & Lachaise, D. (2001). Divergence between *Drosophila santomea* and allopatric or sympatric populations of *D. yakuba* using paralogous amylase genes and migration scenarios along the Cameroon volcanic line. *Molecular Ecology, 20*, 649-660.

Chassagnard, M. T. (1996). Les espèces africaines du sous-genre *Zaprionus s.str.* à six bandes thoraciques (Diptera: Drosophilidae). *Annales de la Société Entomologique de France, 32*, 59-65.

Chassagnard, M. T. & Kraaijeveld, A. R. (1991). The occurrence of *Zaprionus sensu stricto* in the Palearctic region (Diptera: Drosophilidae). *Annales de la Société Entomologique de France, 27*, 495-496.

Chassagnard, M. T. & McEvey, S. F. (1992). The *Zaprionus* of Madagascar, with descriptions of five new species (Diptera: Drosophilidae). *Annales de la Société Entomologique de France, 28*, 317-335.

Chassagnard, M. T. & McEvey, S. F. (1997). Le genre *Phorticella* Duda de la région afrotropicale et de Sumatra. Description de deux nouvelles espèces (Diptera: Drosophilidae). *Annales de la Société Entomologique de France, 33*, 81-87.

Chassagnard, M. T. & Tsacas, L. (1993). Le sous-genre *Zaprionus s.str.* Définition de groupes d'espèces et révision du sous-groupe *vittiger* (Diptera: Drosophilidae). *Annales de la Société Entomologique de France, 29*, 173-194.

Chassagnard, M. T., Tsacas, L. & Lachaise, D. (1997) Drosophilidae (Diptera) of Malawi. *Annals of the Natal Museum, 38*, 61-131.

Cobb, M., Huet, M., Lachaise, D. & Veuille, M. (2000). Fragmented forests, evolving flies: molecular variation in African Populations of *Drosophila teissieri*. *Molecular Ecology, 9*, 1591-1597.

Collart, A. (1937). Les *Zaprionus* du Congo belge (Diptera: Drosophilidae). *Bulletin du Musée Royale d'Histoire Naturelle de Belgique, 13*, 1-15.

Cook, J. M. C. & Rasplus, J. Y. (2003). Mutualists with attitude: coevolving fig wasps and figs. *Trends in Ecology and Evolution, 18*, 241-248.

Coquillett, D. W. (1901). New Diptera from Southern Africa. *Proceedings of the United States National Museum, 24*, 27-32.

Cuenca, A., Asmussen-Lange, C. B. & Borchsenius, F. (2008). A dated phylogeny of the palm tribe Chamaedoreeae supports Eocene dispersal between Africa, North and South America. *Molecular Phylogenetics and Evolution, 46*, 760-775.

Da Lage, J. L., Kergoat, G. J., Maczkowiak, F., Silvain, J. F., Cariou, M. L. & Lachaise, D. (2007). A phylogeny of Drosophilidae using the *Amyrel* gene: questioning the *Drosophila melanogaster* species group boundaries. *Journal of zoological systematics and evolutionary research, 45*, 47-63.

David, J. & Bocquet, C. (1975). Similarities and differences in latitudinal adaptation of two Drosophila sibling species. *Nature, 257*, 588-590.

David, J. R. & Capy, P. (1988). Genetic variation of *Drosophila melanogaster* natural populations. *Trends in Genetics, 4*, 106-111.

David, J. R., Allemand, R., Capy, P., Chakir, M., Gibert, P., Pétavy, G. & Moreteau, B. (2004). Comparative life histories and ecophysiology of *Drosophila melanogaster* and *D. simulans*. *Genetica, 120,* 151-163.

David, J. R., Lemeunier, F., Tsacas, L. & Yassin, A. (2007). The historical discovery of the nine species in the *Drosophila melanogaster* species subgroup. *Genetics, 177,* 1969-1973.

DeSalle, R. (1992). The phylogenetic relationships of flies in the family Drosophilidae deduced from mtDNA sequences. *Molecular Phylogenetics and Evolution, 1,* 31-40.

Drosophila 12 Genomes Consortium (2007). Evolution of genes and genomes on the *Drosophila* phylogeny. *Nature, 450,* 203-218.

Dubois, A. (2007). Phylogeny, taxonomy and nomenclature: the problem of taxonomic categories and of nomenclatural ranks. *Zootaxa, 1519,* 27-68.

Duda, O. (1939). Revision der afrikanischen Drosophiliden (Diptera). I. *Annales Historico-Naturales Musei Nationalis Hungarici, 32,* 1-57.

Duda, O. (1940). Revision der afrikanischen Drosophiliden (Diptera). II. *Annales Historico-Naturales Musei Nationalis Hungarici, 33, 19-53.*

Evenhuis, N. L. (1994). Catalogue of the fossil flies of the world (Insecta: Diptera). Backhuys, Leiden, the Netherlands.

Gao, J. J., Watabe, H. A., Aotsuka, T., Pang, J. F & Zhang, Y. (2007). Molecular phylogeny of the *Drosophila obscura* species group, with emphasis on the Old World species.

Gheerbrant, E. & Rage, J. C. (2006). Paleobiogeography of Africa: How distinct from Gondwana and Laurasia? *Palaeogeography, Palaeoclimatology, Palaeoecology, 241,* 224-246.

Glinka, S., Ometto, L., Mousset, S., Stephan, W. & De Lorenzo, D. (2003). Demography and natural selection have shaped genetic variation in *Drosophila melanogaster*: A multi-locus approach. *Genetics, 165,* 1269-1278.

Grimaldi, D. (1986). The *Chymomyza aldrichii* species-group (Diptera: Drosophilidae): relationships, new Neotropical species, and the evolution of some sexual traits. *Journal of the New York Entomological Society, 94,* 342-371.

Grimaldi, D. (1993). Amber fossil Drosophilidae (Diptera), Part II: Review of the genus *Hyalistata*, new status (Steganinae). *American Museum Novitates, 3084,* 1-15.

Grimaldi, D. & Engel, M. S. (2004). *Evolution of the Insects.* Cambridge University Press.

Grimaldi, D., Ervik, F. & Bernal, R. (2003). Two new Neotropical genera of Drosophilidae (Diptera) visiting palm flowers. *Journal of the Kansas Entomological Society, 76,* 109-124.

Grimaldi, D. A. (1987). Phylogenetics and taxonomy of *Zygothrica* (Diptera: Drosophilidae). *Bulletin of the American Museum of Natural History, 186,* 103-268. Grimaldi, D. A. (1987b) Amber fossil Drosophilidae (Diptera), with particular reference to the Hispaniolan taxa. *American Museum Novitates, 2880,* 1-23.

Grimaldi, D. A. (1988). Relicts in the Drosophilidae (Diptera). In: Liebherr, J. K., (Ed.), *Zoogeography of Caribbean Insects*, (pp. 183-213). Cornell University Press.

Grimaldi, D. A. (1990). A phylogenetic, revised classification of genera in the Drosophilidae (Diptera). *Bulletin of the American Museum of Natural History, 197,* 1-139.

Haag-Liautard, C., Coffey, N., Houle, D., Lynch, M., Charlesworth, B. & Keightley, P. D. (2008) Direct estimation of the mitochondrial mutation rate in *Drosophila melanogaster*. *PLoS Biology, 6,* e204.

Haddrill, P. R., Thornton, K. R., Charlesworth, B. & Andolfatto, P. (2005). Multilocus patterns of nucleotide variability and the demographic and selection history of *Drosophila melanogaster* populations. *Genome Research, 15,* 790-799.

Haffer, J. (1969). Speciation in Amazonian forest birds. *Science, 165,* 131-137.

Harris, N. (2006). The elevation history of the Tibetan Plateau and its implications for the Asian monsoon. *Palaeogeography, Palaeoclimatology, Palaeoecology, 241,* 4-15.

Harry, M., Solignac, M. & Lachaise, D. (1996). Adaptive radiation in the Afrotropical region of the Paleotropical genus *Lissocephala* (Drosophilidae) on the Pantropical genus *Ficus* (Moraceae). *Journal of Biogeography, 23,* 543-552.

Harry, M., Solignac, M. & Lachaise, D. (1998). Molecular evidence for parallel evolution of adaptive syndromes in fig-breeding *Lissocephala* (Drosophilidae). *Molecular Phylogenetics and Evolution, 9,* 542-551.

Hennig, W. (1965). Die Acalyptratae des Baltischen Bernsteins und ihre Bedeutung fur die Erforschung der phylogenetischen Entwicklung dieser Dipteren-Gruppe. *Stuttgarter Beitrage zur Naturkunde, 145,* 1-215.

Hu, Y. G. & Toda, M. J. (2001). Polyphyly of *Lordiphosa* and its relationships in Drosophilinae (Diptera: Drosophilidae). *Systematic Entomology, 26,* 15-31.

Hu, Y. G. & Toda, M. J. (2002). Cladistic analysis of the genus *Dichaetophora* Duda (Diptera: Drosophilidae) and a revised classification. *Insect Systematics & Evolution, 33,* 91-102.

Jacobs, B. F. (2004). Palaeobotanical studies from tropical Africa: relevance to the evolution of forest, woodland and savannah biomes. *Philosophical Transactions of the Royal Society of London (B), 359,* 1573-1583.

Katoh, T., Tamura, K. & Aotsuka, T. (2000). Phylogenetic position of the subgenus *Lordiphosa* of the genus *Drosophila* (Diptera: Drosophilidae) inferred from alcohol dehydrogenase (*Adh*) gene sequences. *Journal of Molecular Evolution, 51,* 122-130.

Kirk-Spriggs, A. H. (2008). A contribution to the knowledge of the immature stages of *Curtonotum* (Diptera: Curtonotidae), from Africa and the Middle East, with a discussion of relationships to other known Ephydroidea larvae. *African Entomology, 16,* 226-243.

Kwiatowski, J., Krawczyk, M., Jaworski, M., Skarecky, D. & Ayala, F.J. (1997). Erratic Evolution of Glycerol-3-Phosphatase Dehydrogenase in *Drosophila, Chymomyza,* and *Ceratitis. Journal of Molecular Evolution, 44,* 9-22.

Lachaise, D. (1975). Les Drosophilidae des savanes préforestières de Lamto (Côte d'Ivoire). III. - Le peuplement du palmier rônier. *Annales de l'Universite d'Abidjan, 8,* 223-280.

Lachaise, D. (1977) Niche separation of African *Lissocephala* within the *Ficus* drosophilid community. *Oecologia, 31,* 201-214.

Lachaise, D. & Chassagnard, M. T. (2001) Seven new montane species of *Drosophila* in the Eastern Arc mountains and Mt Kilimanjaro in Tanzania attesting to past connections between eastern and western African mountains (Diptera: Drosophilidae). *European Journal of Entomology, 98,* 351-366.

Lachaise, D. & McEvey, S. F. (1990). Independent evolution of the same set of characters in fig flies (*Lissocephala,* Drosophilidae). *Evolutionary Ecology, 4,* 358-364.

Lachaise, D. & Silvain, J. F. (2004). How two Afrotropical endemics made two cosmopolitan human commensals: the *Drosophila melanogaster-D. simulans* palaeogeographic riddle. *Genetica, 120,* 17-39.

Lachaise, D. & Tsacas, L. (1983). Breeding-sites in tropical African drosophilids. In: Ashburner, M., Carson, H. L. & Thompson, J. N., (Eds.), *The Genetics and Biology of Drosophila* (vol. 3d, pp. 221-332). Academic Press.

Lachaise, D., Lemeunier, F. & Veuille, M. (1981). Clinal variations in male genitalia in *Drosophila teissieri* Tsacas. *American Naturalist, 117,* 600-608.

Lachaise, D., Tsacas, L. & Couturier, G. (1982). The Drosophilidae associated with tropical African figs. *Evolution, 36,* 141-151.

Lachaise, D., Cariou, M. L., David, J. R., Lemeunier, F., Tsacas, L. & Ashburner, M. (1988). Historical biogeography of the *Drosophila melanogaster* species subgroup. *Evolutionary Biology, 22,* 159-225.

Lachaise, D., Harry, M., Solignac, M., Lemeunier, F., Benassi, V. & Cariou, M. L. (2000). Evolutionary novelties in islands: *Drosophila santomea*, a new melanogaster sister species from Sao Tome. *Proceedings. Royal Society of London. Biological Sciences, 267,* 1487-1495.

Lastovka, P. & Maca, J. (1978). European species of the *Drosophila* subgenus *Lordiphosa* (Diptera, Drosophilidae). *Acta Entomologica Bohemoslovaca, 75,* 404-420.

Linder, H. P. (2001). Plant diversity and endemism in sub-Saharan tropical Africa. *Journal of Biogeography, 28,* 169-182.

Livingstone, D. A. (1975). Late Quaternary climatic change in Africa. *Annual Review of Ecology, Evolution, and Systematics, 6,* 249-280.

Livingstone, D. A. (1982). Quaternary geography of Africa and the refuge theory. In: Prance, G. T. (Ed.), *Biological Diversification in the Tropics* (pp. 523-536). Columbia University Press.

Maczkowiak, F. & Da Lage, J. L. (2006). Origin and evolution of the *Amyrel* gene in the alpha-amylase multigene family of Diptera. *Genetica, 128,* 145-158.

Mallet, J. (2006). What does *Drosophila* genetics tell us about speciation? *Trends in Ecology and Evolution, 21,* 386-393.

Markow, T. A. & O'Grady, P. M. (2007). *Drosophila* biology in the genomic age. *Genetics, 177,* 1269-1276.

Maslin, M. A. & Christenen, B. (2007). Tectonics, orbital forcing, global climate change, and human evolution in Africa: introduction to the African paleoclimate special volume. *Journal of Human Evolution, 53,* 443-464.

McAlpine, J. F. (1989). Phylogeny and classification of the Muscomorpha. In: McAlpine, J.F. et al.,(Eds.), Manual of Nearctic Diptera (vol. 3, pp. 1397-1518). Canada: Research Branch, Agriculture.

McCall, R. A. (1997). Implications of recent geological investigations of the Mozambique Channel for the mammalian colonization of Madagascar. *Proceedings of the Royal Society of London B, 264,* 663-665.

McEvey, S. F., Aulard, S. & Ralisoa-Randrianasolo, O. (1989). An Australian drosophilid (Diptera) on *Eucalyptus* and *Eugenia* (Myrtaceae) flowers in Madagascar. *Journal of the Australian Entomological Society, 28,* 53-54.

Meier, R., Kotrba, M. & Barber, K. (1997). A comparative study of the egg, first-instar, larva, puparium, female reproductive system and natural history of *Curtonotum helvum* (Curtonotidae; Ephydroidea; Diptera). *American Museum Novitates, 3219,* 1-20.

Merçot, H., Defaye, D., Capy, P., Pla, E. & David, J. R. (1994). Alcohol tolerance, ADH activity, and ecological niche of *Drosophila* species. *Evolution, 48,* 746-757.

Mooers, A. O. (2007). The diversity of biodiversity. *Nature, 445,* 717-718.

Morley, R. J. (2003). Interplate dispersal paths for megathermal angiosperms *Perspectives in Plant Ecology, Evolution and Systematics, 6,* 5–20.

Moritz, C. Patton, J. L., Schnieider, C. J. & Smith, T. B. (2000). Diversification of rainforest faunas: an integrated molecular approach. *Annual Review of Ecology, Evoluyion and Systematics, 31,* 533-563.

Nee, S. (2006). Birth-death models in macroevolution. *Annual Review of Ecology, Evoluyion and Systematics, 37,* 1-17.

Nicolson, S. W. (1994). Pollen feeding in the eucalypt nectar fly, *Drosophila flavohirta. Physiological Entomology, 19,* 58-60.

Okada, T. (1985). The genus *Lissocephala* Malloch and an allied new genus of Southeast Asia and New Guinea (Diptera, Drosophilidae). *Kontyu, 53,* 335-345.

Okada, T. (1989). A proposal of establishing tribes for the family Drosophilidae with key to tribes and genera (Diptera). *Zoological Science, 6,* 391-399.

Okada, T. & Carson, H. L. (1983). The genera *Phorticella* Duda and *Zaprionus* Coquillett (Diptera, Drosophilidae) of the Oriental Region and New Guinea. *Kontyu, 51,* 539-553.

O'Grady, P. M. & Kidwell, M. G. (2002). Phylogeny of the subgenus *Sophophora* (Diptera: Drosophilidae) based on combined analysis of nuclear and mitochondrial sequences. *Molecular Phylogenetics and Evolution, 22,* 442-453.

O'Grady, P. & DeSalle, R. (2008). Out of Hawaii: the origin and biogeography of the genus *Scaptomyza* (Diptera: Drosophilidae). *Biology Letters, 4,* 195-199.

O'Grady, P. M., Beardsley, J. W. & Perreira, W. D. (2002). New records for introduced Drosophilidae (Diptera) in Hawaii. *Bishop Museum Occasional Papers, 68,* 34-35.

Pélandakis, M. & Solignac, M. (1993). Molecular phylogeny of *Drosophila* based on ribosomal RNA sequences. *Journal of Molecular Evolution, 37,* 525-543.

Pipkin, S. B. (1965). The influence of adult and larval food habits on population size of Neotropical ground-feeding *Drosophila. American Midland Naturalist, 74,* 1-27.

Plana, V. (2004). Mechanism and tempo of evolution in the African Guineo-Congolian rainforest. *Philosophical Transactions of the Royal Society of London B, 359,* 1585-1594.

Pollock, J. N. (2002). Observations on the biology and anatomy of Curtonotidae (Diptera: Schizophora). *Journal of Natural History, 36,* 1725-1745.

Pool, J. E. & Aquadro, C. F. (2006). History and structure of sub-Saharan populations of *Drosophila melanogaster. Genetics, 174,* 915-929.

Powell, J. R., Sezzi, E., Moriyama, E. N., Gleason, J. M. & Caccone, A. (2003). Analysis of a shift in codon usage in *Drosophila. Journal of Molecular Evolution, 57,* S214-S225.

Rafael, V. (1984). Relations interspécifiques dans le nouveau complexe africain de *Drosophila bakoue* du groupe *melanogaster,* sous-groupe *montium* (Diptera, Drosophilidae). *Bulletin de la Société Zoologique de France, 109,* 179-189.

Remsen, J. & O'Grady, P. (2002). Phylogeny of Drosophilinae (Diptera: Drosophilidae), with comments on combined analysis and character support. *Molecular Phylogenetics and Evolution, 24,* 249-264.

Robe, L. J., Valente, V. L. S., Budnik, M. & Loreto, E. L. S. (2005). Molecular phylogeny of the subgenus Drosophila (Diptera, Drosophilidae) with an emphasis on Neotropical species and groups: A nuclear versus mitochondrial gene approach. *Molecular Phylogenetics and Evolution, 36,* 623-640.

Rønsted, N., Weiblen, G. D., Cook, J. M., Salamin, N., Machado, C. A. & Savolainen, V. (2005). 60 million years of co-divergence in the fig-wasp symbiosis. *Proceedings of the Royal Society of London B, 272*, 2593-2599.

Rønsted, N., Salvo, G. & Savolainen, V. (2007). Biogeographical and phylogenetic origins of African fig species (*Ficus* section *Galoglychia*). *Molecular Phylogenetics and Evolution, 43*, 190-201.

Russo, C. A. M., Takezaki, N. & Nei, M. (1995). Molecular phylogeny and divergence times of drosophilid species. *Molecular Biology and Evolution, 12*, 391-404.

Sanmartín, I. & Ronquist, F. (2004). Southern hemisphere biogeography inferred by event-based models: plants versus animal patterns. *Systematic Biology, 53*, 216-243.

Schofl, G. & Schlötterer, C. (2006). Microsatellite variation and differentiation in African and non-African populations of *Drosophila simulans*. *Molecular Ecology, 15*, 3895-3905.

Séguy, E. (1938). Mission scientifique de l'Omo. Diptera. I. Nematocera et Brachycera. *Mémoires du Museum National d'Histoire Naturelle de Paris, 8*, 319-380.

Sepulchre, P., Ramstein, G., Fluteau, F., Schuster, M., Tiercelin, J. J. & Brunet, M. (2006). Tectonic uplift and Eastern Africa aridification. *Science, 313*, 1419-1423.

Sidorenko, V. S. (2002). Phylogeny of the tribe Steganini Hendel and some related taxa (Diptera, Drosophilidae). *Far Eastern Entomologist, 111*, 1-20.

Stephan, W. & Li, H. (2007). The recent demographic and adaptive history of *Drosophila melanogaster*. *Heredity, 98*, 65-68.

Sultana, F., Hu, Y. G., Toda, M. J., Takenaka, K. & Yafuso, M. (2006). Phylogeny and classification of Colocasiomyia (Diptera, Drosophilidae), and its evolution of pollination mutualism with aroid plants. *Systematic Entomology, 31*, 684-702.

Tajima, F. (1989). Statistical methods to test for nucleotide mutation hypothesis by DNA polymorphism. *Genetics, 123*, 585-595.

Tatarenkov, A., Zurovcova, M. & Ayala. F. J. (2001). *Ddc* and *amd* sequences resolve phylogenetic relationships of *Drosophila*. *Molecular Phylogenetics and Evolution, 20*, 321-325.

Throckmorton, L. H. (1962). The problem of phylogeny in the genus *Drosophila*. *The University of Texas Publication, 6205*, 207-343.

Throckmorton, L. H. (1975). The phylogeny, ecology, and geography of *Drosophila*. In: R. C. King (Ed.), *Handbook of Genetics* (vol. 3, pp. 421-469). Plenum Press.

Tiffney, B. H. & Manchester, S. R. (2001). The use of geological and paleontological evidence in evaluating plant phylogeographic hypotheses in the Northern hemisphere Tertiary. *International Journal of Plant Sciences, 162*, S3-S17.

Toda, M. J. & Peng, T. X. (1992). Some species of the subfamily Steganinae (Diptera: Drosophilidae) from Guangdong Province, southern China. *Annales de la Société Entomologique de France, 28*, 201-213.

Tsacas, L. (1977). Les Curtonotidae (Diptera) de l'Afrique: 1. Le genre *Curtonotum* Macquart. *Annals of the Natal Museum, 23*,145-171.

Tsacas, L. (1990). Drosophilidae de l'Afrique Australe (Diptera). *Annals of the Natal Museum, 31*, 103-161.

Tsacas, L. (2000). *Allopygaea*, nouveau genre afrotropical proche de *Paraleucophenga* Hendel (Diptera: Drosophilidae). *Annales de la Société Entomologique de France, 36*, 121-136.

Tsacas, L. & David, J. (1978). Une septième espèce appartenant au sous-groupe *Drosophila melanogaster* Meigen: *Drosophila orena* spec. nov. du Cameroun. (Diptera: Drosophilidae). *Beitrage zur Entomologie, 28,* 179-182.

Tsacas, L., Lachaise, D. & David, J. R. (1981). Composition and biogeography of the Afrotropical drosophilid fauna. In: M. Ashburner, H. L. Carson & J. N. Thompson (Eds.), *The Genetics and Biology of Drosophila* (vol. 3a, pp. 197-259). Academic Press.

Van Veller, M. G. P., Brooks, D. R. & Zandee, M. (2003) Cladistic and phylogenetic biogeography: the art and the science of discovery. *Journal of Biogeography, 30,* 319-329.

Veuille, M., Baudry, E., Cobb, M., Derome, N. & Gravot, E. (2004). Historicity and the population genetics of *Drosophila melanogaster* and *D. simulans*. *Genetica, 120,* 61-70.

Wheeler, M. R. (1947). The insemination reaction in intraspecific matings of Drosophila. *The University of Texas Publication, 4720,* 78-115.

Wheeler, M. R. (1952). The Drosophilidae of the Nearctic Region exclusive of the genus *Drosophila*. *The University of Texas Publication, 5204,* 162-218.

Wheeler, M. R. (1960). Sternite modification in males of the Drosophilidae (Diptera). *Annals of the Entomological Society of America, 53,* 133-137.

Wiegmann, B. M., Yeates, D. K., Thorne, J. L., Kishino, H. (2003). Time flies, a new molecular time-scale for Brachyceran fly evolution without a clock. *Systematic Biology, 52,* 745-756.

Yassin, A. (2008). Molecular and morphometrical revision of the *Zaprionus tuberculatus* species subgroup (Diptera: Drosophilidae), with descriptions of two cryptic species. *Annals of the Entomological Society of America (in press)*.

Yassin, A., Araripe, L. O., Capy, P., Da Lage, J.-L., Klaczko, L. B., Maisonhaute, C., Ogereau, D. & David, J. R. (2008a). Grafting the molecular phylogenetic tree with morphological branches to reconstruct the evolutionary history of the genus *Zaprionus* (Diptera: Drosophilidae). *Molecular Phylogenetic and Evolution, 47,* 903-915.

Yassin, A., Capy, P., Madi-Ravazzi, L., Ogereau, D. & David, J. R. (2008b). DNA barcode discovers two cryptic species and two independent geographic radiations in the cosmopolitan drosophilid *Zaprionus indianus*. *Molecular Ecology Resources, 8,* 491-501.

Yeates, D. K., Wiegmann, B. M., Courtney, G. W., Meier, R., Lambkin, C. & Pape, T. (2007). Phylogeny and systematics of Diptera: two decades of progress and prospects. *Zootaxa, 1668,* 565-590.

Zhisheng, A, Kutzbach, J. E., Prell, W. L. & Porter, S. C. (2001). Evolution of Asian monsoons and phased uplift of the Himalaya-Tibetan plateau since Late Miocene times. *Nature, 411,* 62-66.

In: Biogeography
Editors: M. Gailis, S. Kalniņš, pp. 137-175

ISBN: 978-1-60741-494-0
© 2010 Nova Science Publishers, Inc.

Chapter 4

HISPANIOLAN SPIDER BIODIVERSITY AND THE IMPORTANCE OF COMBINING NEONTOLOGICAL AND PALAEONTOLOGICAL DATA IN ANALYSES OF HISTORICAL BIOGEOGRAPHY

*David Penney**

Earth, Atmospheric and Environmental Sciences,
The University of Manchester, Manchester, UK

ABSTRACT

Conflicting opinions regarding the relative importance of dispersal versus vicariance in understanding Caribbean biogeography continue to stimulate lively debate, despite recent advances in regional tectonics, geology, palaeogeography and palaeontology. The Greater Antilles are young geographical features, probably mid-Miocene and Hispaniola has a rich terrestrial invertebrate fossil record in the form of Dominican Republic amber inclusions. Spiders are common in Dominican amber and all fossil species belong in extant families. The island is unique in terms of its known spider fauna, in that a similar number of families have described species recorded on Hispaniola from both fossils in amber and from the extant fauna. It is also the region of the world where the amber fauna is most similar to the Recent fauna, and Miocene Hispaniolan spider biodiversity was presumably similar to that at present.

Fossils form empirical data with both phylogenetic and temporal implications and are of paramount importance for understanding historical biogeography. They can play a decisive role in falsifying hypotheses proposed solely on the distributions of extant taxa. It is also possible to generate hypotheses for both fossil and extant faunas that are ultimately falsibiable through the discovery of new Hispaniolan taxa. The amber spider fauna has clear affiliations to the present South American fauna, so how did they colonize the island from the mainland: dispersal or vicariance? The high Miocene biodiversity (including taxa with poor dispersal capabilities) on a geologically young island would seem to add more weight in favour of the vicariance model for explaining on-island spider lineages, although dispersal cannot be excluded for some of the highly dispersive

*Corresponding author: e-mail: david.penney@manchester.ac.uk

taxa. However, a case-by-case approach is required in order to determine the relative contributions of the different models. The future research direction for understanding the biogeographical patterns will lie in the application of cladistic and phylogenetic biogeographical approaches. It is interesting to note that preliminary obersvations based on pedipalp morphology suggest some fossil taxa are more derived than extant forms. The contribution that the Dominican (and Mexican) amber fossil fauna can make to our understanding of the origins of Hispaniolan biodiversity, and Caribbean biogeography in general, should not be underestimated.

Keywords: Amber, Araneae, Caribbean, Dispersal, Fossil, Greater Antilles, Palaeogeography, Palaeontology, Vicariance.

INTRODUCTION

The present rate of global climate change, habitat destruction through anthropogenic and other factors, and the subsequent rate of species extinctions means it is likely that many taxa will disappear before they have been scientifically described. Briggs [1991] likened the current episode of species demise to a present day mass extinction. He suggested that the rate of species loss is greater and more significant than the end-Cretaceous extinction that wiped out the dinosaurs. In order to assess the effects of this current global change it is necessary to inventory as much of the global biota as possible, before components of it disappear forever. However, in addition to recognizing existing biodiversity and biogeographic patterns, it is important that we understand their origins. Thus, we also need to examine past distributions using palaeontological evidence from the fossil record (and consider these in relation to past global climate change over deep time) in order to help us interpret the potential consequences of the current and ongoing episode.

Island faunas pose interesting biogeographical questions because of difficulties associated with immigration and emigration, when compared to continental landmasses. The high biodiversity of the Caribbean, coupled with its complex geological history, has attracted interest in the historical biogeography of the region [Woods and Sergile, 2001], although most studies have focussed on vertebrates [Hedges, 2001]. The Caribbean is considered a biodiversity hotspot and the West Indies is one of the largest and most diverse archipelagos in the world; home to thousands of endemic species. Recent advances in West Indian geology and plate tectonics [Iturralde-Vinent and MacPhee, 1999; Iturralde-Vinent and Lidiak, 2006], increasing availability of fossils (particularly in amber from the Dominican Republic) [Poinar, 1992], and the application of DNA sequencing in the study of evolutionary history [e.g. Ogden and Thorpe, 2002], provides the basis for a new approach to understanding the complex, evolutionary history of the region's biota.

A Catalysis Meeting (held in the Dominican Republic, July 2006) was recently sponsored by the US National Evolutionary Synthesis Center (NESCent) and initiated and organized by Drs R.E. Glor (University of California), R.E. Ricklefs (University of Missouri-St. Louis) and J.B. Losos (Washington University). The primary purpose of the meeting was to facilitate future collaborative, interdisciplinary research by scientists in diverse fields, with the aim of developing a comprehensive new perspective on the evolution of West Indian biodiversity (past and present). Participants included geologists, palaeontologists (including the present author), biogeographers, systematists, and population/evolutionary biologists from six different countries. The following question for the delegates was one of particular interest: *How have historical connections among islands, and to the mainland, contributed to patterns of diversity?*

The origins of the West Indian biota have long been debated and the last century has seen many differences of opinion regarding the various hypotheses provided to explain the distributions observed today, e.g. former landbridge connections, vicariance, and overwater dispersal. One major aim of the NESCent meeting was to produce synergistic collaborations with geological, palaeontological and phylogenetic data (including specialists on numerous, diverse taxonomic groups) combining to result in a fuller understanding of the West Indies' history than would be possible relying on a single type of information. In this paper, I examine the potential role that fossil spiders preserved in Miocene Dominican Republic amber can add to the debate.

FOSSILIFEROUS AMBER FROM THE DOMINICAN REPUBLIC

Amber is a fossil Lagerstätte famous for its exceptional preservation of entombed animal, usually arthropod, inclusions, which are often preserved with life-like fidelity. This is particularly true of Dominican amber, which probably exhibits the best quality of preservation of all know ambers [Grimaldi and Engel, 2005], despite there being more than 160 deposits known worldwide [Martínez-Delclòs *et al.*, 2004]. Although traces of amber are present in various other countries of the Caribbean, e.g., Haiti, Puerto Rico and Jamaica, it occurs in exploitable quantities only in the Dominican Republic [Iturralde-Vinent, 2001]. Dominican amber has been known to Europeans since the end of the fifteenth century [Baroni-Urbani and Saunders, 1980], but was only brought to the attention of scientists in the twentieth [Sanderson and Farr, 1960]. Despite this late onset of palaeontological study, Hispaniola has a rich terrestrial invertebrate fossil record in the form of species described from Dominican Republic amber inclusions.

The amber is found in two main regions of the country: the Cordillera Septentrional in the north, and the Cordillera Oriental in the north-east (Figure 1). The amberiferous region in the north consists of the upper 300m of an Oligocene to Middle Miocene suite of clastic rocks called the La Toca Formation and consists mainly of sandstone containing thin lamellae of lignite, and occasional conglomerate. In the north-east the amber is found embedded in lignite and sandy clay in the 100m thick Yanigua Formation. These two amber deposits are thought to have been deposited in the same sedimentary basin 15–20Ma (but probably closer to 16Ma), prior to their displacement along major faults [Iturralde-Vinent and MacPhee, 1996; see Iturralde-Vinent, 2001 for a comprehensive review of the geology of these deposits]. For a discussion of alternative ages for this amber deposit see Poinar and Poinar [1999].

Poinar [1991] described the extinct tree *Hymenaea protera* (Leguminosae, Caesalpinioideae) as the botanical source of Dominican Republic amber; see also Hueber and Langenheim [1986]. This generic placement was confirmed by Langenheim [1995], through evidence based on resin chemistry and amber plant inclusions. The amber was formed in a tropical climate similar to that in the region today [e.g. Poinar and Poinar, 1999].

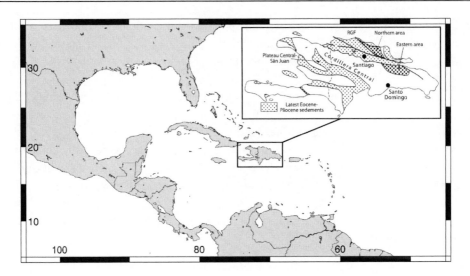

Figure 1. Current geography of Caribbean islands and adjacent continental landmasses, with inset (after Itturalde-Vinent and MacPhee, 1996) showing the location of the main amber mining districts (northern and eastern areas) in Hispaniola.

RESIN AS A TRAP AND THE BIAS OF AMBER PRESERVATION

The preservative and antibiotic qualities of resin, which kill or retard fungal and bacterial growth have been known since antiquity [Poinar and Hess, 1982]. Preservation in amber basically represents an extreme case of mummification with natural embalming by the entrance of plant sap into the organisms' body cavity [Poinar and Hess, 1982]. Despite the exceptional preservation, a major problem faced by palaeontologists is that the fossil record is incomplete and many of the biases with regard to which organisms did, and which did not, get trapped remain unidentified or unquantified. Few authors have investigated the ecology of amber inclusions and quantitative studies of amber palaeobiology are in their infancy. At present, what we do know as a result of quantitative investigations of amber faunas, is that different amber-forming resins appear to have functioned as a trap in the same way [Penney and Langan, 2006], and that they are biased towards sampling active, trunk-dwelling faunas, which are more prone to wandering onto an exposed, sticky resin surface, getting stuck, and which are then entombed by a subsequent resin flow [Penney, 2002—in disagreement with the conclusions of Henwood [1993], which is assessed critically by Penney]. This is easily observed in many specimens as a layering effect in the amber, with inclusions trapped at the interface of the layers. However, organisms were also trapped by becoming engulfed in a single, more rapid flow of less viscous resin [Penney, 2005a].

By far the commonest inclusions encountered in amber are insects, but this is hardly surprising given their global dominance today. Spiders are also common in amber faunas and have a long and diverse geological history extending to the deep Paleozoic, although many Cenozoic spiders belong in extant families [Penney et al., 2003]. Despite this overabundance of insects, our knowledge of fossil spiders in Dominican amber tends to exceed that of insects at ordinal level. For example, combined fossil and Recent Hispaniolan faunal checklists exist for a number of arthropod orders, e.g. Trichoptera [Flint and Pérez-Gelabert, 1999],

Neuroptera [Pérez-Gelabert and Flint, 2000], Blattodea [Gutiérrez and Pérez-Gelabert, 2000], Orthoptera [Pérez-Gelabert, 2001] and Diplopoda [Pérez-Asso and Pérez-Gelabert, 2001], but when compared to spiders [Penney and Pérez-Gelabert, 2002; updated here, see Appendix] these other groups are poorly represented as fossils (Table 1; see also Pérez-Gelabert [1999]). This is not to suggest that these smaller orders cannot play a decisive role in studies of historical biogeography, but the justification for an analysis based on the comparison of extant and fossil spider distributions is clearly evident.

Table 1. Comparison of extant and fossil (*excluding subfossils in copal) species richness values from orders for which data are available (see text for references)

Order	Total spp.	Extant spp.	Fossil spp.	Extant:fossil ratio
Trichoptera	99	77	22	1:0.29
Neuroptera	55	52	3	1:0.06
Blattodea	92	85	7	1:0.08
Orthoptera	111	103	8	1:0.08
Diplopoda	162	146	16	1:0.11
Araneae	467*	313	154*	1:0.49

BIODIVERSITY OF HISPANIOLAN SPIDERS

Hispaniola is unique in terms of its known spider fauna, in that almost the same number of families have described species recorded on Hispaniola from fossils in Miocene Dominican Republic amber (39) and from the extant fauna (42). Furthermore, additional families are also known for the fossil fauna, but have not had formal species described (see later). It is also the region of the world where the amber fauna is most similar to the Recent fauna [Penney, 2005b]. A checklist of known fossil and Recent Hispaniolan spiders was given by Penney and Pérez-Gelabert [2002; with additions by Penney, 2004a]. Since then, a number of authors have provided new records for Hispaniola, including the first identifications of certain families for the island, and as a result of advances in taxonomy and systematics a number of the taxa listed by Penney and Pérez-Gelabert [2002] have undergone name changes or have even been transferred to different families. A complete and updated checklist is provided in the Appendix.

The Extant Fauna

Prior to the twentieth century, only a small number of large, common spiders had been described from Hispaniola. The first important contribution to the knowledge of Hispaniolan spiders was a short paper by Banks [1903], based on specimens collected by R. J. Crew. He recorded 63 species, primarily from the vicinity of Port-au-Prince, Haiti. The principal collectors of Hispaniolan spiders were P. R. Uhler, 1873; W. M. Mann, 1902, 1912–1913; Bates and Darlington, 1934; P. J. Darlington, 1934, 1938; A. Audant and D. Hurst. Uhler's main interest was entomology, although he also collected a number of spiders during his visit to the western part of Haiti, some of which were described over the years by the Lithuanian arachnologist Keyserling. Some of the jumping spiders (Salticidae) collected from Haiti by Mann in 1902 were described by the Peckhams. Bates and Darlington collected a few spiders along with their entomology collections during their visit to Haiti in 1934, and Darlington returned to Hispaniola for more intensive arachnological collecting later that year and also in 1938, during which time he visited eastern and central parts of the island, including some of the highest altitudes in the Caribbean.

Elizabeth Bryant described most of the known Recent spiders from Hispaniola [Bryant, 1943: Salticidae; 1945: Argiopidae = Araneidae, Tetragnathidae, Theridiosomatidae; 1948: numerous families], most of which are held in the Museum of Comparative Zoology, Harvard University. In her final publication on Hispaniolan Araneae, Bryant [1948] considered the total described fauna to consist of 224 species, but recognized that this was probably a gross underestimate of the true total. Since Bryant's synopsis, several authors have reported or described new species for the island [e.g., Muma, 1953; Archer, 1958; Gertsch, 1958; Exline and Levi, 1962; Berman and Levi, 1971; Levi, 1955, 1957, 1959, 1962, 1963ab, 1964, 1977, 1986, 1991, 1993, 1995, 1999, 2001; Chickering, 1967; Gertsch and Ennik, 1983; Dondale and Redner, 1984; Opell, 1979; Maddison, 1986; Smith, 1986; Wunderlich, 1986; Platnick and Shadab, 1974a, 1980, 1989; Richman, 1989]. In more recent years Brescovit [1993, 1999] has described two species of Anyphaenidae, Coyle [1995] two new species of Dipluridae, Edwards and Wolff [1995] reported three new species of jumping spiders, Alayón-García [1992, 1995, 2002, 2004] has described several new spider species, Alayón-García *et al.* [2001] identified the Old World *Cyrtophora citricola* (Araneidae) as a new record for the island, Piel [2001] listed a new Araneidae, Huber [2000] identified one new species in his revision of Neotropical Pholcidae, and Sánchez-Ruíz [2005] described a new caponiid. Rheims and Brescovit [2004] described the first extant Hersiliidae from Hispaniola and Platnick and Penney [2004] identified the first Prodidomidae from from the island, which also represented the first New World record of the genus *Zimiris*. Penney [2004a] provided eight additional new species records for the Island based on a small, opportunistically collected sample of 23 identifiable species. Such a large proportion of new records in such a small sample demonstrates that the extant spider fauna of Hispaniola is poorly known and worthy of further investigation.

The Fossil Fauna

It is only a quarter of a century since Ono [1981] described the first spider preserved in Miocene amber from the Dominican Republic. Subsequently, Dominican amber spiders were described by Schawaller [1981a, 1982, 1984], Wunderlich [1981, 1982, 1986, 1987, 1988, 2004], Reiskind [1989], Wolff [1990] and Penney [2000a, 2001, 2005cd]. Penney [2000b] revised the Dominican amber Anyphaenidae and Penney [in press] revised the Neotropical fossil Hersiliidae. The spiders described as *Mysmena dominicana* Wunderlich, 1998 (Mysmenidae) and *Ceratinopsis deformans* (Wunderlich, 1998) (sub *Grammonota*, Linyphiidae) by Wunderlich [1998] and the specimen reported as Archaeidae [Wunderlich, 1999] from Dominican Republic amber are all subfossils preserved in Madagascan copal [Wunderlich, 2004: 39]. The undescribed specimen identified as Amaurobiidae by Schawaller [1981b] is actually preserved in Baltic amber [Wunderlich, 1988] as is the linyphiid *Custodela lamellata* (Wunderlich, 1988) (sub *Lepthyphantes*) [Wunderlich, 2004: 1342].

Poinar [1992] reported the families Oecobiidae: Urocteinae and Archaeidae as present in Dominican Republic amber, but they are unknown as fossils from this source.

Additional spider taxa reported, but without formally described species, and which are otherwise unrecorded in Dominican amber include: Symphytognathidae *sensu lato* [Schawaller, 1981b], Agelenidae, Philodromidae [Schawaller, 1981b; Wunderlich, 1988], Liocranidae, Pisauridae, Palpimanidae: *Otiothops* sp. [Wunderlich, 1988], Hahniidae [Penney, 1999, based on a juvenile and possibly a misidentification] and Lycosidae [Penney, 2001]. Genera recorded but not identified to species, from families with additional described species in Dominican amber include: Dipluridae: *Ischnothele*?; Oonopidae: *Heteroonops*?, *Opopaea*?; Tetragnathidae: *Leucauge*; Theridiidae: *Craspedisia*; Salticidae: *Descanso* [Wunderlich, 1988], ?*Sarinda* [Wunderlich, 2004] and Salticidae: *Nebridia* [Cutler, 1984]. The spider described as *Anelosimus clypeatus* Wunderlich, 1988 was removed from that genus by Penney [2001] and currently remains unassigned. A complete systematic catalogue, including synonymies, transfers, etc. of fossil spiders preserved in Dominican Republic amber described up until the end of 2005 was provided by Penney [2006a].

COMPARISON OF THE EXTANT AND FOSSIL SPIDER FAUNAS

Quantitative comparisons of inventories of hyperdiverse taxa, such as spiders, from different regions (including through geological time) can be problematic because of undersampling bias. Even intensive inventories of spiders from poorly studied regions are typically composed of many singletons (species represented by only one specimen). Extrapolative richness estimates from such data typically indicate that many species have been missed and comparing overlap between two such inventories can multiply these difficulties, because of uncertainty over whether taxa absent from one site were present and not sampled, or were truly absent. Palaeontological data are scarce by their nature and new species in amber are often described from singletons and only occasionally from a series of specimens, which is usually few in number. Thus, species numerical abundance data are limited in terms of quantitative investigations. Traditionally, analysis of terrestrial arthropod palaeocommunities has been at family level, which is a good predictor of underlying species diversity and is applied extensively as the 'higher taxon approach' by neontologists investigating distribution, ecological correlates, and diversity patterns of extant tropical insects [Labandeira, 2005]. Similar issues relate to qualitative investigations, but not to the same degree and it is reasonably easy to generate falsifiable hypotheses for the fossil and the extant faunas based on family (and also genus) presence and absence data.

There are currently 49 spider families with formal species described from Hispaniola when both Miocene and Recent faunas are considered together (see Appendix). When the families listed without formally described fossil species (see earlier) are included, this value rises to 50, with 32 and 36 shared families respectively. This equates to similarity values for the Miocene and Recent Hispaniolan spider families of 69.6% for families with described species and 72.0% when all fossils are considered. The similarity values for genera are considerably lower at 13.0% for those with described species and 15.1% when those without described species are included; 26 genera are currently considered strictly fossil. However, such genera have, in the past, been subsequently discovered in the extant fauna (see later) or synonymized with recent genera (e.g. Raven [2000: 573] synonymized *Microsteria* Wunderlich, 1988 = *Masteria* L. Koch, 1873 [Dipluridae]), so this value is considered an upper limit for the taxa described to date. The species similarity is 0% because all described Dominican amber fossil and sub-fossil spiders are considered extinct [Penney, 2006a].

For families recorded from 15 or more species interesting similarities and dissimilarities exist between the Miocene and Recent faunas (Appendix). The frequency of genera and species is similar between the Miocene amber and Recent faunas for Pholcidae and Theridiidae, they are distinctly dissimilar for Tetragnathidae, Araneidae and Salticidae and are also slightly dissimilar for Anyphaenidae. These similarities may be an indication that those families were as diverse in the Miocene as they are at present and the dissimilarities may reflect the opposite. However, it must be remembered that the Recent fauna is poorly known and some of the dissimilarities may be purely the result of taphonomic biases associated with the amber fauna, i.e., there may be some mechanism related either directly to the resin formation or the ecology of these spider families that reduced their chances of getting caught and preserved as amber inclusions [e.g., Penney, 2002; Penney and Langan, 2006]. In this case, it may be argued that these families may also have been as diverse on Hispaniola in the Miocene as they are today, but that they are under-represented as amber inclusions.

Araneidae and Tetragnathidae are usually relatively large, conspicuous, aerial web-spinning spiders. Therefore, they are often common in collections of Recent spiders because they are easily seen by collectors. In addition, their large size may facilitate their escape from the sticky resin trap, preventing them occurring frequently as amber inclusions. The jumping spiders (Salticidae) form the largest extant spider family, with more than 5,035 species in 553 genera [Platnick, 2006] and have a global distribution (excluding Antarctica). However, salticids appear to be geologically young (see known fossil range in Figure 2). The active predatory behavior of salticids predisposes them to entrapment in resin [Penney, 2002] and they are currently unknown from Cretaceous ambers, so it may be that Salticidae is a recently evolved family [e.g., García-Villafuerte and Penney, 2003] and this may explain why there are considerably more species known from the Recent fauna. However, this 'over-representation' of salticids for the extant fauna may also be due to their diurnal, relatively rapid, start–stop, hunting behaviour, which makes them conspicuous to collectors. Alternatively, the Salticidae are a notoriously difficult family from a taxonomic perspective and many taxonomists are reluctant to study palaeontological material. Except for the few well-resolved groups, fossil salticid taxonomy is beyond the scope of the current author, so it may well be that fossil amber salticids exist, but that they await formal description.

The only obvious case that does not fit the above scenarios is the family Dictynidae. These appear to have been considerably more diverse on Hispaniola in the Miocene than they are at present, with 15 known fossil species and only one Recent example (although the author has collected a second extant species but currently only from a single female, making an identification difficult at this stage). All the fossil dictynids appear to be valid species because Wunderlich [1988] based his diagnoses on distinct differences in male pedipalp morphology, except for *P. singularis*, which is known from only one female. All known Dictynidae amber fossils belong to strictly fossil genera (Appendix), so it may be the case that they were more diverse in the past, at least on Hispaniola. However, the Recent Hispaniolan spider fauna is not known well enough to support this conclusion at present, and future collections of extant material may reveal more species of Dictynidae and maybe also examples of these genera currently known only from fossils.

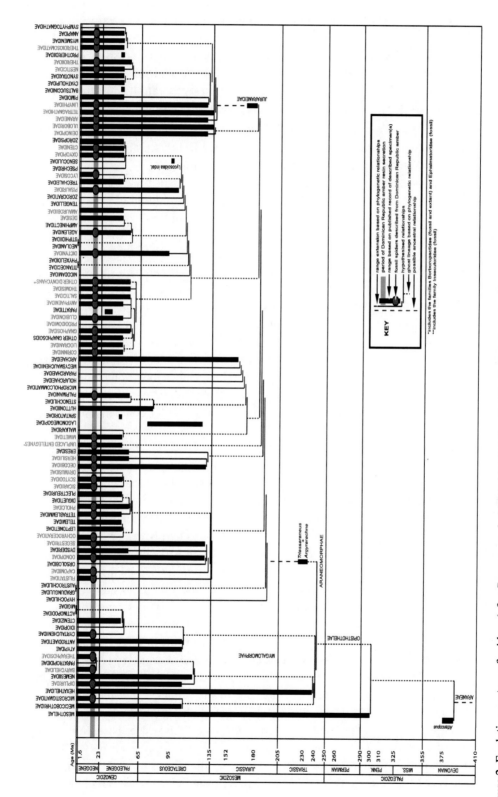

Figure 2. Evolutionary tree of spiders (after Penney *et al.*, 2003), highlighting the fossil (red circles) and Recent (green text) Hispaniolan faunas.

When Bryant [1948] published her synopsis of the Recent Hispaniolan spider fauna she listed 224 species for the island. Since then few authors have focused specifically on this fauna, which now numbers 313 species (Appendix). As a comparison to Hispaniola Bryant [1948] noted that 280 extant spider species occured in Cuba. In contrast to Hispaniola, arachnological studies have progessed with regard to Cuban spiders and Alayón-García [2000] listed 567 species (more than twice the number cited by Bryant). Thus, the currently known spider fauna for Hispaniola is probably a gross under-representation of the true total. Although Hispaniola is smaller than Cuba ($76,071km^2$ c.f. $114,525km^2$) it has a considerably greater altitudinal variation with a maximum of 3,174m c.f. 1,999m for Cuba [Flint and Pérez-Gelabert, 1999] providing a broad range of habitat variation.

There is general agreement that there was a very substantial rise in global biodiversity through the Cenozoic and especially the Neogene (i.e. the last 23 million years) [Crame and Rosen, 2002]. This was a time of crucial plate tectonic movement that led to climate change and essentially gave the tropics their modern form [Crame and Rosen, 2002]. Given the young age of Dominican amber (15–20Ma) and the similarity of the extant and amber spider faunas (Figure 2, Appendix), the *a priori* assumption is that the Hispaniolan spider fauna had a similar diversity in the Miocene to that observed/expected today. However, our knowledge of Cenozoic Hispaniolan biodiversity will always be far from complete because not all Miocene species will have been preserved in amber, and of those that were, only a relatively small proportion will be found as fossils. Within the amber forest, resin secretion was restricted to a single tree species and would therefore be selective in the organisms it trapped, maybe temporally (e.g. seasonal) as well as spatially. Thus, amber inclusions will represent only a subsample of the total fauna present at the time. Nonetheless, given the similarity of both faunas (Figure 2; Appendix) they do afford an excellent basis for comparison in studies of historical biogeography.

It should be noted that some differences between the Hispaniolan fossil and Recent faunas do occur in other groups. For example, certain extant genera of Hymenoptera, Isoptera, Lepidoptera and Coleoptera were present on Hispaniola in the Miocene but are absent from the island today [Wilson, 1985; Nagel, 1997; Poinar and Poinar, 1999; Hall *et al.*, 2004]. Many insects are more susceptible to extinction than spiders because they form specialized associations with other organisms, such as plants (as herbivores or pollinators), or as parasites, whereas most spiders are generalist predators and if one prey item becomes extinct they can easily switch to another source [Penney *et al.*, 2003; Penney, 2004b]. In addition, extinction resistance in spiders may be facilitated by their ability to survive extended periods of starvation, or other adverse conditions, by entering a state of metabolic torpor.

Poinar [1999] proposed that the demise of certain groups known from Dominican amber but absent from Hispaniola (and other islands in the Greater Antilles) today resulted from a cool period associated with increased ariditiy during the Plio-Pleistocene. However, evidence from sedimentary and geomorphological data, alluvial terraces and albedo reflectivity indices suggest that the Dominican Republic was not drastically affected by the Pleistocene glaciations [Schubert, 1988]. Curtis *et al.* [2001] proposed that both the continental and insular Caribbean regions were cool and arid during the Late Pleistocene, based on oxygen isotope analysis of carbonate shells from lake sediment cores. However, their only Antillean sample was from the south-western tip of Haiti and their results [Curtis *et al.*, 2001: Figure 2] show considerably less variation in this sample than those from the mainland, and indeed, not a great deal of aridity in comparison to that at present, although they support their ideas with previously published palynological data. Their conclusion that analyses of West Indian biogeography must be cognizant of Pleistocene and Holocene climatic changes is certainly a valid one.

THE IMPORTANCE OF AMBER FOSSILS IN STUDIES OF HISTORICAL BIOGEOGRAPHY

Fossils are of paramount importance in studies of historical biogeography. They can play a decisive part in the falsification of hypotheses proposed solely on the distributions of extant taxa [e.g., Eskov, 1990]. As an arachnological example, the current Gondwanan distribution of the extant spider family Archaeidae supports the theory of mobilistic biogeography, i.e., that the fragmentation of Gondwanaland and the subsequent continental drift can explain the current distribution of this family, with extant fauna restricted to South Africa and Madagascar (an apochoric distribution; e.g., Kraus [1978]). However, because fossils of this family are highly diverse in Baltic amber [Wunderlich, 2004], and also occur in Tertiary French amber [Penney, 2006b], Cretaceous Burmese amber [Penney, 2003] and in sediments from the Jurassic of Kazakhstan [Eskov, 1987], the palaeontological data demonstrate that their current distribution is plesiochoric, and so a different explanation is required. Another arachnological example concerns the family Huttonniidae, with extant species restricted to New Zealand, but with fossils described in Cretaceous amber from Canada [Penney and Selden, 2006].

The theory of ousted relicts [e.g., Eskov and Golovatch, 1986] proposes that austral disjunctions result from a formerly pancontinental distribution followed by the extinction of 'intermediate links' from the northern continents. There is a considerable amount of palaeontological data, in the form of northern hemisphere fossil representatives of Recent austral taxa of numerous different animal groups in support of this theory [e.g., Eskov, 1987: table 2; Eskov, 1992: table 1], and this tends to be the rule rather than the exception. However, the above authors did not consider the phylogenetic placement of the fossil taxa, and from the viewpoint of competing theories in historical biogeography, these amber fossils are only important if they are not plesiomorphic or the sister group to the extant taxa [Grimaldi, 1992].

Additional problems that may occur when generating biogeographical hypotheses without considering fossils concern non-indigenous taxa. For example, one might be tempted to explain Hispaniolan spider families known in the Recent fauna from only synanthropic or widespread continental species as recent introductions or colonizers (e.g., the families Oecobiidae and Filistatidae: see Appendix). The only extant oecobiid on Hispaniola is the synanthropic *Oecobius concinnus* Simon, 1893, which is distributed from USA to Brazil, the West Indies, the Galapagos Islands and introduced elsewhere [Santos and Gonzaga, 2003] and the only extant filistatid on the island is *Kukulcania hibernalis* (Hentz, 1842), which has a widespread distribution in the New World, so is also presumably an introduced species. However, the possibility that these families are recent introductions can be exluded when the fossil record is taken into consideration because both are described from fossil species in amber [Wunderlich, 1988; Penney, 2005d], so were already present on Hispaniola in the Miocene.

The main models for explaining historical biogeography (dispersal and vicariance) differ with regard to the age of barriers and disjunctions, thus both appear open to falsification through statements about the age of one or the other [Platnick and Nelson, 1978]. The minimum age of a taxon is that of the oldest fossil attributed to it or its sister taxon. A vicariance correlation can be falsified if one of the disjunct taxa is older than the barrier and a dispersal correlation can be falsified if one of the taxa is the same age as, or older, than the barrier. However, both models include protection from falsification based on palaeontological evidence through rejection of a particular barrier in favour of an older barrier, which can be postulated if not demonstrated [Platnick and Nelson, 1978].

It is imperative that palaeontological data are considered when generating hypotheses to explain the biogeographic patterns observed today because fossils form empirical data with both phylogenetic and temporal implications. As pointed out by Platnick and Nelson [1978], the kinds of biogeographic analyses found relegated to the back pages of systematic revisions cannot be justified, yet they continue to be published without taking into consideration palaeontological or geological data.

GREATER ANTILLEAN ORIGINS, PALAEOGEOGRAPHY AND CONSEQUENCES FOR BIOGEOGRAPHY

Prior to plate tectonics providing a mechanism for vicariance in the latter part of the twentieth century, the 'land bridge' was the primary alternative mechanism to dispersal in biogeography [Hedges, 2001]. Indeed, Petrunkevitch [1928] considered the Greater Antillean spider fauna to represent an eastern outgrowth of the Central American fauna by way of a presumed earlier land connection and subsequent continent–island vicariance. However, such a land connection appears never to have existed [Ross and Scotese, 1988; Iturralde-Vinent and MacPhee, 1999]. The most recent hypothesis proposes that during the Eocene–Oligocene transition, the developing northern Greater Antilles and northwestern South America were briefly (33–35Ma) connected by a landspan (a subaerial connection between a continent and one or more off-shelf islands) centered on the emergent Aves Ridge, but the massive uplift that apparently permitted these connections was finished by 32Ma [Iturralde-Vinent and MacPhee, 1999]. However, this idea is not accepted by all workers, e.g., Hedges [2001] criticised the GAARlandia hypothesis of Iturralde-Vinent and MacPhee [1999] on a point by point basis, and favoured a dispersal model as the primary means for explaining the origins of West Indian biodiversity. The criticims of Hedges [2001] were refuted convincingly by MacPhee and Iturralde-Vinent [2005], and so the debate continues.

Nonetheless, the Greater Antilles in their current guise appear to be relatively young geographical features, probably no older than the middle Miocene [Iturralde-Vinent and MacPhee, 1999], rather than having evolved from the Proto-Greater Antillean and subsequently the Greater Antillean landmass formed on the west of the Proto-Caribbean region during the late Lower Cretaceous as proposed by Ross and Scotese [1988]. Therefore, all on-island lineages forming the Recent fauna must be younger than Middle Eocene [Iturralde-Vinent and MacPhee, 1999]. It is likely, and even probable, that Caribbean islands existed during the Mesozoic and that they doubtless had a biota as well, but so far as can be determined, none of the terrestrial environments that may have supported such life continued into the Cenozoic [MacPhee and Iturralde-Vinent, 2000]. The Proto-Antilles were located only one to three crater diameters away [Hedges, 2001] from the Chicxulub impact site in Yucutan at the K/T extinction event (65Ma). Whatever land masses were emergent in this region at the time would have had all flora and fauna anihilated as a result of associated mega-tsunamis and hypercanes (giant hurricanes) and the knock-on consequences in terms of climate and environmental change (see Robertson *et al.* [2004] for additional local effects). During the period of amber-forming resin secretion (15–20Ma; Iturralde-Vinent and MacPhee [1996]) Hispaniola was a distinct island. There may have been a connection to Puerto Rico via a narrow neck of land, but this is not certain [Iturralde-Vinent and MacPhee, 1999]. Since then, tectonic movements of the Caribbean Plate have carried Hispaniola slowly eastwards to its current position.

BIODIVERSITY PREDICTIONS AND ORIGINS OF HISPANIOLAN SPIDERS

It is not necessarily possible to predict the presence of undiscovered fossil spiders from families currently unknown from the amber fauna, based on what we observe in the Recent fauna. However, we can hypothesise their existence on Hispaniola during the time of resin secretion during the Miocene and that they evaded capture because of their habitat preference or ecological niche [Penney, 1999]. On the other hand, families present in the amber fauna but unrecorded from the Recent fauna can be predicted to occur on the island. For example, based on presence and absence data of endemic and non-endemic spider species in the fossil and extant Hispaniolan faunas, Penney [1999] made the following predictions: the families Cyrtaucheniidae, Microstigmatidae, Nemesiidae, Ochyroceratidae, Tetrablemmidae, Palpimanidae, Hersiliidae, Symphytognathidae *s.l.*, Anapidae, Mysmenidae and Hahniidae would form components of the Recent fauna; Filistatidae and Desidae colonized Hispaniola following the Miocene amber formation; Drymusidae, Amaurobiidae and Deinopidae were present on Hispaniola during the Tertiary, but avoided capture in resin or have yet to be found in the amber; and Scytodidae, Oecobiidae, Uloboridae, Dictynidae and Clubionidae have colonized Hispaniola since the Miocene amber formation but these families, which were present on Hispaniola during the period of amber formation, also contain undiscovered endemic species.

All these hypotheses are falsifiable or supportable through future discoveries of new Hisaniolan taxa. Indeed, extant Hersiliidae have been described from the island by Rheims and Brescovit [2004] and a new extant species of the genus *Arachnolithulus* Wunderlich, 1988 (Ochyroceratidae) originally known only from fossil species has also recently been found on Hispaniola [Hormiga, pers. comm. 2005, in prep.]. The idea that Filistatidae first colonized the island post-Miocene has been falsified by the discovery of fossil species in amber by Penney [2005d]. The remaining hypotheses still stand, except possibly for the presence of Hahniidae which was based on a dubious identification [Penney and Pérez-Gelabert, 2002]. The same basic principles may be applied to presence and absence data for genera to generate higher resolution hypotheses, but caution should be exercised because higher taxonomic levels will be more susceptible to extinction than lower ones, unless of course they are monotypic.

It is clearly evident that many different types of organismal associations have a long geological history and many were already well established by the Mioecene, such as, plant–pollinator, plant–herbivore, predator–prey, host–parasite relationships, phoresy, commensalism, etc. Thus, it is also possible to predict the presence of other taxa from expected associations with observed taxa. This is possibly a more realistic proposition for the extant fauna, but nonetheless, fossil evidence supports this idea for the Miocene fauna also. For example, Wunderlich [1988] described the extant spider genus *Ischnothele* Ausserer, 1875 (Dipluridae) and Penney [2000a] described a new species of Mysmenidae from the extant genus *Mysmenopsis* Simon, 1897, both from Dominican amber. Of the 17 extant species of *Mysmenopsis* known at the time, Platnick and Shadab [1978] reported five of them as kleptoparasites in the webs of diplurid spiders. The two Recent species (*M. monticola* and *M. furtiva*) to which the fossil species is most closely related were described by Coyle and Meigs [1989] from Jamaica; they reported a probable kleptoparasitic mode of life because they were collected from the webs of *Ischnothele* spp., and Coyle *et al.* [1991] provided direct evidence for the kleptoparasitic and commensualistic nature of *M. furtiva* in the web of its diplurid host. Based on the idea of behavioural fixity/uniformitariansim [Boucot, 1989], which infers that extinct organisms behaved in a similar manner to their Recent relatives at genus and often family level, there is no reason to believe that the fossil species behaved any differently; because it is morphologically similar to the Jamaican species (with only minor differences in genitalic structure and tibial spine numbers). Thus, the presence of Dipluridae in the above example could have been used to predict the presence of Mysmenidae on Hispaniola during the Miocene, but even more so vice versa.

Patterns of endemism in other groups have been invoked in discussions of West Indian biogeography [e.g. Genaro and Tejuca, 2001; Hedges, 2001], but it is unclear at present what this adds to the debate on the grand scale. The problem lies in the range of variation between groups, for example, regional endemism in vertebrates ranges from 35% (Aves) to 99% (Amphibia) with an average of 74% [Hedges, 2001: table 1] and in Cuba insect ordinal endemism may range from 0% to >90% with an average range of 40–60% [Genaro and Tejuca, 2001]. The latter authors note a lack of information for Hispaniolan extant insect biodiversity. In their paper on Hispaniolan spiders, Penney and Pérez-Gelabert [2002] listed 116 of the 296 extant species (39%) as endemic. However, given the poorly known nature of the extant fauna and the fact that many of the so-called endemics were described by Bryant (see earlier), who has subsequently had many of her species synonymized by other authors, the accuracy of this value as a true representation is highly dubious. Until the existing taxa are revised and the Recent fauna is better known, little can be derived from general studies of spider endemism with regard to Caribbean (especially Hispaniolan) biogeography.

Once such work is undertaken, interesting patterns can be expected and particularly so when the extant taxa are compared with closely related fossil species. For example, *Trachelas* (Corinnidae) spiders from North and Central America were revised by Platnick and Shadab [1974ab] and placed into four species groups separated on the grounds of genitalia, somatic morphology and geography. The *bicolor* group is the only one found in the West Indies. Penney [2001] described the new fossil species *Trachelas poinari* from Dominican amber, which also belongs in this group and is most closely related to *T. borinquensis* Gertsch, 1942 (Recent, endemic to Puerto Rico), and the only other *Trachelas* species to possess a retrolateral patellar apophysis. This may represent an example of island–island vicariance following the separation of Puerto Rico from the northern Hispaniola block, presumably sometime within the last 14Ma [e.g. Iturralde-Vinent and MacPhee, 1999].

As discussed above, the competing arguments regarding dispersal versus vicariance in understanding Caribbean biogeography continue to stimulate lively debate and conflicting opinions. In both cases a barrier is involved. In the vicariance model, dispersal takes place in the absence of a barrier (the appearance of the barrier fragments the range of an ancestral species), whereas in the dispersalist model, dispersal occurs across an existing barrier [Platnick and Nelson, 1978]. The dispersal capabilities of organisms vary and this is clearly going to have important consequences for those with poor disperal abilities when trying to cross a barrier (not to mention the additional problems of finding the right habitat and establishing a founder population once it/they arrive). This is particularly true of spiders, in which some species can disperse great distances at high altitudes through ballooning (for a review of spider families that balloon see Bell *et al.* [2005]), whereas some fossorial, burrow-dwelling species may disperse only very short distances from the maternal burrow. For example, this concept is evident in the extremely isolated Hawaiian Islands, where the few spider families that have colonized the archipelago, have been from highly dispersive taxa and appear to have done so repeatedly [Gillespie *et al.*, 1998]. Consequently, one might expect a relatively low familial diversity for islands on which dispersal is the primary mode of immigration. This is not the case for Hispaniola now, or back in the Miocene [Penney and Pérez-Gelabert, 2002; Penney, 2005b; Appendix], and the significance of this observation increases when one considers the relatively young age of the island. This would seem to add more weight in favour of the vicariance model for explaining on-island spider lineages, although dispersal cannot be ruled out for some of the highly dispersive taxa.

Evidence for links with the South American fauna can be found in the Recent distributions of numerous extant spider genera known as fossils in Dominican amber (see Appendix). For example, *Masteria* L. Koch, 1873 (Dipluridae), *Psalistops* Simon, 1889 (Barychelidae), *Misionella* Ramírez and Grismado, 1997 (Filistatidae), *Coryssocnemis* Simon, 1893 and *Modisimus* Simon, 1893 (Pholcidae), *Nops* MacLeay, 1839 (Caponiidae), *Monoblemma* Gertsch, 1941 (Tetrablemmidae), *Stenoonops* Simon, 1891 (Oonopidae), *Mysmenopsis* Simon, 1897 (Mysmenidae), *Azilia* Keyserling, 1881 and *Cyrtognatha* Keyserling, 1881 (Tetragnathidae), *Anyphaenoides* Berland, 1913 and *Lupettiana* Brescovit, 1996 (Anyphaenidae), *Corinna* C. L. Koch, 1841 (Corinnidae), *Pseudosparianthis* Simon, 1887 (Sparassidae), *Corythalia* C. L. Koch, 1850 and *Lyssomanes* Hentz, 1845 (Salticidae) and *Molinaranea* Mello-Leitão, 1940 [Penney, in prep.] (Araneidae) have extant species restricted to, or primarily distributed within, South America, the Greater and Lesser Antilles and Central America (Panama). The remaining genera are more widespread and some include cosmopolitan taxa, but all are well represented in South America or the Neotropics. None of the extant spider genera recorded from Miocene Dominican amber present any reasoning to detract from a South American origin for this fauna.

The future research direction for understanding these biogeographical patterns will lie in the application of cladistic and phylogenetic approaches to biogeography. Cladistic biogeography emphasizes the search for congruent biogeographic patterns using cladograms, disregarding both dispersal and viariance explanations *a priori*. It assumes correspondence between taxonomic relationships and area relationships is biogeographically informative (see Reiskind [1989] for an example concerning Caribbean spiders including a species in

Dominican amber) and that general area cladograms can be derived from individual area cladograms of different taxa [Morrone and Crisci, 1995]. Phylogenetic biogeography suggests that if different monophyletic groups show the same biogeographic pattern, then they probably share the same biogeographic history and infers that primitive members of a taxon are found closer to its centre of origin than more apomorphic ones, which are found at the periphery [Morrone and Crisci, 1995]. Indeed, it is interesting to note that preliminary obersvations based on pedipalp morphology of Miocene Dominican amber fossils, e.g., *Misionella* (Filistatidae) and *Molinaranea* (Araneidae; Penney [in prep.]), when compared to extant taxa restricted to South America, suggest that the amber species are more derived and thus further removed from the centre of origin, which in this case would be South America. Clearly, there is great potential for additional research in this area.

MEXICAN AMBER AS AN ADDITIONAL COMPARATIVE PALAEONTOLOGICAL RESOURCE

Another potentially informative and comparative palaeontological resource for understanding the historical biogeography of Caribbean spiders (and other groups) is amber from Chiapas, Mexico. However, the Early Miocene–Late Oligocene age of this amber [e.g., Poinar, 1992] is not as well constrained as previously thought, and it may actually be the same age as Dominican amber [Solórzano Kraemer and Rust, in prep.; Solórzano Kraemer, pers. comm. 2006], which would obviously have interesting implications for direct comparisons with the fossil Hispaniolan fauna. The degree of preservation is not quite as good as that in Dominican amber, but spiders are relatively common as inclusions. Unfortunately, only a few spiders have been described from this deposit [Petrunkevitch, 1963, 1971; García-Villafuerte and Vera, 2002; Garcia-Villafuerte and Penney, 2003; Penney, in press; García-Villafuerte, in prep.; Dunlop *et al.*, in prep.] but the taxonomy of those described by Petrunkevitch is poor and his taxa are in need of revision [Penney, in press]. Penney [in press] revised the Neotropical fossil Hersiliidae, which included fossil taxa in Mexican amber. A number of genera (including extant and fossil) were synonymized and a number of extant species had to be transferred to a previously strictly fossil genus. This highlights the importance of palaeontologists considering Recent taxa when describing fossils, and likewise the need for neontologists to consider fossil taxa. Without such, cross-disciplinary awareness and consideration, which unfortunately is often lacking by palaeontologists and neontologists alike, comparisons of fossil and extant assemblages will be pointless and uniformative at best, but misleading at worst.

The potential importance of Mexican amber for understanding the origins of Hispaniolan spiders derives from the fact that the Yucatan Peninsula did not have land connections to South America (post K/T and prior to the Mexican amber formation). Thus, any similarities of Mexican amber spiders to the South American fauna must be a result of dispersal, despite the recent Panamanian isthmus formation. Thus, a comparative analysis of Mexican and Dominican amber spiders with proximate extant faunas will aid our understanding of the relative contributions of vicariance and dispersal for explaining the origins of current Caribbean biogeography.

CONCLUSION

Fossils form empirical data with both phylogenetic and temporal implications and are of paramount importance for understanding historical biogeography. They can play a decisive role in falsifying hypotheses proposed solely on the distributions of extant taxa. It is also possible to generate hypotheses for both fossil and extant faunas that are ultimately falsibiable through the discovery of new Hispaniolan taxa. The amber spider fauna has clear affinities to the present South American fauna, so how did they colonize the island from the mainland: dispersal or vicariance? The high Miocene biodiversity (including taxa with poor dispersal capabilities) on a geologically young island would seem to add more weight in favour of the vicariance model for explaining on-island spider lineages, although dispersal cannot be excluded for some of the highly dispersive taxa. However, a case-by-case approach is required in order to determine the relative contributions of the different models. The future research direction for understanding the biogeographical patterns will lie in the application of cladistic and phylogenetic biogeographical approaches and it is interesting to note that preliminary obersvations based on pedipalp morphology suggest some fossil taxa are more derived than extant forms.

Both the fossil and Recent Hispaniolan spider faunas are worthy of further investigation, particularly because of their similarity, which means they can be used as directly comparable, complementary data sets to address various questions relating to Caribbean biogeography. Clearly, the greater the completeness of the faunal inventories and the larger the datasets in such analyses the more reliable the conclusions. The contribution that the Dominican Republic (and Mexican) amber fossil fauna can make to our understanding of the origins of Caribbean biogeography should not be underestimated.

APPENDIX: CHECKLIST OF HISPANIOLAN SPIDER SPECIES

Families are listed in systematic order following Platnick [2006]. Only full species are included in the list. Species from Navassa, a tiny island (5.2km^2) 64km west of Hispaniola are not included; a list of the 58 known spider species was provided by Alayón-García [2001]. In numerous old arachnological works the dates printed on the journals do not actually represent the correct publication year. All publication dates for the Recent fauna and generic placements follow Platnick [2006]. Names in square brackets are nomina dubia; * = fossil genus; † = fossil species; [†] = subfossil species. This list is based primarily on that of Penney and Pérez-Gelabert [2002] and Penney [2006a] with emendations and updates as required. References for the extant taxa are not cited but these can be located easily using the bibliography in Platnick [2006]. In total, the Hispaniolan spider fauna currently comprises 470 named species (excluding Theridiidae: gen. indet. *clypeatus* [Wunderlich, 1988)). The fossil fauna consists of 154 named species in 83 genera and 39 families. None of the families, 31% of the genera and all the species are extinct. Three subfossil species have been described from copal. The extant fauna consists of 313 named species, 175 genera and 42 families.

Order Araneae Suborder Opisthothelae Infraorder Mygalomorphae

Family dipluridae

Ischnothele jeremie Coyle, 1995
Ischnothele garcia Coyle, 1995
†*Masteria sexoculata* (Wunderlich, 1988)

Family cyrtaucheniidae

†*Bolostromus destructus* Wunderlich, 1988

Family microstigmatidae

*†*Parvomygale distincta* Wunderlich, 2004

Family barychelidae

Psalistops fulvus Bryant, 1948
†*Psalistops hispaniolensis* Wunderlich, 1988
Psalistops maculosus Bryant, 1948
Trichopelma nitidum Simon, 1888

Family theraphosidae

Citharacanthus spinicrus (Latreille, 1819)
Cyrtopholis agilis Pocock, 1903
Cyrtopholis cursor (Ausserer, 1875)
Holothele sericea (Simon, 1903)
*†*Ischnocolinopsis acutus* Wunderlich, 1988
Phormictopus cancerides (Latreille, 1806)
Infraorder Araneomorphae

Family filistatidae

Kukulcania hibernalis (Hentz, 1842)
†*Misionella didicostae* Penney, 2005

Family sicariidae

†*Loxosceles aculicaput* Wunderlich, 2004
Loxosceles caribbaea Gertsch, 1958
Loxosceles cubana Gerstch, 1958
†*Loxosceles defecta* Wunderlich, 1988
†*Loxosceles deformis* Wunderlich, 1988

Loxosceles taino Gerstch and Ennik, 1983

Family scytodidae

Scytodes fusca Walckenaer, 1837
Scytodes longipes Lucas, 1844
†*Scytodes piliformis* Wunderlich, 1988
†*Scytodes planithorax* Wunderlich, 1988
†*Scytodes stridulans* Wunderlich, 1988

Family drymusidae

Drymusa simoni Bryant, 1948

Family ochyroceratidae

*†*Arachnolithulus longipes* Wunderlich, 2004
*†*Arachnolithulus pygmaeus* Wunderlich, 1988

Family pholcidae

Artema atlanta Walckenaer, 1837
†*Coryssocnemis velteni* Wunderlich, 2004
Leptopholcus hispaniola Huber, 2000
Micropholcus fauroti (Simon, 1887)
Modisimus femoratus Bryant, 1948
†*Modisimus calcar* Wunderlich, 1988
†*Modisimus calcaroides* Wunderlich, 1988
†*Modisimus crassifemoralis* Wunderlich, 1988
Modisimus fuscus Bryant, 1948
Modisimus glaucus Simon, 1893
†*Modisimus oculatus* Wunderlich, 1988
†*Modisimus tuberosus* Wunderlich, 1988
Modisimus vittatus Bryant, 1948
†*Pholcophora brevipes* Wunderlich, 1988
†*Pholcophora gracilis* Wunderlich, 1988
†*Pholcophora longicornis* Wunderlich, 1988
Physocyclus globosus (Taczanowski, 1874)
Tainonia serripes (Simon, 1893)
*†*Serratochorus pygmaeus* Wunderlich, 1988

Family caponiidae

Caponina darlingtoni Bryant, 1948
Nops blandus (Bryant, 1942)

Nops ernestoi Sánchez-Ruíz, 2005
Nops gertschi Chickering, 1967
†*Nops lobatus* Wunderlich, 1988

Family tetrablemmidae
†*Monoblemma? spinosum* Wunderlich, 1988

Family segestriidae
Ariadna multispinosa Bryant, 1948
†*Ariadna paucispinosa* Wunderlich, 1988

Family oonopidae

*†*Fossilopaea sulci* Wunderlich, 1988
Ischnothyreus peltifer (Simon, 1891)
†*Oonops seldeni* Penney, 2000
Oonops validus Bryant, 1948
†*Orchestina dominicana* Wunderlich, 1981
†*Orchestina tibialis* Wunderlich, 1988
†*Stenoonops incertus* (Wunderlich, 1988)
Triaeris stenaspis Simon, 1891

Family mimetidae

†*Mimetus? bituberculatus* Wunderlich, 1988
Mimetus hispaniolae Bryant, 1948
Mimetus syllepsicus Hentz, 1832

Family oecobiidae

Oecobius concinnus Simon, 1893
†*Oecobius piliformis* Wunderlich, 1988

Family hersiliidae

†*Prototama maior* (Wunderlich, 1988)
†*Prototama media* (Wunderlich, 1988)
†*Prototama minor* (Wunderlich, 1987)
Yabisi guaba Rheims and Brescovit, 2004

Family deinopidae

Deinopis lamia MacLeay, 1839

Family uloboridae

†*Miagrammopes dominicanus* Wunderlich, 2004
Miagrammopes latens Bryant, 1936
Philoponella semiplumosa (Simon, 1893)
Uloborus glomosus (Walckenaer, 1842)
Uloborus trilineatus Keyserling, 1883
Zosis geniculata (Oliver, 1789)

Family nesticidae

Eidmannella pallida (Emerton, 1875)
*†*Hispanonesticus latopalpus* Wunderlich, 1986
Nesticus maculatus Bryant, 1948

Family theridiidae

†*Achaearanea extincta* Wunderlich, 1988
Anelosimus studiosus (Hentz, 1850)
†*Argyrodes crassipatellaris* Wunderlich, 1988
Argyrodes elevatus Taczanowski, 1873
Argyrodes nephilae Taczanowski, 1873
†*Argyrodes parvipatellaris* Wunderlich, 1988
Ariamnes haitensis (Exline and Levi, 1962)
†*Chrosiothes biconigerus* Wunderlich, 1988
†*Chrosiothes curvispinosus* Wunderlich, 1988
†*Chrosiothes emulgatus* Wunderlich, 1988
†*Chrosiothes longispinosus* Wunderlich, 1988
†*Chrosiothes monoceros* Wunderlich, 1988
†*Chrosiothes tumulus* Wunderlich, 1988
†*Chrosiothes unicornis* Wunderlich, 1988
Chrysso albomaculata O.P.-Cambridge, 1882
†*Chrysso? conspicua* Wunderlich, 1988
†*Chrysso? dubia* Wunderlich, 1988
Chrysso pulcherrima (Mello-Leitão, 1917)
Coleosoma floridanum Banks, 1900
*†*Cornutidion elongatum* Wunderlich, 1988
Craspedisia spatulata Bryant, 1948
Dipoena dominicana Wunderlich, 1986
†*Dipoenata altioculata* Wunderlich, 1988
†*Dipoenata cala* Wunderlich, 1988
†*Dipoenata clypeata* Wunderlich, 1988
†*Dipoenata globulus* Wunderlich, 1988
Dipoenata morosa (Bryant, 1948)
[†]Dipoenata praedominicana (Wunderlich, 1986)

†Dipoenata stipes Wunderlich, 1988
†Dipoenata yolandae Wunderlich, 1988
[†]Episinus antecognatus Wunderlich, 1986
†Episinus brevipalpus Wunderlich, 1988
†Episinus cornutus Wunderlich, 1988
Episinus dominicus Levi, 1955
Episinus gratiosus Bryant, 1940
†Episinus praecognatus Wunderlich, 1982
†Episinus tuberosus Wunderlich, 1988
Faiditus americanus (Taczanowski, 1874)
Faiditus caudatus (Taczanowski, 1874)
Faiditus darlingtoni (Exline and Levi, 1962)
†Lasaeola puta Wunderlich, 1988
†Lasaeola pristina (Wunderlich, 1988)
†Lasaeola vicina (Wunderlich, 1982)
†Lasaeola vicinoides Wunderlich, 1988
Latrodectus geometricus C.L. Koch, 1841
Latrodectus mactans (Fabricius, 1775)
Neospintharus furcatus (O.P.-Cambridge, 1894)
Nesticodes rufipes (Lucas, 1846)
Spintharus flavidus Hentz, 1850
†Spintharus longisoma Wunderlich, 1988
Steatoda erigoniformis (O.P.-Cambridge, 1872)
Steatoda grossa (C. L. Koch, 1838)
†Stemmops incertus Wunderlich, 1988
†Stemmops prominens Wunderlich, 1988
†Styposis pholcoides Wunderlich, 1988
Theridion antillanum Simon, 1894
Theridion australe Banks, 1899
†Theridion contrarium Wunderlich, 1988
Theridion dilucidum Simon, 1897
†Theridion erectoides Wunderlich, 1988
†Theridion erectum Wunderlich, 1988
[Theridion fuesslini Simon, 1894]
Theridion hassleri Levi, 1963
†Theridion inversum Wunderlich, 1988
Theridion melanostictum O.P.-Cambridge, 1876
Theridion positivum Chamberlin, 1924
†Theridion variosoma Wunderlich, 1988
†Theridion wunderlichi Penney, 2001
Theridula gonygaster (Simon, 1873)
Thymoites banksi (Bryant, 1948)
Thymoites guanicae (Petrunkevitch, 1930)
Thymoites pallidus (Emerton, 1913)
Tidarren sisyphoides (Walckenaer, 1842)
Wamba congener O.P.-Cambridge, 1896

Family theridiosomatidae

[*Allototua guttata* Bryant, 1945]
Ogulnius fulvus Bryant, 1945
Ogulnius latus Bryant, 1948
*†*Palaeoepeirotypus iuvenis* Wunderlich, 1988
*†*Palaeoepeirotypus iuvenoides* Wunderlich, 1988
†*Theridiosoma incompletum* Wunderlich, 1988
Wendilgarda clara Keyserling, 1886

Family anapidae

*†*Palaeoanapis nana* Wunderlich, 1988

Family mysmenidae

*†*Dominicanopsis grimaldii* Wunderlich, 2004
†*Mysmenopsis lissycoleyae* Penney, 2000

Family linyphiidae

Ceraticelus paludigenus Crosby and Bishop, 1925
Eperigone serrata Ivie and Barrows, 1935
†*Floricomus fossilis* Penney, 2005
Florinda coccinea (Hentz, 1850)
Frontinella bella Bryant, 1948
Frontinella communis (Hentz, 1850)
Grammonota calcarata Bryant, 1948
Lomaita darlingtoni Bryant, 1948
†*Meioneta bigibber* (Wunderlich, 1988)
†*Meioneta fastigata* (Wunderlich, 1988)
†*Meioneta separata* (Wunderlich, 1988)
†*Selenyphantes flagellifera* (Wunderlich, 1986)
Tutaibo anglicanus (Hentz, 1850)

Family tetragnathidae

Agriognatha argyra Bryant, 1945
Agriognatha espanola Bryant, 1945
Agriognatha rucilla Bryant, 1945
Antillognatha lucida Bryant, 1945
†*Azilia hispaniolensis* Wunderlich, 1988
Azilia montana Bryant, 1940?
Chrysometa bigibbosa (Keyserling, 1864)
Chrysometa conspersa (Bryant, 1945)

Chrysometa cornuta (Bryant, 1945)
Chrysometa maculata (Bryant, 1945)
Chrysometa obscura (Bryant, 1945)
Chrysometa sabana Levi, 1986
†*Cyrtognatha weitschati* Wunderlich, 1988
Glenognatha mira Bryant, 1945
Hispanognatha guttata Bryant, 1945
†*Homalometa fossilis* Wunderlich, 1988
Leucauge argyra (Walckenaer, 1842)
Leucauge regnyi (Simon, 1897)
Leucauge venusta (Walckenaer, 1842)
Leucage venustella Strand, 1916
†*Nephila breviembolus* Wunderlich, 1986
Nephila clavipes (Linnaeus, 1767)
†*Nephila dommeli* Wunderlich, 1982
†*Nephila furca* Wunderlich, 1986
†*Nephila longembolus* Wunderlich, 1986
†*Nephila tenuis* Wunderlich, 1986
Tetragnatha elongata Walckenaer, 1842
Tetragnatha nitens (Audouin, 1826)
Tetragnatha orizaba (Banks, 1898)
Tetragnatha pallescens F.O.P.-Cambridge, 1903
†*Tetragnatha pristina* Schawaller, 1982
Tetragnatha tenuissima O.P.-Cambridge, 1889

Family araneidae

Acacesia hamata (Hentz, 1847)
Acanthepeira stellata (Walckenaer, 1805)
Aculepeira busu Levi, 1991
Aculepeira visite Levi, 1991
Alcimosphenus licinus Simon, 1895
Allocyclosa bifurca (McCook, 1887)
†Araneometa excelsa Wunderlich, 1988
†Araneometa herrlingi Wunderlich, 1988
†Araneometa spirembolus Wunderlich, 1988
Araneus bryantae Brignoli, 1983
Araneus elizabethae Levi, 1991
Araneus hispaniola (Bryant, 1945)
Araneus hotteiensis (Bryant, 1945)
†*Araneus? nanus* Wunderlich, 1988
[*Araneus perplexus* (Walckenaer, 1842)]
Argiope argentata (Fabricius, 1775)
Argiope trifasciata (Forskål, 1775)
Cyclosa berlandi Levi, 1999
Cyclosa bifurcata (Walckenaer, 1842)

Cyclosa caroli (Hentz, 1850)
Cyclosa haiti Levi, 1999
Cyclosa turbinata (Walckenaer, 1842)
Cyclosa walckenaeri (O.P.-Cambridge, 1889)
Cyrtophora citricola (Forskål, 1775)
†*Enacrosoma verrucosa* (Wunderlich, 1988)
Eriophora ravilla (C.L. Koch, 1844)
Eustala bisetosa Bryant, 1945
Eustala delasmata Bryant, 1945
Eustala devia (Gerstch and Mulaik, 1936)
Eustala fuscovittata (Keyserling, 1864)
Eustala perdita Bryant, 1945
Eustala vegeta (Keyserling, 1865)
*†*Fossilaraneus incertus* Wunderlich, 1988
Gasteracantha cancriformis (Linnaeus, 1758)
Gea heptagon (Hentz, 1850)
Kapogea sellata (Simon, 1895)
Larinia minor (Bryant, 1945)
Mangora fascialata Franganillo, 1936
Mecynogea martiana (Archer, 1958)
Metazygia cienaga Levi, 1995
Metazygia crewi (Banks, 1903)
Metazygia dubia (Keyserling, 1864)
Metazygia gregalis (O.P.-Cambridge, 1889)
Metepeira compsa (Chamberlin, 1916)
Metepeira datona Chamberlin and Ivie, 1942
Metepeira jamaicensis Archer, 1958
Metepeira triangularis (Franganillo, 1930)
Metepeira vigilax (Keyserling, 1893)
Micrathena forcipata (Thorell, 1859)
Micrathena militaris (Fabricius, 1775)
Micrathena similis Bryant, 1945
Neoscona arabesca (Walckenaer, 1842)
Neoscona marcanoi Levi, 1993
Neoscona moreli (Vinson, 1863)
Neoscona nautica (L. Koch, 1875)
Neoscona oaxacensis (Keyserling, 1864)
Ocrepeira darlingtoni (Bryant, 1945)
Ocrepeira serrallesi (Bryant, 1947)
Parawixia tredecimnotata F.O.P.-Cambridge, 1904
*†*Pycnosinga fossilis* Wunderlich, 1988
Verrucosa arenata (Walckenaer, 1842)
Wagneriana tauricornis (O.P.-Cambridge, 1889)
Wagneriana undecimtuberculata (Keyserling, 1865)
Wagneriana vegas Levi, 1991
Witica crassicaudus (Keyserling, 1865)

Family lycosidae

Agalenocosa bryantae (Roewer, 1951)
Arctosa inconspicua (Bryant, 1948)
Hogna reducta (Bryant, 1942)
Hogna tantilla (Bryant, 1948)
Pardosa hamifera F.O.P.-Cambridge, 1902
Pardosa portoricensis Banks, 1902
Pirata sedentarius Montgomery, 1904
Trochosa reichardtiana Strand, 1916

Family pisauridae

Thaumasia annecta Bryant, 1948
Thaumasia marginella (C.L. Koch, 1847)
Tinus connexus (Bryant, 1940)

Family oxyopidae

Hamataliwa communicans (Chamberlin, 1925)
Hamataliwa haytiana (Chamberlin, 1925)
Hamataliwa nigritarsa Bryant, 1948
Hamataliwa rana (Simon, 1897)
Oxyopes crewi Bryant, 1948
†*Oxyopes defectus* Wunderlich, 1988
Peucetia viridans (Hentz, 1832)

Family zoridae

Odo abudi Alayón-García, 2002

Family ctenidae

Ctenus avidus Bryant, 1948
Ctenus darlingtoni Bryant, 1948
Ctenus haitiensis Strand, 1909
Ctenus haina Alayón-García, 2004
Ctenus hiemalis Bryant, 1948
Ctenus insulanus Bryant, 1948
Ctenus jaragua Alayón-García, 2004
Ctenus manni Bryant, 1948
Ctenus monticola Bryant, 1948
Ctenus naranjo Alayón-García, 2004
Cupiennius salei (Keyserling, 1877)
*†*Nanoctenus longipes* Wunderlich, 1988

Trujillina hursti (Bryant, 1948)
Trujillina spinipes Bryant, 1948

Family desidae

Paratheuma insulana (Banks, 1902)

Family dictynidae

Emblyna altamira (Gertsch and Davis, 1942)
*†*Hispaniolyna hirsuta* Wunderlich, 1988
*†*Hispaniolyna magna* Wunderlich, 1988
*†*Palaeodictyna intermedia* Wunderlich, 1988
*†*Palaeodictyna longispina* Wunderlich, 1988
*†*Palaeodictyna? singularis* Wunderlich, 1988
*†*Palaeodictyna spiculum* Wunderlich, 1988
*†*Palaeodictyna termitophila* Wunderlich, 1988
*†*Palaeodictyna unispina* Wunderlich, 1988
*†*Palaeolathys? circumductus* Wunderlich, 1988
*[†]*Palaeolathys copalis* Wunderlich, 1986
*†*Palaeolathys quadruplex* Wunderlich, 1988
*†*Palaeolathys similis* Wunderlich, 1988
*†*Palaeolathys spinosa* Wunderlich, 1986
*†*Succinya longembolus* Wunderlich, 1988
*†*Succinya pulcher* Wunderlich, 1988
*†*Succinya spinipalpus* Wunderlich, 1988

Family amaurobiidae

Neowadotes casabito Alayón-García, 1995
Tugana crassa (Bryant, 1948)
Tugana infumata (Bryant, 1948)
Retiro gratus (Bryant, 1948)

Family miturgidae

Cheiracanthium inclusum (Hentz, 1847)
†*Strotarchus heidti* Wunderlich, 1988
Teminius hirsutus (Petrunkevitch, 1925)
Teminius insularis (Lucas, 1857)
Teminius monticola (Bryant, 1948)

Family anyphaenidae

Anyphaena dominicana Roewer, 1951

Anyphaena modesta Bryant, 1948
Anyphaena pusilla Bryant, 1948
†*Anyphaenoides bulla* (Wunderlich, 1988)
Hibana tenuis (L. Koch, 1866)
Hibana velox (Becker, 1879)
Lupettiana levii Brescovit, 1999
†*Lupettiana ligula* (Wunderlich, 1988)
Lupettiana parvula (Banks, 1903)
Lupettiana spinosa (Bryant, 1948)
Thaloe ennery Brescovit, 1993
Thaloe remotus (Bryant, 1948)
Wulfila fasciculus (Bryant, 1948)
Wulfila fragilis (Bryant, 1948)
Wulfila gracilipes (Banks, 1903)
†*Wulfila spinipes* Wunderlich, 1988

Family Liocranidae?

Lausus pulchellus Bryant, 1948 (*I. sedis*)

Family clubionidae

Elaver implicata (Gertsch, 1941)
†*Elaver nutua* (Wunderlich, 1988)

Family corinnidae

†*Castianeira tenebricosa* Wunderlich, 1988
*†*Chemmisomma dubia* Wunderlich, 1988
†*Corinna flageliformis* Wunderlich, 1988
Corinna parvula Bryant, 1940
Corinna toussainti Bryant, 1948
Creugas gulosus Thorell, 1878
†*Megalostrata grandis* Wunderlich, 1988
Phrurolithus spinosus Bryant, 1948
Trachelas bicolor Keyserling, 1887
Trachelas dilatus Platnick and Shadab, 1974
Trachelas erectus Platnick and Shadab, 1974
†*Trachelas poinari* Penney, 2001
Trachelas tomaculus Platnick and Shadab, 1974
Xeropigo tridentiger (O.P.-Cambridge, 1869)

Family trochanteriidae

*†*Veterator angustus* Wunderlich, 1988

*†*Veterator ascutum* Wunderlich, 1988
*†*Veterator incompletus* Wunderlich, 1982
*†*Veterator longipes* Wunderlich, 1988
*†*Veterator loricatus* Wunderlich, 1988
*†*Veterator porrectus* Wunderlich, 1988
*†*Veterator viduus* Wunderlich, 1988

Family prodidomidae

Zimiris doriai Simon, 1882

Family gnaphosidae

Cesonia nadleri Platnick and Shadab, 1980
*†*Drassyllinus aliter* Wunderlich, 1988
Gnaphosa sericata (L. Koch, 1866)
Sergiolus magnus (Bryant, 1948)

Family selenopidae

Selenops bani Alayón-García, 1992
†*Selenops beynai* Schawaller, 1984
†*Selenops dominicanus* Wunderlich, 2004
Selenops insularis Keyserling, 1881
Selenops marcanoi Alayón-García, 1992
Selenops pensilis Muma, 1953
Selenops phaselus Muma, 1953

Family sparassidae

Heteropoda venatoria (Linnaeus, 1767)
Olios antiguensis (Keyserling, 1880)
†*Pseudosparianthis pfeifferi* (Wunderlich, 1988)
Stasina saetosa Bryant, 1948

Family philodromidae

Apollophanes punctatus (Bryant, 1948)

Family thomisidae

*†*Heterotmarus altus* Wunderlich, 1988
Isaloides toussainti Banks, 1903
*†*Komisumena rosae* Ono, 1981
Misumenops asperatus (Hentz, 1847)

Misumenops bellulus (Banks, 1896)
Misumenops californicus (Banks, 1896)
Misumenops celer (Hentz, 1847)
Rejanellus pallescens (Bryant, 1940)
Rejanellus venustus (Bryant, 1948)
Tobias taczanowskii Roewer, 1951

Family salticidae

Agobardus anormalis Keyserling, 1885
Agobardus anormalis montanus Bryant, 1943
Agobardus brevitarsus Bryant, 1943
Agobardus obscurus Bryant, 1943
Agobardus perpilosus Bryant, 1943
Anasaitis morgani (Peckham and Peckham, 1901)
Antillattus gracilis Bryant, 1943
Antillattus placidus Bryant, 1943
Bythocrotus cephalotes (Simon, 1888)
Cobanus cambridgei (Bryant, 1943)
Commoris modesta Bryant, 1943
Compsodecta haytiensis (Banks, 1903)
Compsodecta peckhami Bryant, 1943
Corythalia banksi Roewer, 1951
Corythalia elegantissima (Simon, 1888)
Corythalia locuples (Simon, 1888)
     †Corythalia ocululiter Wunderlich, 1988
     †*Corythalia pilosa* Wunderlich, 1982
     †*Corythalia scissa* Wunderlich, 1988
*†*Descangeles pygmaeus* Wunderlich, 1988
Descanso formosus Bryant, 1943
Descanso magnus Bryant, 1943
Descanso montanus Bryant, 1943
Dinattus erebus Bryant, 1943
Dinattus heros Bryant, 1943
Dinattus minor Bryant, 1943
Eris flava (Peckham and Peckham, 1888)
Habronattus brunneus (Peckham and Peckham, 1901)
Habronattus paratus (Peckham and Peckham, 1896)
Hasarius adansoni (Audouin, 1826)
Hentzia antillana Bryant, 1940
Hentzia mandibularis (Bryant, 1943)
Hentzia vittata (Keyserling, 1885)
Hentzia zombia Richman, 1989
Icius separatus Banks, 1903
Jollas armatus (Bryant, 1943)
Jollas crassus (Bryant, 1943)

Lyssomanes antillanus Peckham and Wheeler, 1889
Lyssomanes perplexa Peckham and Peckham, 1901
†*Lyssomanes pristinus* Wunderlich, 1986
†*Lyssomanes pulcher* Wunderlich, 1988
Lyssomanes viridis (Walckenaer, 1837)?
Maeotella perplexa (Peckham and Peckham, 1901)
Menemerus bivittatus (Dufour, 1831)
Metacyrba pictipes Banks, 1903
Metacyrba taeniola (Hentz, 1846)
Myrmarachne parallela (Fabricius, 1798)
Nebridia manni Bryant, 1943
Nebridia mendica Bryant, 1943
Paraphidippus aurantius (Lucas, 1833)
Parathiodina compta Bryant, 1943
Pelegrina proxima (Peckham and Peckham, 1901)
Pensacola darlingtoni Bryant, 1943
Pensacola electa Bryant, 1943
Pensacola maxillosa Bryant, 1943
Pensacola montana Bryant, 1943
Pensacola peckhami Bryant, 1943
*†*Pensacolatus coxalis* Wunderlich, 1988
*†*Pensacolatus spinipes* Wunderlich, 1988
*†*Pensacolatus tibialis* Wunderlich, 2004
Phidippus regius C.L. Koch, 1846
*†*Phlegrata pala* Wunderlich, 1988

ACKNOWLEDGMENTS

I thank F. Columbus (Nova Science Publishers) for the invitation to submit to this volume, D. E. Pérez-Gelabert (Smithsonian Institution, Washington DC), M. Solórzano Kraemer (Institut für Paläontologie, Bonn) for information, N. Platnick (American Museum of Natural History, New York), H. Levi (Museum of Comparative Zoology, Harvard) and G. Hormiga (George Washington University, Washington DC) for personal communications.

REFERENCES

Alayón-García, G. (1992). La familia Selenopidae (Arachnida, Araneae) en República Dominicana. *Poeyana*, *419*, 1–10.
Alayón-García, G. (1995). Nuevo género de Agelenidae (Arachnida, Araneae) de República Dominicana. *Poeyana*, *450*, 1–8.
Alayón-García, G. (2000). Las arañas endémicas de Cuba (Arachnida, Araneae). *Rev. Ibér. Aracnol.*, *2*, 1–48.

Alayón-García, G. (2001). Lista preliminar de las arañas (Araneae) de la isla de Navassa. *Cocuyo*, *10*, 18–22.

Alayón-García, G. (2002). Nueva especie de *Odo* Keyserling (Araneae, Zoridae) de República Dominicana. *Rev. Ibérica Aracnol.*, *5*, 29–32.

Alayón-García, G. (2004). Notas sobre la familia Ctenidae (Arachnida, Araneae) en la Hispaniola, con la descripción de tres nuevas especies. *Rev. Ibér. Aracnol.*, *9*, 277–283.

Alayón-García, G., Armas, L. F. De & Antún, A. J. A. (2001). Presencia de *Cyrtophora citricola* Forskål, 1775) (Araneae, Araneidae) en las Antillas. *Rev. Ibér. Aracnol.*, *4*, 9–10.

Archer, A. F. (1958). Studies in the orb-weaving spiders. 4. *Amer. Mus. Novit.*, *1922*, 1–21.

Banks, N. (1903). A list of the Arachnida from Haiti, with descriptions of new species. *Proc. Acad. Nat. Sci. Philadelphia, 55*, 340–345.

Baroni-Urbani, C. & Saunders, J. B. (1980). The fauna of the Dominican Republic amber, the present status of knowledge. In *Proceedings of the Ninth Caribbean Geological Conference* (Santo Domingo; August 1980) (pp. 213–223).

Bell, J. R., Bohan, D. A., Shaw, E. M. & Weyman, G. S. (2005). Ballooning dispersal using silk, world fauna, phylogenies, genetics and models. *Bull. Entomol. Res., 95*, 69–114.

Berman, J. D. & Levi, H. W. (1971). The orb–weaver genus *Neoscona* (Araneidae). *Bull. Mus. Comp. Zool., 141*, 465–500.

Boucot, A. J. (1989). *Evolutionary palaeobiology of behaviour and coevolution.* Amsterdam: Elsevier.

Brescovit, A. D. (1993). *Thaloe* and *Bromelina*, novos géneros de aranhas neotropicais da la familia Anyphaenidae (Arachnida, Araneae). *Rev. Bras. Entomol., 37*, 693–703.

Brescovit, A. D. (1999). Revisao das aranhas do género *Lupettiana* Brescovit (Araneae, Anyphaenidae, Anyphaeninae). *Rev. Bras. Zool., 16 (Supl. 2)*, 63–76.

Briggs, J. C. (1991). A Cretaceous–Tertiary mass extinction? Were most of Earth's species killed off? *BioScience, 41*, 619–624.

Bryant, E. B. (1943). The salticid spiders of Hispaniola. *Bull. Mus. Comp. Zool., 92*, 445–522.

Bryant, E. B. (1945). The Argiopidae of Hispaniola. *Bull. Mus. Comp. Zool., 95*, 357–418.

Bryant, E. B. (1948). The spiders of Hispaniola. *Bull. Mus. Comp. Zool., 100*, 332–447.

Chickering, A. M. (1967). The genus *Nops* (Araneae, Caponiidae) in Panama and the West Indies. *Breviora, 274*, 1–19.

Coyle, F. A. (1995). A revision of the funnelweb mygalomorph spider subfamily Ischnothelinae (Araneae, Dipluridae). *Bull. Amer. Mus. Nat. Hist., 226*, 1–133.

Coyle, F. A. & Meigs, T. E. (1989). Two new species of kleptoparasitic *Mysmenopsis* (Araneae, Mysmenidae) from Jamaica. *J. Arachnol., 17*, 59–70.

Coyle, F. A., O'Shields, T. C. & Perlmutter, D. G. (1991). Observations on the behaviour of the kleptoparasitic spider, *Mysmenopsis furtiva* (Araneae, Mysmenidae*). J. Arachnol., 19*, 62–66.

Crame, J. A. & Rosen, B. R. (2002). Cenozoic palaeogeography and the rise of modern biodiversity patterns. In Crame, J. A. & Owen, A. W. (Eds.), *Palaeobiogeography and Biodiversity Change, the Ordovician and Mesozoic–Cenozoic Radiations* (Geological Society Special Publication No. 194, pp. 153–168). London: The Geological Society.

Curtis, J. H., Brenner, M. & Hodell, D. A. (2001). Climate change in the circum-Caribbean (Late Pleistocene to present) and implications for regional biogeography. In Woods,

C. A. & Sergile, F. E. (Eds.). *Biogeography of the West Indies: patterns and perspectives* (second edition, pp. 35–54). Boca Raton: CRC Press.

Cutler, B. (1984). *Late Oligocene amber salticids from the Dominican Republic.* Peckhamia, *2*, 45–46.

Dondale, C. D. & Redner, J. H. (1984). Revision of the *milvina* group of the wolf spider genus *Pardosa* (Araneae, Lycosidae). *Psyche, 91*, 67–117.

Edwards, G. B. & Wolff, R. J. (1995). A list of the jumping spiders (Salticidae) of the islands of the Caribbean region. *Peckhamia, 3*, 27–60.

Eskov, K. Y. (1987). A new archaeid spider (Chelicerata, Araneae) from the Jurassic of Kazakhstan, with notes on the so-called "Gondwanan" ranges of Recent taxa. *N. Jb. Geol. Paläont., Abh., 175*, 81–106.

Eskov, K. Y. (1990). Spider palaeontology, present trends and future expectations. *Acta Zool. Fennica, 190*, 123–127.

Eskov, K. Y. (1992). Archaeid spiders from Eocene Baltic amber (Chelicerata, Araneida, Archaeidae) with remarks on the so-called "Gondwanan" ranges of Recent taxa. *N. Jb. Geol. Paläont., Abh., 185*, 311–328.

Eskov, K. Y. & Golovatch, S. I., (1986). On the origin of trans-Pacific disjunctions. *Zool. Jb. Syst., 113*, 265–285.

Exline, H. & Levi, H. W. (1962). American spiders of the genus *Argyrodes* (Araneae, Theridiidae). *Bull. Mus. Comp. Zool., 127*, 75–204.

Flint, O. S. Jr. & Pérez-Gelabert, D. E. (1999). Checklist of the Caddisflies (Trichoptera) of Hispaniola. *Nov. Carib., 1*, 33–46.

García-Villafuerte, M. A. & Vera, V. F. (2002). Arañas fosiles incluidas en ambar del Oligoceno-Mioceno de Simojovel, Chiapas, México. In Entomología Mexicana 2002 Vol. 1. 50 aniversario de la Sociedad Mexicana de Entomología (pp. 20–25). Guanajuato, Gto, México: E.G.V. a. A.E.M. e. J.R. Napoles.

García-Villafuerte, M. A. & Penney, D. (2003). *Lyssomanes* (Araneae, Salticidae) in Oligocene–Miocene Chiapas amber. *J. Arachnol., 33*, 400–404.

Genaro, J. A. & Tejuca, A. E. (2001). Patterns of endemism and biogeography of Cuban insects. In C. A. Woods and F. E. Sergile (Eds.). *Biogeography of the West Indies: patterns and perspectives* (second edition, pp. 77–83). Boca Raton: CRC Press.

Gertsch, W. J. (1958). The spider genus *Loxosceles* in North America, Central America, and the West Indies. *Amer. Mus. Novit., 1907*, 1–46.

Gertsch, W. J. & Ennik, F. (1983). The spider genus *Loxosceles* in North America, Central America and the West Indies (Araneae, Loxoscelidae). *Bull. Amer. Mus. Nat. Hist., 175*, 264–360.

Gillespie, R. G., Rivera, M. & Garb, J. (1998). Sun, surf and spiders, taxonomy and phylogeography of Hawaiian Araneae. In P. A. Selden (Ed.) *Proceedings of the 17th European Colloquium of Arachnology,* Edinburgh 1997 (pp. 41–51). Burham Beeches: British Arachnological Society.

Grimaldi, D. A. (1992). Vicariance biogeography, geographic extinctions, and the North American Oligocene tsetse flies. In Novacek, M. J. & Wheeler, Q. D. (Eds.) *Extinction and phylogeny* (pp. 178–204). New York: Columbia University Press.

Grimaldi, D. & Engel, M. S. (2005). *Evolution of the insects.* New York: Cambridge University Press.

Gutiérrez, E. & Pérez-Gelabert, D. E. (2000). Annotated checklist of Hispaniolan cockroaches. *Trans. Amer. Entomol. Soc., 126*, 423–445.

Hall, J. P. W., Robbins, R. K. & Harvey, D. J. (2004). Extinction and biogeography in the Caribbean, new evidence from a fossil riodinid butterfly in Dominican amber. *Proc. R. Soc. Lond. B, 271*, 789–801.

Hedges, S. B. (2001). Biogeography of the West Indies: an overview. In C. A. Woods and F. E. Sergile (Eds.). *Biogeography of the West Indies: patterns and perspectives* (second edition, pp. 15–33). Boca Raton: CRC Press.

Henwood, A. (1993). *Ecology and taphonomy of Dominican Republic amber and its inclusions. Lethaia. 26*, 237–245.

Huber, B. A. (2000). New World pholcid spiders (Araneae, Pholcidae), A revision at generic level. *Bull. Amer. Mus. Nat. Hist., 254*, 1–348.

Hueber, F. M. & Langenheim, J. H. (1986). Dominican amber tree had African ancestors. *Geotimes*, 31, 8–10.

Iturralde-Vinent, M. A. (2001). Geology of the amber-bearing deposits of the Greater Antilles. *Carib. J. Sci., 37*, 141–167.

Iturralde-Vinent, M. A. & Lidiak, E. G. (Eds.) (2006). Caribbean Plate Tectonics: Stratigraphic, Magmatic, Metamorphic and Tectonic Events (UNESCO/IUGS IGCP Project 433). *Geol. Acta, 4 (1–2)*, 1–341.

Iturralde-Vinent, M. A. & MacPhee, R. D. E. (1996). Age and palaeogeographical origin of Dominican amber. *Science, 273*, 1850–1852.

Iturralde-Vinent, M. A. & MacPhee, R. D. E. (1999). Paleogeography of the Caribbean region, implications for Cenozoic biogeography. Bull. *Amer. Mus. Nat. Hist., 238*, 1–95.

Kraus, O. (1978). Zoogeography and plate tectonics, introduction to a general discussion. *Abh. Vehr. Naturwiss. Ver. Hamburg, 21/22*, 33–41.

Labandeira, C. C. (2005). The fossil record of insect extinction, new approaches and future directions. *Amer. Entomol., 51*, 14–29.

Langenheim, J. H. (1995). Biology of amber-producing trees, focus on case studies of *Hymenaea* and *Agathis. Proc. ACS Symp. Ser., 617*, 1–31.

Levi, H. W. (1955). The spider genera *Episinus* and *Spintharus* from North America, Central America and the West Indies (Araneae, Theridiidae). *J. New York Entomol. Soc., 62*, 65–90.

Levi, H. W. (1957). The spider genera *Chrysso* and *Tidarren* in America (Araneae, Theridiidae). *J. New York Entomol. Soc., 63*, 59–81.

Levi, H. W. (1959). The spider genera *Achaearanea, Theridion* and *Sphyrotinus* from Mexico, Central America, and the West Indies (Araneae, Theridiidae). *Bull. Mus. Comp. Zool., 121*, 55–163.

Levi, H. W. (1962). More American spiders of the genus *Chrysso* (Araneae, Theridiidae). *Psyche, 69*, 209–237.

Levi, H. W. (1963a). The American spider genera *Spintharus* and *Thwaitesia. Psyche, 70*, 223–234.

Levi, H. W. (1963b). The American spiders of the genus *Theridion* (Araneae, Theridiidae). *Bull. Mus. Comp. Zool., 129*, 483–592.

Levi, H. W. (1964). The spider genus *Thymoites* in America (Araneae, Theridiidae). *Bull. Mus. Comp. Zool., 130*, 445–471.

Levi, H. W. (1977). The American orb-weaver genera *Cyclosa*, *Metazygia* and *Eustala* North of Mexico (Araneae, Araneidae). *Bull. Mus. Comp. Zool.*, *148*, 61–127.

Levi, H. W. (1986). The Neotropical orb-weaver genera *Chrysometa* and *Homalometa* (Araneae, Tetragnathidae). *Bull. Mus. Comp. Zool.*, *151*, 91–215.

Levi, H. W. (1991). The Neotropical orb-weaver genera *Edricus* and *Wagneriana* (Araneae, Araneidae). *Bull. Mus. Comp. Zool.*, *152*, 363–415.

Levi, H. W. (1993). American *Neoscona* and corrections to previous revisions of Neotropical orb-weavers (Araneae, Araneidae). *Psyche*, *99*, 221–239.

Levi, H. W. (1995). The Neotropical orb-weaver genus *Metazygia* (Araneae, Araneidae). *Bull. Mus. Comp. Zool.*, *154*, 63–151.

Levi, H. W. (1999). The Neotropical and Mexican orb weavers of the genera *Cyclosa* and *Allocyclosa* (Araneae, Araneidae). *Bull. Mus. Comp. Zool.*, *155*, 299–379.

Levi, H. W. (2001). The orb weavers of the genera *Molinaranea* and *Nicolepeira*, a new species of *Parawixia*, and comments on orb weavers of temperate South America (Araneae, Araneidae). *Bull. Mus. Comp. Zool.*, *155*, 445–475.

MacPhee, R. D. E & Iturralde-Vinent, M. A. (2000). A short history of Greater Antillean land mammals: biogeography, paleogeography, radiations, and extinctions. *Tropics*, *10*, 145–154.

MacPhee, R. D. E & Iturralde-Vinent, M. A. (2005). The interpretation of Caribbean palaeogeography, reply to Hedges. In Alcover, J. A. & Bover, P. (Eds.) Proceedings of the international symposium "insular vertebrate evolution, the palaeontological approach". Monograf. *Soc. d'Hist. Nat. Balears*, *12*, 175–184.

Maddison, W. (1986). Distinguishing the jumping spiders *Eris militaris* and *Eris flava* in North America (Araneae, Salticidae). *Psyche*, *93*, 141–149.

Martínez-Delclòs, X., Briggs, D. E. G. & Peñalver, E. (2004). Taphonomy of insects in carbonates and amber. Palaeogeog. Palaeoclimatol. *Palaeoecol.*, *203*, 19–64.

Morrone, J. J. & Crisci, J. V. (1995). Historical biogeography, introduction to methods. *Ann. Rev. Ecol. Syst.*, *26*, 371–401.

Muma, M. H. (1953). A study of the spider family Selenopidae in North America, Central America, and the West Indies. *Amer. Mus. Novit.*, *1629*, 1–55.

Nagel, P. (1997). New fossil paussids from Dominican amber with notes on the phylogenetic systematics of the paussine complex (Coleoptera, Carabidae). *Syst. Entomol.*, *22*, 345–362.

Ogden, R. & Thorpe, R. S. (2002). Molecular evidence for ecological speciation in tropical habitats. *Proc. Nat. Acad. Sci.*, USA, *99*, 13612–13615.

Ono, H. (1981). First record of a crab spider (Thomisidae) from Dominican amber (Amber Collection Stuttgart, Arachnida, Araneae). *Stuttgarter Beitr. Naturk.* (B), *73*, 1–13.

Opell, B. D. (1979). Revision of the genera and tropical American species of the spider family Uloboridae. *Bull. Mus. Comp. Zool.*, *148*, 443–549.

Penney, D. (1999). Hypotheses for the Recent Hispaniolan spider fauna based on the Dominican Republic amber spider fauna. *J. Arachnol.*, *27*, 64–70.

Penney, D. (2000a). Miocene spiders in Dominican amber (Oonopidae, Mysmenidae). *Palaeontology*, *43*, 343–357.

Penney, D. (2000b). Anyphaenidae in Miocene Dominican Republic amber (Arthropoda, Araneae). *J. Arachnol.*, *28*, 223–226.

Penney, D. (2001). Advances in the taxonomy of spiders in Miocene amber from the Dominican Republic (Arthropoda, Araneae). *Palaeontology, 44*, 987–1009.

Penney, D. (2002). Paleoecology of Dominican amber preservation—spider (Araneae) inclusions demonstrate a bias for active, trunk–dwelling faunas. *Paleobiology, 28*, 389–398.

Penney, D. (2003). *Afrarchaea grimaldii*, a new species of Archaeidae (Araneae) in Cretaceous Burmese amber. *J. Arachnol., 31*, 122–130.

Penney, D. (2004a). New extant and fossil Dominican Republic spider records, with two new synonymies and comments on taphonomic bias of amber preservation. *Rev. Ibér. Aracnol., 9*, 183–190.

Penney, D. (2004b). Does the fossil record of spiders track that of their principal prey, the insects? Trans. *Royal Soc. Edinb., Earth Sci., 94*, 275–281.

Penney, D. (2005a). Fossil blood droplets in Miocene Dominican amber yield clues to speed and direction of resin secretion. *Palaeontology, 48*, 925–927.

Penney, D. (2005b). Importance of Dominican Republic Amber for determining taxonomic bias of fossil resin preservation—A case study of spiders. *Palaeogeog. Palaeoclimatol. Palaeoecol., 223*, 1–8.

Penney, D. (2005c). First Caribbean *Floricomus* (Araneae, Linyphiidae), a new species in Miocene Dominican Republic amber. A new synonymy for the extant American fauna. *Geol. Acta*, 3, 59–64.

Penney, D. (2005d). First fossil Filistatidae, A new species of *Misionella* in Miocene amber from the Dominican Republic. *J. Arachnol., 33*, 93–100.

Penney, D. (2006a, for 2005). An annotated systematic catalogue, including synonymies and transfers, of Miocene Dominican Republic amber spiders described up until 2005. *Rev. Ibér. Aracnol.*, 12, in press.

Penney, D. (2006b). A new fossil oonopid spider, in lowermost Eocene amber from the Paris Basin, with comments on the fossil spider assemblage. *Afr. Invert.*, 47, in press.

Penney, D. (in press). Tertiary Neotropical Hersiliidae (Arthropoda, Araneae) with new combinations for the extant fauna and comments on historical biogeography of the family. *Palaeontology*.

Penney, D. & Langan, A. M. L. (2006). Comparing amber fossils across the Cenozoic. *Biol. Lett.*, doi:10.1098/rsbl.2006.0442

Penney, D. & Pérez-Gelabert, D. E. (2002). Comparison of the Recent and Miocene Hispaniolan spider faunas. *Rev. Ibér. Aracnol., 6*, 203–223.

Penney, D. & Selden, P. A. (2006). First fossil Huttoniidae (Araneae), in Late Cretaceous Canadian Cedar and Grassy Lake ambers. *Cret. Res., 27*, 442-446.

Penney, D., Wheater, C. P. & Selden, P. A. (2003). Resistance of spiders to Cretaceous–Tertiary extinction events. *Evolution, 57*, 2599–2607.

Pérez-Asso, A. R. & Pérez-Gelabert, D. E. (2001). Checklist of the millipedes (Diplopoda) of Hispaniola. *Bol. S.E.A., 28*, 67–80.

Pérez-Gelabert, D. E. (1999). Catálogo systemático y bibliografía de la biota fósil en ámbar de la República Dominicana. *Hispaniolana, 1*, 1–65.

Pérez-Gelabert, D. E. (2001). Preliminary checklist of the Orthoptera (Saltatoria) of Hispaniola. *J. Orthoptera Res., 10*, 63–74.

Pérez-Gelabert, D. E. & Flint, O. S. Jr. (2000). Annotated list of the Neuroptera of Hispaniola, with new faunistic records of some species. *J. Neuropterol., 3*, 9–23.

Petrunkevitch, A. (1928). The Antillean spider fauna—a study in geographic isolation. *Science, 68*, 650.

Petrunkevitch, A. (1963). Chiapas amber spiders. *Uni. Calif. Pubs. Entomol., 31*, 1–40.

Petrunkevitch, A. (1971). Chiapas amber spiders, 2. *Uni. Calif. Pubs. Entomol., 63*, 1–44.

Piel, W. H. (2001). The systematics of Neotropical orb-weaving spiders in the genus *Metepeira* (Araneae, Araneidae). *Bull. Mus. Comp. Zool., 157*, 1–92.

Platnick, N. I. (2006). The world spider catalogue (version 6.5). Online at, http://research.amnh.org/entomology/spiders/catalog/INTRO2.html.

Platnick, N. I. & Nelson, G. (1978). A method of analysis for historical biogeography. *Syst. Zool., 27*, 1–16.

Platnick, N. I. & Penney, D. (2004). A revision of the widespread spider genus *Zimiris* (Araneae, Prodidomidae). *Amer. Mus. Novitat., 3450*, 1–12.

Platnick, N. I. & Shadab, M. U. (1974a). A revision of the *bispinosus* and *bicolor* groups of the spider genus *Trachelas* in North and Central America and the West Indies. *Amer. Mus. Novit., 2560*, 1–34.

Platnick, N. I. & Shadab, M. U. (1974b). A revision of the *tranquillus* and *speciosus* groups of the spider genus *Trachelas* in North and Central America and the West Indies. *Amer. Mus. Novit., 2553*, 1–34.

Platnick, N. I. & Shadab, M. U. (1978). A review of the spider genus *Mysmenopsis* (Araneae, Mysmenidae). *Amer. Mus. Novit., 2661*, 1–22.

Platnick, N. I. & Shadab, M. U. (1980). A revision of the spider genus *Cesonia* (Araneae, Gnaphosidae). *Bull. Amer. Mus. Nat. Hist., 165*, 335–386.

Platnick, N. I. & Shadab, M. U. (1989). A review of the spider genus *Teminius* (Araneae, Miturgidae). *Amer. Mus. Novit., 2963*, 1–12.

Poinar, G. O. Jr. (1991). *Hymenaea protera* sp.n. (Leguminosae, Caesalpinioideae) from Dominican amber has African affinities. *Experientia, 47*,1075–1082.

Poinar, G. O. Jr. (1992). Life in amber. California: Stanford University Press.

Poinar, G. O. Jr. & Hess, R. (1982). Ultrastructure of 40-million-year-old insect tissue. *Science, 215*, 1241–1242.

Poinar, G. O. Jr. (1999). Extinction of tropical insect lineages in Dominican amber from Plio-Pleistocene cooling events. *Russian Entomol. J., 8*, 1–4.

Poinar, G. O. Jr. & Poinar, R. (1999). *The amber forest: a reconstruction of a vanished world.* New Jersey: Princeton University Press.

Raven, R. J. (2000). Taxonomica Araneae I, Barychelidae, Theraphosidae, Nemesiidae and Dipluridae (Araneae). *Mem. Qld Mus., 45*, 569–575.

Reiskind, J. (1989). The potential of amber fossils in the study of the biogeography of spiders in the Caribbean with the description of a new species of *Lyssomanes* from Dominican amber (Araneae, Salticidae). In C. A. Woods (Ed.) *Biogeography of the West Indies: past, present and future* (pp. 217–227). Gainseville: Sandhill Crane Press.

Rheims, C. A. & Brescovit, A. D. (2004). Revision and cladistic analysis of the spider family Hersiliidae (Arachnida, Araneae) with emphasis on Neotropical and Nearctic species. *Insect Syst. Evol., 35*, 189–239.

Richman, D. B. (1989). A revision of the genus *Hentzia* (Araneae, Salticidae). *J. Arachnol., 17*, 285–344.

Robertson, D. S., McKenna, M. C., Toon, O. B., Hope, S., and Lillegraven, J. A. (2004). Survival in the first hours of the Cenozoic. *GSA Bull., 116*, 760–768.

Ross, M. I. & Scotese, C. R. (1988). A hierarchical tectonic model of the Gulf of Mexico and Caribbean region. *Tectonophysics*, *155*, 139–168.

Sánchez-Ruíz, A. (2005). Una nueva especie de *Nops* MacLeay, 1839 (Araneae, Caponiidae) de República Dominicana. *Rev. Ibér. Aracnol.*, *11*, 23–27.

Sanderson, M. W. & Farr, T. H. (1960). Amber with insect and plant inclusions from the Dominican Republic. *Science, 131*, 1313.

Santos, A. J. & Gonzaga, M. O. (2003). On the spider genus *Oecobius* Lucas, 1846 in South America (Araneae, Oecobiidae). *J. Nat. Hist.*, *37*, 239–252.

Schawaller, W. (1981a). The spider family Hersiliidae in Dominican amber (Amber Collection Stuttgart, Arachnida, Araneae). *Stuttgarter Beitr. Naturk.* (B), *79*, 1–10.

Schawaller, W. (1981b). Survey of the spider–families in Dominican amber and other Tertiary resins (Amber Collection Stuttgart, Arachnidae, Araneae). *Stuttgarter Beitr. Naturk.* (B), *77*, 1–10.

Schawaller, W. (1982). Spinnen der Familien Tetragnathidae, Uloboridae und Dipluridae in Dominikanischem Bernstein und allgemeine Gesichtspunkte (Arachnida, Araneae). Stuttgarter Beitr. *Naturk.* (B), *89*, 1–19.

Schawaller, W. (1984). The family Selenopidae in Dominican amber (Arachnida, Araneae). Stuttgarter Beitr. *Naturk.* (B), *103*, 1–8.

Schubert, C. (1988). Climatic changes during the last glacial maximum in northern South America and the Caribbean, a review. *Interciencia*, *13*, 128–137.

Smith, A. M. (1986). *The tarantula classification and identification guide.* London: Fitzgerald Publishing.

Wilson, E. O. (1985). Invasion and extinction in the West Indian ant fauna, evidence from Dominican amber. *Science*, *229*, 265–267.

Wolff, R. J. (1990). A new species of *Thiodina* (Araneae, Salticidae) from Dominican amber. *Acta Zool. Fennica, 190*, 405–408.

Woods, C. A. and Sergile, F. E. (Eds.) (2001). Biogeography of the West Indies: patterns and perspectives (second edition). *Boca Raton:* CRC Press.

Wunderlich, J. (1981). Fossile Zwergsechsaugenspinnen (Oonopidae) der Gattung *Orchestina* Simon, 1882 in Bernstein mit Ammerkungen zur Sexual-biologie (Arachnidae, Araneae). *Mitt. Geol.-Paläont. Inst. Univ. Hamburg, 51*, 83–113.

Wunderlich, J. (1982). Die häufigsten Spinnen (Araneae) des Dominikanischen Bernsteins. *Neue Entomol. Nachr., 1*, 26–45.

Wunderlich, J. (1986). Spinnenfauna Gestern und Heute, 1: Fossile Spinnen in Bernstein und Ihre Heute Lebenden Verwandten. Wiesbaden: Erich Bauer Verlag bei Quelle and Meyer.

Wunderlich, J. (1987). *Tama minor*, n. sp., eine fossile Spinnerart der Familie Hersiliidae in Dominikanischem Bernstein (Arachnida, Araneae). *Entomol. Z.*, *97*, 93–96.

Wunderlich, J. (1988). Die Fossilen Spinnen im Dominikanischem Bernstein. *Beitr. Araneol., 2*, 1–378.

Wunderlich, J. (1998). Beschreibung der ersten fossilen Spinnen der Unterfamilien Mysmeninae (Anapidae) und Erigoninae (Linyphiidae) im Dominikanischen Bernstein (Arachnida, Araneae). *Entomol. Z., 108*, 363–367.

Wunderlich, J. (1999). Two subfamilies of spiders (Araneae, Linyphiidae, Erigoninae and Anapidae, Mysmeninae) new to Dominican amber – or falsificated amber? *Est. Mus. Cienc. Nat. Àlava, 14 (Num. Espec. 2)*, 167–172.

Wunderlich, J. (Ed.) (2004). Fossil spiders in amber and copal. *Beitr. Araneol., 3ab*, 1–1908.

In: Biogeography
Editors: M. Gailis, S. Kalninš, pp. 175-203

ISBN: 978-1-60741-494-0
© 2010 Nova Science Publishers, Inc.

Chapter 5

BIRD FAUNA ALONG A RURAL-URBAN GRADIENT AT REGIONAL AND LOCAL SCALE IN COMUNIDAD VALENCIANA (SPAIN)

*Enrique Murgui**

Grupo para el Estudio de las Aves. G.V. Marqués del Turia, 28, 46005 Valencia (Spain).

ABSTRACT

The distribution and abundance of bird species along a rural-urban gradient placed in Comunidad Valenciana (Spain) was examined. At a regional scale (2800 km^2) I used data on presence-absence of breeding bird species in 10 x 10 km squares in order to determine the patterns of bird species' richness along the rural-urban gradient. Such squares were grouped into four sectors according to a decreasing degree of urbanization and they represent: Mediterranean transitional woodland shrub, mostly woody crops, mostly permanently irrigated land, and mostly urban areas. At a local scale (Valencia, 76 Km^2) I used the abundance of wintering and breeding bird species recorded in 118 squares of 700 x 700 m. Such squares were grouped into four classes depending on the cover of rural and urban landscapes: permanently irrigated land, mixed, mostly urban, and urban landscape. At regional scale, breeding bird richness did not show significant changes along the gradient, whereas data at the local scale revealed a decreasing number of bird species with increasing urbanization. Patterns at the regional scale were presumably related with: (i) the spatial variation of bird richness which increased from inland to the coastal fringe; (ii) a greater diversity of habitats because of development; (iii) a moderate level of urbanization; (iv) the use of a grain large enough to sample different habitats many of them sufficiently large to fulfil the requirements of many species. At the local scale, the following factors were apparently important: (i) a higher degree of urbanization; (ii) the smaller grain employed which sampled a less diverse set of habitats, and the extent of them was in many cases too small to fulfil habitat requirements of sensitive species; and (iii) the fact that urban parks were rarely suitable to open-field bird species of the surrounding habitat. The proportion of threatened species showed some variation along the gradient depending on the spatial scale and criterion employed, but overall the number of threatened species increased with increasing urbanization. The

---

* Corresponding author: enmurpe@alumni.uv.es

study area is a very developed region, where conservation strategies can not be designed without a full consideration of the intricate set of socio-economic processes which have driven most of the changes in the landscape. Currently, efforts should be orientated to conserve the biological communities and the ecological role of the agricultural land which still constitutes the landscape matrix at many spatial scales and administrative boundaries.

INTRODUCTION

About 50 % of the ice-free land surface has been substantially transformed by humans (Vitousek et al., 1997; Foley, 2005) and such land-cover changes are the primary driver of the decline of ecosystems services and global biodiversity (Sala et al., 2000; Stuart-Chapin III et al., 2000; Baillie et al., 2004). One of the most severe cases of land conversion is urbanization (Vitousek et al., 1997; Miller and Hobbs, 2002), where the former landscape is replaced to a greater or lesser degree by man-made structures (buildings, roads, paved areas...); in extreme cases, as occurs in compact cities, all the landscape is dominated by such structures, leaving only patches of the former landscape, and "ex novo" habitats like urban parks and gardens.

Urbanization is not only an intense and damaging modification of landscapes, it is a widespread one. Since the 19th century, urbanization has showed an increasing growth globally both in terms of human population and in land cover (Pacione, 2001). Currently almost half of the world's population resides in urban areas, reaching 80% in some regions, and the population growth expected for the world in the next thirty years mainly will be concentrated in urban areas (UN, 2005). From an environmental point of view, it is a cause of concern that most of the future increase in the human population and its concomitant land-cover changes will take place in tropical or subtropical regions (UN, 2005), which typically exhibit high levels of biodiversity (Brown and Lomolino, 1998).

Although ecological studies conducted in urban areas have a long tradition (for examples see Gilbert, 1989) they are relatively few compared with those carried out in natural areas (Miller and Hobbs, 2002). However, probably as a consequence of a greater awareness of the deleterious influence of urbanization on global biodiversity and environmental processes, in the last decade the interest in the urban ecosystem has increased noticeably, and urban ecology is being developed in a new framework that integrates social and ecological principles (e.g. Grimm et al., 2000).

One widespread approach in studying the effects of urbanization on ecological systems is gradient analyses in which the changes in the variable of interest are quantified in a range of land-use types from rural (or wild land landscapes) to the urban ones (McDonnell and Pickett, 1990). Such an approach has been employed in numerous studies addressing the effects of urbanization on bird fauna and on other taxonomic groups (see Marzluff *et al.*, 2001; Chace and Walsh, 2006; Garden *et al.*, 2006).

The response of bird fauna to urbanization seems little consistent, and different studies have reported maximum density and bird species richness at the undeveloped sectors of the gradient, at intermediate sectors, and even at the most urbanized ones (for review see Marzluff, 2001). Among the reasons behind such different patterns (see Discussion) an important one could be the effect of spatial scale because it has been long recognized that ecosystem processes change if the spatial resolution of analyses is changed (e.g. Wiens, 1989;

Whittaker et al., 2001). For instance, studies analyzing the coincidence of people (which in many regions could be considered a proxy of urbanization degree) and biodiversity patterns have reported positive and negative correlations, and these differences are largely due to the grain and extent of studies (see Pautasso, 2007).

Following a general trend in Europe (EEA, 2006b) Spain has experienced one of the greatest increases in artificial surfaces (about 30 %) in the 1990–2000 period at expenses mainly of agricultural land and pastures (OSE, 2007). This phenomenon is particularly acute on the Mediterranean coast where in places like the Comunidad Valenciana region the increase has been about 50% (OSE, 2007). In Comunidad Valenciana, increasing urbanization has been the outcome of the growing of compact cities, and the development at unprecedented rates of urban sprawl as occurs in other zones of southern Europe (EEA, 2006c). However, the effect of such processes on biodiversity is little known.

This study sought to determine the extent to which bird communities and populations in Comunidad Valenciana are influenced by urbanization. To accomplish this objective I quantified changes on bird abundance and diversity along a gradient from rural-wildland to urban areas at regional (2800 km^2) and local (76 km^2) scales. Additionally, for the local scale, the data allows us to explore whether changes along the urban gradient are influenced by seasonality (Murgui, 2007a).

STUDY AREA

For the purpose of this study, I used data about two areas of very different extent in the Comunidad Valenciana region (Spain): part of the municipality of Valencia (76 km^2), and a larger area (2800 km^2) which includes the former one (Figure 1).

The municipality of Valencia comprises 134.6 Km^2 and it can be divided into two well-differentiated parts (Figure 1): the Albufera Natural Park (a large wetland) to the south, and the urban and rural area to the north. Fieldwork was carried out in this northern area of approximately 76 km^2 hereafter referred to as Valencia (for details see Murgui, 2005).

The urban area of Valencia is inhabited by about 800,000 people. It is a dense and heterogeneous urban fabric, comprising primarily block buildings, most of them having five or more stories. Buildings are interrupted by four other habitat types: urban parks (Murgui, 2007a), wooded streets (Murgui, 2007b), residential gardens, and patches of derelict land, i.e. non developed spaces dominated by spontaneous ruderal vegetation (Carretero and Aguilella, 1995; Murgui, 2005). Valencia also includes a harbour zone.

The rural area surrounding the city (Figure 2) is composed mostly by horticultural fields of small size (< 1 ha) with row crop cultivates (2298 ha) along with orange groves (571 ha) which are more abundant in the north (Díaz and Galiana, 1996). This area also includes some "natural" habitats: a small (69 ha) Aleppo Pine *Pinus halepensis* wood with Mediterranean shrub, and the old course of the river Turia that was redirected outside the city after flooding in 1957 by means of a large channel. This later includes a small (1 ha) wetland composed of Reed *Phragmites communis* and Reedmace *Typha dominguensis* (Murgui, 2005).

The 2800 km^2 area constitutes 12 % of the Comunidad Valenciana territory (Figure 1), and it encompasses the most representative habitats of the region. From the coastal fringe inland, we could distinguish (Figure 2): wetlands (the Albufera Natural Park) and rice fields, irrigated land (horticultural row crops and orange groves), non-irrigated land dominated by woody crops both perennial (carob and olive groves) and deciduous ones (almond groves and vineyards), and Mediterranean pinewoods and shrublands, including Cork Oak *Quercus suber* forest, the latter in the Sierra Calderona Natural Park in the northwest.

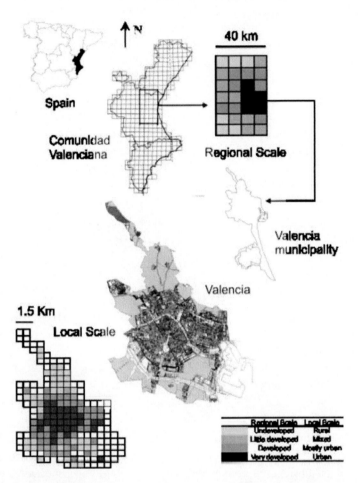

Figure 1. Study area. The gradient at regional scale included 28 squares of 10 x 10 Km, the gradient at local scale included 118 squares of 700x 700 Km. For criteria under classification of sectors of the gradient see Landscape Measurement.

BIRD CENSUSES

In Valencia, the study area was divided into 210 squares of 49 ha (700 m x 700 m). At the periphery of the study area were 31 irregularly bounded squares that were eliminated due to their small size, yielding a total of 197 squares (Murgui, 2009) . Fieldwork was carried out during the years 1997 and 1998. The first year I surveyed half of the 197 squares on two occasions, once during the winter (between December 1st and February 28th) and once during the breeding period (March 15th–July 15th). The second year the remaining squares were surveyed using the same protocol. Each square was surveyed by walking in the morning between 8.00 and 11.00 a.m. on days when the weather was good (i.e. avoiding rainy and windy days). One hour per visit was spent in each 49 ha square (for census effort see Roberge and Svensson, 2003) and a proportional time in those of lesser size recording all birds seen or heard in the different habitats of the square except those overflying that did not use the square (for similar census methods e.g. Jokimäki and Kaisanlathi-Jokimäki, 2003).

Figure 2. Main habitat types at the regional gradient. From top down, and from left to right: (a) rice fields (photo J.A. Peris), (b) row crops, (c) orange groves, (d) woody crops (carob, olive and almond trees), (e) shrubland, (f) woodland (photo J.A.Peris).

Data on birds of the coarse landscape were obtained from the Atlas of the breeding birds of Spain (Martí and Del Moral, 2003) which incorporates information on the distribution and abundance of bird species in 10 x 10 km UTM squares. I extracted data on presence-absence of the species in the 28 UTM squares corresponding to the 2800 km^2. One hundred and forty species were recorded in this area, which represents the 73 % of the breeding species of the whole Comunidad Valenciana region. The 140 species were classified into different groups upon taxonomy and habitat use according with Diaz et al. (1996) and Telleria et al. (1999).

Conservation Status

Bird species richness can be a misleading indicator of biodiversity, especially in studies including urban areas which can hold many common species, or even many introduced ones as occurs in Valencia (Murgui, 2001; Murgui and Valentín, 2003). Therefore, I also used the number of threatened bird species across sectors of the gradients to measure the effects of urbanization on biodiversity. Three basic criteria constitute the core of the threatened classification systems (e.g. IUCN see Collar, 1999): range size, population size, and population trend of species. Data on birds of Valencia allowed using the two former in order to establish different conservation categories at the local scale. Threatened species were classified as "Endangered" or "Vulnerable". The endangered ones fulfilled any of the following criteria: (i) recorded in less of 11 squares; (ii) population below 50 individuals. Vulnerable species were defined by any of the following criteria: (i) population size between 51 and 500 individuals; (ii) species with population size > 500 but recorded in less of 100 squares.

To assess the conservation status of breeding bird species in Comunidad Valenciana I used data about the range size of species (assuming that a small range size often corresponds to a small population size, see Gaston et al., 2000) because I am not aware of reliable population estimates. Comunidad Valenciana consists in 295 10 X 10 Km squares (Urios et al., 1991). Threatened species include the Endangered ones (recorded in < 10% of the 295 squares), Vulnerable (11-20 %) and Rare (21-50%). At continental level I used the categories proposed by Birdlife (BirdLife, 2004). Threatened species were those considered as Endangered, Vulnerable, Rare, and Declining.

LANDSCAPE MEASUREMENTS

In Valencia, the patches of the different habitat types (see Study Area) in each square were located on a 1:4000 map with the aid of 1:3500 aerial photographs (Murgui, 2009). Each patch of habitat was characterized according to patch area, perimeter, and distance from the nearest patch of the same habitat (Murgui, 2009). Using such measurements I obtained the amount of each habitat type per square.

In order to avoid the influence of the size of the squares on bird species richness only the 131 squares with a size of 49 ha were used for analyses. From these I eliminated the squares with wetland sites (n = 13) because this habitat and most of the species inhabiting it were absent of the rest of the squares. The remaining 118 squares (Figure 1) were grouped into four categories according with a decreasing proportion of cultivated land cover (Table 1).

Table 1. Percentage of landcover of the different habitat types across the sectors of the rural-urban gradient

	n	Percentage ± S.D. of landcover				
		Cultivated	Derelict	Parks	Gardens	Built-up
Rural	38	83 ± 12	3 ± 5	0.1 ± 0.4	0.3 ± 0.7	12 ± 13
Mixed	9	52 ± 5	2 ± 1	1 ± 2	1 ± 2	23 ± 13
Mostly urban	39	18 ± 19	14 ± 15	5 ± 7	2 ± 4	60 ± 22
Urban	32	0	6 ± 9	9 ± 12	1 ± 2	82 ± 13

At the regional scale, using 1:50.000 maps and aerial photographs the percentage of urban land cover was obtained from the 100 1x1 km squares in each of the 28 squares measuring 10 x 10 Km. Urban land-cover ranged between 10 and 4165 ha, and the urban gradient (Figure 1) was defined as follows:

1. Undeveloped (n = 6 squares): < 100 ha of urban land-cover per square (mean ± S.E. = 35.00 ± 206.64 ha).
2. Little developed (n = 9): urban landcover ranging between 101 and 400 ha (248.33 ± 168.72 ha).
3. Developed (n = 8): urban landcover ranging between 401 and 1.000 ha (566.25 ± 178.95 ha).
4. Very developed (n = 5): urban landcover ranging between 1.001 and 4165 ha (1994.00 ± 226.36 ha).

Although, this is admittedly a rough measurement of urbanization I think that it is enough to the purposes of this paper. The classification of squares showed a spatial pattern coherent with the current situation in Comunidad Valenciana i.e. a decreasing level of development with increasing distance from the coastal fringe (Fig 1). The undeveloped sector coincides with a rugged landscape where woodland and scrubland predominates; the little developed sector is dominated by woody crops with remnants of wooded areas; in the developed sector irrigated land is the main land cover, and finally in the very developed sector irrigated land is often substituted by urban habitats. Graphical examples of these habitats are provided in Figure 2.

DATA ANALYSES

At both spatial scales using the bird species richness per square, I calculated the mean number of species at each category of the gradients as defined above. The proportion of threatened bird species per square was also obtained, and then I calculated the mean proportion of such species at each sector of the gradients. For Valencia, I calculated also the mean number of birds of each individual species at each sector of the gradient. Statistical differences in the species richness, proportion of threatened species, and abundance of bird species across categories were tested by means of unifactorial ANOVA test. Previously the normality of the variables was checked with the one sample Kolmogorov-Smirnov Test. As some parameters differed significantly from normality the data were normalized when appropriate using the square-root (x+3/4), log (x+1) and arcsin (x) transformations (Zar, 1996).

RESULTS

Bird Community aong the Urban Gradient

At the regional scale (i.e. 2800 km^2) mean number of breeding bird species recorded in each sector of the gradient slightly increased with increasing urbanization (Figure 3) but no statistical difference across sectors was detected (Unifactorial ANOVA Test $F_{3,24} = 2.58$, $P = 0.07$). Note however that the standard deviation was greater in the undeveloped sector, suggesting that the different habitat types of the squares included in the sector may play a role in results. At local scale (Valencia, 76 km^2) there was a statistically significant decreasing number of bird species with increasing urbanization (Figure 3) both in winter (Unifactorial ANOVA Test Winter $F_{3,114} = 15.22$, $P < 0.0001$) and in the breeding season ($F_{3,114} = 12.19$, $P < 0.0001$). In both seasons, species richness was significantly greater in the Rural and Mixed than in the Mostly Urban and Urban sectors (Figure 3).

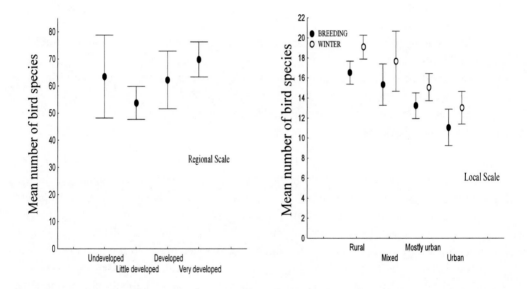

Figure 3. Variation of mean (and 95 % CI) bird species richness per square along an urban gradient at two different scales regional (2800 km^2 in Comunidad Valenciana region) and local (76 km^2 in the municipality of Valencia)

At the regional scale, the proportion of bird species of the different groups experienced some changes along the gradient according with the availability of habitats (Table 2). In overall, the diurnal raptors (e.g. Golden Eagle *Aquila chrysaetos*) and birds associated to cliffs (e.g. Black Weather *Oenanthe leucura*) became rarer along the gradient, whereas aquatic (e.g. Grey Heron *Ardea cinerea*, Mallard *Anas platyrhynchos*, Reed Warbler *Acrocephalus scirpaceus*), and exotic bird species (e.g. Monk Parakeet *Myiopsitta monachus*) became commoner. The rest of groups showed a similar proportion of species along the gradient.

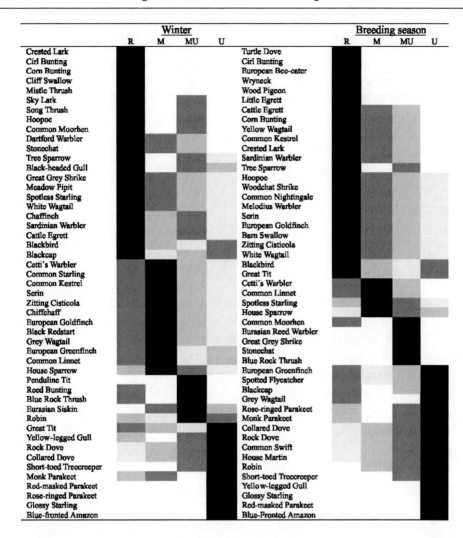

Figure 4. Variation in the density of bird species along the rural-urban gradient at a local scale (Valencia). Decreasing grey tone indicates a decreasing density. R: rural; M: mixed; MU: mostly urban; U: urban.

Table 2. Mean percentage ± standard error of species of different bird groups along the sectors of the urban gradient at regional scale.

Bird group	Sector of the gradient			
	Undeveloped	Little developed	Developed	Very Developed
Diurnal raptors	3.93 ± 0.98	0.93 ± 0.80	2.88 ± 0.85	1.23 ± 1.08
Nocturnal raptors	5.07 ± 0.49	5.62 ± 0.40	5.26 ± 0.43	4.31 ± 0.54
Cliff birds	9.41 ± 1.10	3.05 ± 0.90	3.88 ± 0.95	2.21 ± 1.27
Waterfowl	0.95 ± 2.86	3.79 ± 2.33	7.31 ± 2.48	10.69 ± 3.13
Aquatic passerines	0.91 ± 1.43	4.51 ± 1.17	3.73 ± 1.24	6.21 ± 1.51
Woodland birds	5.37 ± 0.41	5.89 ± 0.33	5.41 ± 0.35	4.87 ± 0.45
Farmland birds	48.95 ± 2.78	53.43 ± 2.27	47.47 ± 2.41	46.91 ± 3.05
Aerial insectivores	5.37 ± 0.41	5.89 ± 0.33	5.41 ± 0.35	4.87 ± 0.45
Exotic species	0	0.58 ± 0.29	0.66 ± 0.31	1.96 ± 0.39

Table 3. Proportion of species threatened (± S.E.) according with different criteria [local (Valencia), regional (Comunidad Valenciana), and continental (Europe)] along different sectors of the urban gradients at local (Rural to Urban) and regional (Undeveloped to Very developed) scales. F shows the results of a one way ANOVA test between the proportions of threatened species. * $P< 0.05$, ***: $P<0.001$

	Local scale			Regional scale	
Sector of the gradient	Local Threat	Regional Threat	European Threat	Regional Thr.	European Thr.
Rural/Undeveloped	45.93 ± 2.26	6.93 ± 1.08	17.11 ± 1.42	15.76 ± 4.03	37.21 ± 1.52
Mixed/ Little developed	37.65 ± 4.64	7.67 ± 2.33	20.56 ± 2.92	15.91 ± 3.29	37.57 ± 1.24
Mostly urban/ Developed	28.02 ± 2.23	11.32 ± 1.07	15.45 ± 1.41	21.98 ± 3.49	36.05 ± 1.32
Urban/ Very Developed	21.28 ± 2.46	14.19 ± 1.18	12.85 ± 1.55	30.84 ± 4.41	33.56 ± 1.67
	$F_{3,114}= 20.42$***	$F_{3,114}= 7.55$***	$F_{3,114}= 2.37$ n.s.	$F_{3,24}= 2.99$*	$F_{3,24}= 1.37$ n.s.

At the local scale, in Valencia, the mean abundance per square of bird species (for scientific names see Appendix 1) showed some variation across the gradient, and for many species differences were statistically significant in both seasons (see Appendix 2). A summary of such variation is depicted in Figure 4. From winter to the breeding season the most noticeable patterns were (i) a decrease in the number of species (17 to 9 species) whose abundance peaked at the intermediate sectors of the gradient mainly due to the decrease of species peaking in the "Mixed" sector (12 to 4 species) ; (ii) an increase in the number of species (32 to 40 species) whose abundance peaked at the extremes of the gradient due mainly to an increase in the number of the species peaking in the most urbanized sector of the gradient (10 to 16 species).

Variation in the Number of Threatened Species aong the Rural-Urban Gradient

When the proportion of threatened species (according with different criteria) along the gradient at different spatial scales was compared (Table 3) some patterns were apparent. First, the proportion of species threatened at continental scale (Europe) decreased with increasing urbanization at any spatial scale, but this decrease was small (about 4 %) and not statistically significant. Secondly, the proportion of species threatened in a regional context (Comunidad Valenciana) showed a statistically significant increase with increasing urbanization irrespectively of the scale of the analyses. By last, the opposite pattern was detected for the proportion of species threatened in the local context (Valencia) which significantly decreased with increasing urbanization.

BIRD FAUNA AT THE LOCAL URBAN GRADIENT

At the local scale, the number of bird species decreased with an increasing level of urbanization, and most of the species reached maximum densities at the two less urbanized sectors of the gradient. This pattern is in accordance with many other studies about the effects of urbanization on birds (Clergeau et al., 1998; Palomino and Carrascal, 2006) although other authors have reported bird species richness, and bird abundance of individual species peaking at intermediate levels of development (Blair, 1996) or even at the most developed level (Rosenberg et al., 1987). Provide that grain and extent of studies were similar, such conflicting patterns can arise, as it has been pointed out by Hennings and Hedge (2003) because differences in the number and typology of the categories included in the gradient; additionally, not only the proportion of a particular habitat may vary across the gradient but its structure and degree of fragmentation (Porter et al., 2001) thus hampering comparisons.

In Valencia the decrease of bird richness and abundance along the urban gradient is probably related with two factors. First, most of the bird species recorded in the cultivated land of Valencia are open field species (e.g. Crested Lark *Galerida cristata*, Skylark *Alauda arvensis*) or shrubland species (e.g. Stonechat *Saxicola rubetra*, Zitting Cisticola *Cisticola juncidis*) which use a diverse set of habitats (row crops, ploughed fields, weedy stubbles...) absent in the urbanized landscape, excepting the patches of derelict land, some of them very similar to weedy stubbles (Carretero and Aguilella, 1995; Murgui, 2006). Secondly, urban parks of Valencia are designed like a wooded habitat which would be suitable mainly for the scarce number of woodland species (e.g. Great Tit *Parus major*) in the cultivated landscape. Therefore, there is a loss of species from the rural to the urbanized landscape that the presence of urban parks can not compensate, as is reflected by the decreasing of bird richness in spite of an increasing cover of urban parks along the urban gradient.

The relatively low number of woodland bird species using urban parks is not only the consequence of the scarcity of such species in the surrounding landscape. The small size of urban parks of Valencia, its sparse tree cover, and the virtual lack of a shrub layer (Murgui, 2007a) probably limits (trough the shortage of resources) its ability in attracting some of the woodland or shrubland bird species of the periphery. Such effect would be particularly pronounced during the breeding season, greatly limiting the urban populations of species that easily use the orange groves of the cultivated landscape for breeding (e.g., Nightingale *Luscinia megarhynchos* or Melodius Warbler *Hippolais polyglotta*) or woodland/shrubland species (e.g., Short-toed Treecreeper *Certhia brachydactyla* or Sardinian Warbler *Sylvia melanocephala*) relatively common in forested areas not too distant from Valencia (Murgui, 1996, 1998). Note, however, that the role of urban parks in increasing the biodiversity of the study area is not expendable. Populations of very scarce breeding bird populations in Valencia like Robin *Erithacus rubecula*, Blackcap *Sylvia atricapilla*, and Short-toed Treecreeper, and even more widespread species like Great Tit, Spotted Flycatcher *Muscicapa striata*, and Greenfinch *Carduelis chloris* (Murgui, 2005) reached their maximum densities in the more urbanized sector of the gradient through their occupancy of urban parks.

By last, the low suitability of urban parks and other urban habitats would be reduced by the effect of human disturbance, especially for bird species associated to open spaces which are usually very sensitive to human activity (Fernández-Juricic et al., 2001) thus limiting its presence and abundance even in *a priori* suitable urban habitats (e.g. lawn in urban parks).

Little evidence was found to support an effect of seasonality in the distribution of bird fauna along the rural-urban gradient, probably reflecting the small seasonal changes of the habitat structure, and of food availability in urban areas (Shochat *et al.*, 2006; Murgui, 2007a). The most noticeable effect of seasonality was a higher number of species in all sectors of the gradient in winter than during the breeding season (for a similar pattern see Clergeau et al., 1998) due to (i) the arrival of wintering species and populations from more northerly countries, and from other parts of the Iberian Peninsula (Tellería, 2004) which exploit the abundance of weeds (e.g., genus *Diplotaxis, Sysimbrium, Sonchus*) associated to cultivated fields and to derelict land (Carretero and Aguilella, 1995), and the large amount of fruits provided by a number of native (e.g. *Phoenix spp, Pistacia lentiscus*) and exotic (e.g. *Ligustrum japonica, Cotoneaster horizontalis*) plant fruit species of urban parks (Debusche and Isenmann, 1990; Murgui, 2007a); (ii) the low ability of many habitats in fulfilling the nesting requirements of many potentially breeding species, discussed previously.

BIRDS AT REGIONAL URBAN GRADIENT

Oppositely to the pattern recorded at local scale, breeding bird species richness tended to increase with level of urbanization at regional scale, a result that fits well with those of Pautasso (2007) who notes that in studies using a sampling unit above 1 Km^2 a positive correlation between human presence and biodiversity is usually found, and the opposite trend occurs at finer scales. However, is noteworthy that González-Taboada et al. (2007) using data from the same source than in the present study i.e. the Atlas of the breeding birds in Spain (Martí and Del Moral, 2003) but with a much larger sample size (which included the 4994 10 x 10 km squares corresponding to Spain) obtained a negative correlation between bird species richness and urbanization degree at the 10 x 10 km squares.

The absence of a negative effect of urbanization at regional scale could be the consequence of three factors (i) an intrinsically richer bird species pool in areas presently urbanized; (ii) an increase in bird richness with the development of such areas; (iii) the use of a spatial scale enough large to capture spatial variation of bird richness. We will consider in detail these factors in the following paragraphs.

At a regional scale, squares placed near the coast formerly were wetlands and forested lowlands (Costa, 1999) two very productive habitat types which probably showed high biodiversity (for the relationships between productivity and biodiversity, see Hawkins, 2003; Currie, 2004). As early as in the first century BC the wetlands and lowlands became to be transformed in cultivated land (Ribera, 2003), and with the course of time in urban areas. Therefore, there is a coincidence of high biodiversity with people (Baillie et al., 2004) a pattern repeatedly documented (e.g. Balmford et al., 2001; Araujo, 2003). Differences in bird richness between the coastal fringe and the Mediterranean forest inland probably are exacerbated by the fact that the forest bird communities in the Mediterranean region constitutes an impoverished pool of the northern Palearctic forest bird communities (Ramírez and Tellería, 2003).

Although the lack of data do not allows a precise balance between extinctions and colonisations in the study area during historical times, is quite possible that the land transformation of the primeval habitats into an agricultural land mosaic increased bird richness (Covas and Blondel, 1998). This would be the case of tree nesting species which benefited from orchards, and of birds associated to cliffs (Rock Dove *Columba livia*, Common swift *Apus apus*, House Martin *Delichon urbica*, Barn Swallow *Hirundo rustica*, even Blue Rock Thrush *Monticola solitarius*) which readily adopt man made structures to nest.

By last, using a grain of 10 x 10 km the percentage of developed land probably is at most 60 % in our data which means that another landscape types are sampled in each square, and that their size may be enough large to fulfil the habitat requirements of many species (Pautasso, 2007). In short, at regional scale the size of the squares is sufficiently large to capture the *per se* bird-rich species pool inhabiting the areas that currently are urbanized. Oppositely, at the local scale not only a more reduced collection of habitats per square is usually sampled but their absolute sizes are considerably smaller thus limiting the presence of some wide-range and sensitive species. Therefore, an increasing in the urban cover is negative to bird richness.

THREATENED BIRDS ALONG THE URBAN GRADIENT: A MATTER OF SPATIAL SCALE AND CRITERION

Our results indicating decreasing or increasing bird species richness with urbanization could be considered trivial as long as such species were considered not threatened (though this is a debatable issue, as we discuss later). However, when attention was focused in species which exhibited some degree of threat under different criteria, results showed that the number of such species was affected by urbanization, and some times in a quite unexpected way.

When the conservation status of bird in Europe was considered (BirdLife, 2004) the proportion of species threatened was little affected by urbanization both at regional and at local scale. Such result mostly reflects the underlying criterion in the classification of BirdLife which give most importance to declining species associated to farmland in Central Europe and UK (e.g. Stonechat or Crested Lark). Such species are (still) widespread and common in the study area, and in Spain; oppositely, relatively uncommon species in the study area (like for instance the Robin *Erithacus rubecula* or the Great Spotted Woodpecker *Dendrocopos major*) are considered in a secure conservation status in Europe (discussed in Carrascal and Lobo, 2003). Therefore, as most of the threatened species in Europe are relatively common in many of the habitats in the study area, its proportion along the gradient remained fairly constant.

At the other extreme, when conservation status criteria obtained from local data was applied at a local scale i.e. Valencia, the proportion of threatened species decreased along the urban gradient. As occurs in other studies conducted in urban areas the distribution of abundance in the bird community is much skewed with a few widespread bird species showing very high abundances (e.g. House Sparrow *Passer domesticus*, Collared Dove *Streptopelia decaocto*, Common Swift *Apus apus*) and the rest exhibiting medium or low abundances and restricted ranges (Murgui, 2005). As abundant species were associated to the urban environment, and the rarer species to the surrounding cultivated land, is not therefore strange than the proportion of threatened species were significantly higher in the less urbanized sectors of the gradient.

When the threat status of birds in Comunidad Valenciana was used to assess the variation along the gradient then the proportion of threatened species *increased* with urbanization at both local and regional scale. Such counterintuitive result is the outcome of two mechanisms: (i) the spatial distribution of some bird groups in Comunidad Valenciana; (ii) the use of spatial range as the only available information to assess threat status.

At the regional scale (i.e. 2800 km^2) the threatened birds recorded belong to mainly diurnal raptors, bird species associated to cliffs, waterfowl, and a group of woodland/farmland birds. The former groups are widespread in the less developed sectors of the gradient (inland areas) where their preferred habitats exist, whereas wildfowl are associated with the most developed sectors (near the coastal fringe) where wetlands are placed. Some threatened woodland birds occur inland (e.g. Greater Spotted Woodpecker or Nuthatch *Sitta europaea*), but the bulk of threatened species are wildfowl. Additionally, other species like Robin and Blackcap, and the Collared Dove are common in developed areas. Robin and Blackcap are species adapted to mesic conditions, and limited in termomediterranean areas to patches of moist habitats like humid forests, urban parks and gardens, and river banks (Carbonell, 2003; Purroy, 2003) i.e. habitats more frequent near the coastal fringe; the Collared Dove is a recent addition to the bird fauna of Comunidad Valenciana (first records in 1984 see Díaz et al., 1996) which is associated to a relatively few number of urban settings (Gámez, 2003). As a result of these patterns the mean proportion of threatened species slightly increased with the development at the regional scale.

What is the effect of the distribution of threatened birds in Comunidad Valenciana on the proportion of threatened species along the rural-urban gradient at the local scale? Raptors, birds of cliffs and wildfowl are very uncommon in all sectors of the gradient, and thus the proportion of threatened birds per square is determined by species (Blackcap, Robin, Collared Dove) associated mainly to urban areas. Therefore, the mean proportion of these birds increases with the urbanization, much more considering that the mean number of species decreases with urbanization.

The results outlined above are partially the consequence of an assessment of the conservation status of birds in Comunidad Valenciana using the spatial range of species but in absence of data about population sizes. Such incomplete knowledge leads to classify as equally threatened in Comunidad Valenciana the Bonelli´s Eagle *Hieraaetus fasciatus* with a population size of c. 90 breeding pairs (Real, 2003) and the Collared Dove which has a population size of c. 2000 individuals only in Valencia (Murgui, 2005). Unfortunately, data about population sizes is known only for some raptors and wetland species, and thus a full correction of such distortions is not possible.

CONSERVATION STRATEGIES OF BIRD FAUNA IN A HIGHLY ANTROPIZED REGION

Centuries ago primeval wetlands and lowlands of the study area were transformed in rice fields and irrigated land, and Mediterranean woodlands in woody crops. Later, especially during 1950-1970 a great extent of the irrigated land mostly was heavily urbanized. In the last four decades, urbanization has continued at expenses not only of irrigated land, but of inland wood crops and shrubland, adopting in some places the form of urban sprawl (OSE, 2007). In such scenario, ecological phenomena can not be understand, and conservation strategies can not be designed without a full consideration of the intricate set of socio-economical processes which have driven most of the changes in the landscape. Therefore, a detailed analysis of conservation strategies is out of the possibilities of this chapter. Rather I will concentrate in a few questions which could be useful in order to put conservation strategies into perspective.

1. Should Conservation Strategies be Conducted at Local Scales, and Should be Allocated Resources to the Conservation of Species which are Common at Regional or Continental Scales?

In our study, urbanization affected more seriously biodiversity (i.e. the proportion of threatened species increased) at decreasing spatial scales. In other words, many species threatened in Valencia were not threatened in a wider area surrounding it, in Comunidad Valenciana region or in Europe. Is therefore reasonable to consider whether the limited resources for conservation should allocated in such species or in other ones with higher probability of extinction, for instance Bonelli´s Eagle which is considered Endangered in Europe (Tucker and Heath, 1994). We could even leave aside this kind of dilemmas, and simply ask whether it is necessary to invest in conserving the scarce (in Valencia) Linnet *Carduelis cannabina* when may have lots in a near village; even more: does make sense take care of the apparently abundant House Sparrow?

Hunter and Hutchinson (1994) have provided compelling arguments for conserving taxa that are not currently at risk of extinction. We will focus on two of them. First, conservation efforts directed towards populations of species which are still in a secure status is the best way to prevent that they become endangered in the future (Hunter Jr and Hutchinson, 1994). Additionally, such species might be secured for relatively little cost (Possingham et al., 2002). As a paradigmatic example, three decades ago no one had thought that the House Sparrow would be a declining species in many European cities (Shaw *et al.*, 2008), and that the species would be listed a as species of European conservation concern (BirdLife, 2004). Secondly, ecosystem functioning, and conservation of biodiversity are linked to the diversity of populations at local scale (Hunter Jr and Hutchinson, 1994; Luck et al., 2003). For instance, conservation of the Bonelli´s Eagle will be unsuccessful if we fail to maintain healthy populations of preys like Rabbit *Oryctulagus cuniculus* and Red-legged Partridge *Alectoris rufa* (Ontiveros et al., 2004) which were common in the past, and such populations must be enough abundant at local scale (i.e. at the home range area of eagles): for the conservation of Bonelli´s Eagle in Comunidad Valenciana it does not matter whether grey partridges are abundant in Scotland.

2. Which Landscapes Should be Conserved?

In Comunidad Valenciana region, protected areas (PA) fall mainly on two categories: protected wetlands along the coastal fringe, and reserves placed on woody or scrubby habitats. Such locations have been chosen under different criteria, but it is not known whether they adequately represent the biodiversity of the region or whether they are suitable to promote the persistence of biodiversity and ecosystem processes (see Pressey et al., 2007). Maiorano et al (2006) in a recent GAP and irreplaceability analyses conducted over terrestrial vertebrates from Italy (a Mediterrranean country similar in many aspects with Comunidad Valenciana) have reported that many species of interest are concentrated in the Mediterranean coastal areas where existing PAs are scarcer and smaller, and human pressure is higher. A study undertaken in Comunidad Valenciana could give similar results leading to the conclusion that the agricultural landscape has been clearly ignored in conservation strategies.

However, the role of agricultural or cultural landscapes in conserving biodiversity is far from negligible (Daily, 2001; Vandermeer and Perfecto, 2007) especially in regions like the Mediterranean basin where human activity during thousand of years have configured a complex and ecologically rich cultural landscape (Covas and Blondel, 1998).

Results of our study would support such view as was indicated by the very high species richness in squares where agricultural land predominated. However, neither the irrigated land nor wood crops (about 350.000 ha according with MMARM, 2008) are protected *per se*. Recently one of the most valuable landscapes the Huerta from Valencia (one of the cultural landscapes identified in Europe see EEA, 1995) is intended to be protected through a PAT (Plan de Acción Territorial). Still in its first stage, such plan constitutes a diagnostic of the current state of the Huerta, and seeks to establish some guidelines over the urban and infrastructure development in about 12.000 ha of Huerta (CMAUH, 2008).

3. Which Actions Should be Undertaken to Conserve the Agricultural Landscape and the Bird Fauna it Harbours?

Currently the agricultural land *still* constitutes the landscape matrix at many spatial scales and administrative boundaries thus holding many different and sometimes conflicting land uses (for an area showing a very similar situation see Martínez-Fernández et al., 2000). Therefore, developing conservation strategies is a very complex issue. From a socio-economic perspective a quite likely fate for many farms in the next decades will be their transformation in urban or industrial land, or its abandonment mainly due to three factors (for a similar case see Saunders and Briggs, 2002): (i) Often fields are of very small size (< 1 ha in the Huerta see Díaz and Galiana, 1996) which means that in spite of the high productivity of the irrigated land (Caballer, 1994), farms often are not economically viable, especially in a context of decreasing commodity prices and increasing production costs (EEA, 2006a). Such situation is more serious in the less productive wooded crops placed inland. (ii) Agricultural land (especially the Huerta) is placed in a region with very competing land uses where land development often is much more cost-effective for farmers than cultivating land; (iii) Average age of farmers is high, and farmer's descendants are not prone to continue their parents job (Cabrejas and García, 1997).

In the scenario depicted above to stop the loss of agricultural land will be possible only if governments at different levels (city councils, regional and national governments, CEE) undertake a decided and imaginative effort. One important obstacle is that landscape planning is dictated by administrative boundaries (see Saunders and Briggs, 2002) largely through land use plans (PGOUs Plan General de Ordenación Urbana) elaborated and periodically revised by city councils. The main outcome of such planning is that the municipalities exhibit in their peripheries small-size (and often temporary until the next PGOU revision) patches of agricultural land surrounded by built-up areas or infrastructures. To reverse this situation only will be possible if planning is coordinated at coarser spatial scales (Gaja and Boira, 1994), and promotes the establishment of larger agricultural areas shared by different municipalities.

Ultimately, conservation strategies should try to preserve two functions of the agricultural matrix: (i) to provide habitat requirements of species inhabiting the agricultural landscape; (ii) to allow the movement of organisms inhabiting the agricultural land, and between other habitats placed inside the landscape matrix. Such functions would be better accomplished by larger agricultural areas, but not only the size but the quality of habitat or its spatial arrangement can strongly influence its permeability to movement or colonization by organisms. Currently, agricultural areas (especially in the irrigated land) are crossed by railways, roads and motorways, and many times cultivated fields are beside buildings and industrial sites. Such circumstances promote habitat loss, a high degree of fragmentation, and degradation of the habitat thus impeding the establishment of sensitive species, and limiting the population size of many other ones through different mechanisms like higher disturbance, decreasing feeding opportunities, barriers to movement and so on (Saunders et al., 1991; Murcia, 1995; Fahrig, 2003). Degradation of the farming landscape has occurred not only through increasing development but from changes in the agricultural system itself. Following a widespread trend (Matson et al., 1997; for an example see Nakamura and Short, 2001) the cultivated landscape of the Huerta has suffered from agricultural intensification which among other changes (e.g. phenomena of water pollution and salinisation) has produced the removal of almost any patch of seminatural habitat (Maron and Fitzsimons, 2007 see Fig 2b); additionally dirt-sided irrigation ditches have been rebuilt as sheer-sided concrete channels (Figure 5). Such features have resulted in the lost or extreme simplification of plant and animal communities, and a serious reduction of the feeding and nesting opportunities for many bird species (Donald *et al.*, 2001).

Figure 5. Sheer-sided concrete irrigation channel (left) and dirt-sided irrigation ditches (right) showing a herbaceous layer and some small trees (genus *Salix* and *Ulmus*).

It is important to point out that efforts towards maintain areas as larger as possible is not an excuse to adopt a cavalier attitude to smaller patches of agricultural land which many times appear encapsulated in the urban areas (for the role of small habitat patches in conserving biodiversity see e.g. Fischer and Lindenmayer, 2002). Animals operate at different spatial scales, and therefore small pieces of agricultural land which could be not suitable for a Common Kestrel *Falco tinnunculus* would be enough large for serins *Serinus serinus*, and much more for many invertebrate species (Hostetler, 1999); in addition, small agricultural patches represent a link with the rural past thus providing invaluable educational and cultural benefits (Kendle and Forbes, 1997).

CONCLUSION

Increasing urbanization in the study area has resulted in dramatic changes in the landscape, and it has exerted a deleterious influence on many species and ecological processes. Fortunately urbanization at the regional scale is still not so widespread to impede the existence of a diversified bird fauna, but prospects indicate that such situation could seriously deteriorate in the near future if urban growth is not properly limited and directed. However it could be a mistake to think about urban areas as the source of all ecological harm. Urban areas could help in conserving biodiversity in at least three ways. First, it is not uncommon that many cities harbour rich faunas and floras (some examples cited in Miller and Hobbs, 2002), and sometimes threatened species (like the Peregrine Falcon *Falco peregrinus* see Crick et al., 2003); secondly many urban habitats like urban parks and gardens (Savard et al., 2000; Fernández-Juricic and Jokimäki, 2001; Gaston et al., 2005), the roofs of buildings (Oberndorfer et al., 2007), and the wooded streets (Murgui, 2007b) could be designed and managed in order to attract species inhabiting the landscape surrounding cities. Finally, and perhaps this is the most important issue, most people live today in cities and such number will increase in the next decades. Therefore, as it has been repeatedly stated (Kendle and Forbes, 1997; McKinney, 2002; Dunn *et al.*, 2006) decisions on the conservation policies at every scale will depend not on rural or wild land dwellers but on people inhabiting cities because of their increasing political influence in democracies. If through a daily contact with nature (Turner et al., 2004; Miller, 2005) citizens are more sensitive to environmental issues, it is likely that they would care for nature conservation not only in their own backyard but in a global context.

Like it or not, "*the fight of the future of the countryside...is taking place in the towns*" (Kendle and Forbes, 1997). As it was quoted by the poet Johann Hölderlin in his novel *Hyperion*, we hope that "*where the danger grows, grows as well what can save us*".

Appendix 1. Scientific Names

Barn Swallow *Hirundo rustica*	Hoopoe *Upupa epops*
Blackbird *Turdus merula*	House Martin *Delichon urbica*
Blackcap *Sylvia atricapilla*	House Sparrow *Passer domesticus*
Black-headed Gull *Larus ridibundus*	Little Egrett *Egretta garzetta*
Black Redstart *Phoenicuros ochruros*	Meadow Pipit *Anthus pratensis*
Blue-fronted Amazon *Amazona aestiva*	Melodius Warbler *Hippolais polyglotta*
Blue Rock Thrush *Monticola solitarius*	Mistle Thrush *Turdus viscivorus*
Cattle Egrett *Bubulcus ibis*	Monk Parakeet *Myiopsitta monachus*
Cetti´s Warbler *Cettia cetti*	Penduline Tit *Remiz pendulinus*
Chaffinch *Fringilla coelebs*	Red-masked Parakeet *Aratinga erythrogenis*
Chiffchaff *Phylloscopus collybita*	Reed Bunting *Emberiza schoeniclus*
Cirl Bunting *Emberiza schoeniclus*	Robin *Erithacus rubecula*
Cliff Swallow *Ptyonoprogne rupestris*	Rock Dove *Columba livia*
Collared Dove *Streptopelia decaocto*	Rose-ringed Parakeet *Psittacula krameri*
Common Linnet *Carduelis cannabina*	Sardinian Warbler *Sylvia melanocephala*
Common Kestrel *Falco tinnunculus*	Serin *Serinus serinus*
Common Moorhen *Gallinula chloropus*	Short-toed Treecreeper *Certhia brachydactyla*
Common Nightingale *Luscinia megarhynchos*	Sky Lark *Alauda arvensis*
Common Starling *Sturnus vulgaris*	Song Thrush *Turdus philomelos*
Common Swift *Apus apus*	Spotless Starling *Sturnus unicolor*
Corn Bunting *Miliaria calandra*	Spotted Flycatcher *Muscicapa striata*
Crested Lark *Galerida cristata*	Stonechat *Saxicola torquata*
Dartford Warbler *Sylvia undata*	Tree Sparrow *Passer montanus*
Eurasian Reed Warbler *Acrocephalus palustris*	Turtle Dove *Streptopelia turtur*
Eurasian Siskin *Carduelis spinus*	White Wagtail *Motacilla alba*
European Bee-eater *Merops apiaster*	Woodchat Shrike *Lanius senator*
European Goldfinch *Carduelis carduelis*	Wood Pigeon *Columba palumbus*
Glossy Starling *Lamprotornis chaelybaeus*	Wryneck *Jynx torquilla*
Great Grey Shrike *Lanius meridionalis*	Yellow-legged Gull *Larus cachinnans*
Great Tit *Parus major*	Yellow Wagtail *Motacilla flava*
Grey Wagtail *Motacilla cinerea*	Zitting Cisticola *Cisticola juncidis*

Appendix 2. Mean bird abundance per square along the rural-urban gradient. *P*: probability of ANOVA test (when >2 sectors are compared) or Student T test (2 sites compared)

Breeding season	Rural	Mixed	MUrb.	Urban	P		Rural	Mixed	MUrb.	Urban	P
Turtle Dove	0.45	0.00	0.00	0.00		Serin	22.45	20.78	7.10	4.28	***
Cirl Bunting	0.26	0.00	0.00	0.00		European Goldfinch	12.32	21.11	9.18	5.00	***
European Bee-eater	0.18	0.00	0.00	0.00		Spotless Starling	8.92	14.44	12.33	5.75	n.s.
Wryneck	0.13	0.00	0.00	0.00		Barn Swallow	7.21	6.77	3.91	2.30	**
Wood Pigeon	0.03	0.00	0.00	0.00		White Wagtail	1.18	1.00	0.74	0.31	*
Cattle Egrett	0.71	0.33	0.44	0.00	n.s.	Blackbird	7.82	2.78	2.59	6.03	**
Corn Bunting	0.42	0.11	0.03	0.00	*	Spotted Flycatcher	1.11	1.56	1.10	2.06	n.s.
Yellow Wagtail	0.39	0.11	0.05	0.00	n.s.	Great Tit	1.66	1.33	0.74	1.25	*
Sardinian Warbler	0.29	0.11	0.21	0.00	n.s.	European Greenfinch	1.95	1.67	1.59	2.00	n.s.
Crested Lark	0.13	0.11	0.10	0.00	n.s.	House Sparrow	70.68	127.89	84.59	75.28	**
Common Linnet	0.03	0.33	0.03	0.00	n.s.	Collared Dove	1.26	3.56	12.38	34.00	***
Cetti´s Warbler	0.05	0.11	0.05	0.00	n.s.	Rock Dove	1.63	6.78	10.54	56.59	***
Tree Sparrow	1.84	0.00	1.21	0.00	n.s.	Common Swift	15.15	30.71	25.84	45.16	**
Common Moorhen	0.08	0.00	0.21	0.00	n.s.	House Martin	2.55	6.77	9.25	7.71	**
Eurasian Reed Warbler	0.00	0.00	0.10	0.00		Monk Parakeet	0.11	0.00	0.13	1.28	n.s.
Stonechat	0.00	0.00	0.08	0.00		Robin	0.00	0.11	0.15	0.47	n.s.
Woodchat Shrike	0.66	0.44	0.13	0.03	***	Blackcap	0.21	0.00	0.31	0.31	***
Common Nightingale	2.47	0.78	0.49	0.09	***	Yellow Wagtail	0.08	0.00	0.00	0.28	n.s.
Melodius Warbler	1.68	0.89	0.21	0.09	***	Short-toed Treecreeper	0.00	0.00	0.03	0.16	n.s.
Zitting Cisticola	3.82	5.78	2.51	0.34	***	Yellow-legged Gull	0.00	0.00	0.00	0.41	
Hoopoe	1.36	1.11	0.23	0.16	***	Glossy Starling	0.00	0.00	0.00	0.16	
Great Grey Shrike	0	0	0.03	0		Rose-ringed Parakeet	0.11	0.00	0.21	0.38	n.s.
						Red-masked Parakeet	0.00	0.00	0.00	1.03	
						Blue-Fronted Amazon	0.00	0.00	0.00	0.09	
						Blue Rock Thrush	0.00	0.00	0.03	0.00	

*P < 0.05 ** P < 0.01 *** P < 0.001

Winter	Rural	Mixed	MUrb.	Urban	*P*		Rural	Mixed	MUrb.	Urban	*P*
Crested Lark	0.74	0.00	0.00	0.00		Black Redstart	7.42	8.78	6.82	3.53	***
Cirl Bunting	0.13	0.00	0.00	0.00		Zitting Cisticola	2.32	2.44	1.95	0.22	***
Corn Bunting	0.05	0.00	0.00	0.00		Grey Wagtail	1.42	1.78	0.82	0.41	**
Mistle Thrush	0.03	0.00	0.00	0.00		Chaffinch	3.16	1.11	1.15	1.19	***
Song Thrush	0.32	0.00	0.21	0.00	n.s.	Sardinian Warbler	3.11	1.56	2.08	1.00	**
Hoopoe	0.21	0.00	0.03	0.00	*	European Greenfinch	1.74	2.11	1.05	1.66	n.s.
Common Moorhen	0.08	0.00	0.05	0.00	n.s.	Blackbird	6.13	2.78	2.05	4.34	*
Reed Bunting	0.05	0.00	0.15	0.00	n.s.	Black-headed Gull	1.37	0.00	0.69	0.69	n.s.
Stonechat	1.16	1.56	0.69	0.00	*	Yellow-legged Gull	0.79	0.00	1.28	1.44	n.s.
Dartford Warbler	0.05	0.11	0.08	0.00	n.s.	Great Tit	1.37	1.22	0.85	1.72	n.s.
Great Grey Shrike	0.16	0.11	0.08	0.03	n.s.	Blackcap	4.29	1.11	1.56	4.09	**
Cetti's Warbler	0.05	0.22	0.03	0.03	*	Robin	2.61	2.11	3.08	3.00	n.s.
Cattle Egret	3.39	1.56	2.31	0.06	*	Eurasian Siskin	0.00	0.44	1.21	0.25	n.s.
Tree Sparrow	1.61	0.00	0.23	0.09	*	House Sparrow	77.68	193.22	99.82	76.16	***
Common Kestrel	0.42	0.44	0.26	0.09	*	Rock Dove	3.37	5.00	8.28	70.63	***
Meadow Pipit	3.42	3.22	2.13	0.16	***	Collared Dove	1.39	2.33	13.10	30.50	***
Common Linnet	0.45	2.22	0.13	0.16	n.s.	Short-toed Treecreeper	0.00	0.00	0.03	0.09	n.s.
Serin	24.61	42.56	17.82	5.84	***	Red-masked Parakeet	0.00	0.00	0.00	0.97	
Chiffchaff	18.58	42.11	14.64	5.50	***	Rose-ringed Parakeet	0.00	0.00	0.00	0.31	
European Goldfinch	17.05	45.89	14.92	5.28	***	Glossy Starling	0.00	0.00	0.00	0.09	
Spotless Starling	15.18	12.56	12.51	6.38	**	Blue-fronted Amazon	0.00	0.00	0.00	0.06	
Penduline Tit	0	0	0.11	0		Blue Rock Thrush	0.03	0.00	0.03	0.00	
Cliff Swallow	0.16	0	0	0							

*P < 0.05 ** P < 0.01 *** P < 0.001

REFERENCES

Araujo, M., 2003. The coincidence of people and biodiversity in Europe. Global Ecology and Biogeography 12, 5-12.

Baillie, J., Hilton-Taylor, C., Stuart, S. (Eds.), 2004. 2004 IUCN Red List of Threatened Species. A Global Species Assessment. IUCN, Gland, Switzerland.

Balmford, A., Moore, J.L., Brooks, T., Burgess, N., Hansen, L.A., Williams, P., 2001. Conservation conflicts across Africa. Science 291, 2616-2619.

BirdLife, 2004. Birds in the European Union: a status assessment. BirdLife International, Wageningen, The Netherlands.

Blair, R., 1996. Land use and avian species diversity along an urban gradient. Ecological Applications 6, 506-519.

Brown, J., Lomolino, M., 1998. Biogeography. Sinauer Associates, Sunderland.

Caballer, V., 1994. Viabilidad económica de la huerta de Valencia. In: Salvador, P. (Ed.), Seminario Internacional sobre la Huerta de Valencia. Ayuntamiento de Valencia, Valencia, pp. 155-176.

Cabrejas, M., García, E., 1997. València, l´Albufera, l´Horta: Medi ambient i conflicte social. Universitat de València, València.

Carbonell, R., 2003. Curruca Capirotada Sylvia atricapilla. In: Martí, R., Del Moral, J. (Eds.), Atlas de las Aves reproductoras de España. Dirección General de Conservación de la Naturaleza-Sociedad Española de Ornitología, Madrid, pp. 484-485.

Carrascal, L., Lobo, J., 2003. Respuestas a viejas preguntas con nuevos datos: estudio de los patrones de distribución de la avifauna española y consecuencias para su conservación. In: Martí, R., del Moral, J. (Eds.), Atlas de las Aves Reproductoras de España. Dirección General de Conservación de la Naturaleza

Sociedad Española de Ornitología, Madrid, pp. 651-668.

Carretero, J.L., Aguilella, A., 1995. Flora y vegetación nitrófilas del término municipal de la ciudad de Valencia. Ajuntament de València, Valencia.

Chace, J.F., Walsh, J.J., 2006. Urban effects on native avifauna: a review. Landscape and Urban Planning 74, 46-49.

Clergeau, P., Jean-Pierre,L, Mennechez, G., Falardeau, G., 1998. Bird abundance and diversity along an urban-rural gradient: a comparative study between two cities on different continents. The Condor 100, 413-425.

CMAUH, 2008. Plan de Acción Territorial de Protección de la Huerta de Valencia. Conselleria de Medi Ambient i Aigua, Urbanisme i Habitatge. Generalitat Valenciana, Valencia.

Collar, N., 1999. Risk Indicators and Status Assessment in Birds. In: del Hoyo, J., Elliot, A., Sargatal, J. (Eds.), Handbook of the Birds of the World. Vol.5. Barn-owls to Hummingbirds. Lynx Edicions, Barcelona, pp. 13-32.

Costa, M., 1999. La vegetación y el paisaje en las tierras valencianas. Editorial Rueda, Madrid.

Covas, R., Blondel, J., 1998. Biogeography and history of the Mediterranean bird fauna. Ibis 140, 395-407.

Crick, H., Banks, A., Coombes, R., 2003. Findings of the National Peregrine Survey 2002. BTO News 248, 8-9.

Currie, D.J., 2004. Predictions and tests of climate-based hypotheses of broad-scale variation in taxonomic richness. Ecology Letters 7, 1121-1134.

Daily, G., 2001. Ecological Forecasts: Countryside. Nature 411, 245.

Debusche, M., Isenmann, P., 1990. Introduced and cultivated fleshy-fruited plants: consequences of a mutualistic Mediterranean plant-bird system. In: di Castri, F., Hansen, A., Debussche, M. (Eds.), Biological invasions in Europe and the Mediterranean Basin. Kluwer Academic Publishers.

Díaz, M., Asensio, B., Tellería, J., 1996. Aves Ibéricas I. No paseriformes. J.M. Reyero. Editor, Madrid.

Díaz, M., Galiana, F., 1996. Estudio paisajístico de la huerta de Valencia. Ajuntament de València, Valencia.

Donald, P., Green, R.E., Heath, M., 2001. Agricultural intensification and the collapse of Europe´s farmland birds. Proc. R. Soc. London (B) 268, 25-29.

Dunn, R., Gavin, M., Sánchez, M., Solomon, J., 2006. The Pigeon Paradox: Dependence of Global Conservatio on Urban Nature. Conservation Biology 20, 1814-1816.

EEA, 1995. Europe´s Environment. The DOBRIS Assessment. European Environmental Agency.

EEA, 2006a. Integration of environment into EU agriculture policy — the IRENA indicator-based assessment report. European Environmental Agency.

EEA, 2006b. Land accounts for Europe 1990–2000. European Environmental Agency.

EEA, 2006c. Urban sprawl in Europe. European Environmental Agency.

Fahrig, L., 2003. Effects of habitat fragmentation on biodiversity. Annual Review of Ecology and Evolutionary Systems . 34, 487-515.

Fernández-Juricic, E., Jimenez, M., Lucas, E., 2001. Alert distances as an alternative measure of bird tolerance to human disturbances: implications for park design. Environmental Conservation 28, 263-269.

Fernández-Juricic, E., Jokimäki, J., 2001. A habitat island approach to conserving birds in urban landscapes: cases studies from southern and northern Europe. Biodiversity and Conservation 10, 2023-2043.

Fischer, J., Lindenmayer, D., 2002. Small patches can be valuable for biodiversity conservation: two case studies on birds in Southeastern Australia. Biological Conservation 106, 129-136.

Foley, J.A., 2005. Global consequences of land use. Science 309, 570-574.

Gaja, F., Boira, J., 1994. Planeamiento y realidad urbana en la ciudad de Valencia (1939-1989). Cuadernos de Geografía 55, 63-89.

Gámez, I., 2003. Tórtola Turca Streptopelia decaocto. In: Martí, R., Del Moral, J. (Eds.), Atlas de las aves reproductoras de España. Dirección General de Conservación de la Naturaleza-Sociedad Española de Ornitología, Madrid, pp. 304-305.

Garden, J., McAlpine, C., Pereson, A., Jones, D., Possingham, H., 2006. Review of the ecology of Australian urban fauna: a focus on spatially explicit processes. Austral Ecology 31, 126-148.

Gaston, K., Blackburn, T., Greenwood, J., Gregory, R., Quinn, R., Lawton, J., 2000. Abundance-occupancy relationships. Journal of Applied Ecology 37 (S1), 39-59.

Gaston, K., Smith, R., Thompson, K., Warren, P., 2005. Urban domestic gardens (II): experimental tests of methods for increasing biodiversity. Biodiversity and Conservation 14, 395-413.

Gilbert, O., 1989. The Ecology of Urban Habitats. Chapman and Hall, London.

González-Taboada, F., Nores, C., Alvarez, M., 2007. Breeding bird species richness in Spain: assessing diversity hypothesis at various scales. ECOGRAPHY 30, 241-250.

Grimm, N., Grove, J., STA., Reman, C.L., 2000. Integrated approaches to long-term studies of urban ecological systems. BioScience 50, 571-584.

Hawkins, B.A., 2003. Energy, water, and broad-scale geographic patterns of species richness. Ecology 84, 3105-3117.

Hennings, L., Edge, W., 2003. Riparian bird community structure in Portland, Oregon: habitat, urbanization, and spatial scale patterns. The Condor 105, 288-302.

Hostetler, M., 1999. Scale, birds, and human decisions: a potential for integrative research in urban ecosystems. Landscape and Urban Planning 45, 15-19.

Hunter Jr, M., Hutchinson, A., 1994. The virtues and shortcomings of parochialism: conserving species that are locally rare, but globally common. Conservation Biology 8, 1163-1165.

Jokimäki, J., Kaisanlathi-Jokimäki, M., 2003. Spatial similarity of urban bird communities: a multiscale approach. Journal of Biogeography 30, 1183-1193.

Kendle, T., Forbes, S., 1997. Urban Nature Conservation. Landscape Management in the Urban Countryside. E & FN Spon, Oxford.

Luck, G., Daily, G., Ehrlich, P., 2003. Population diversity and ecosystem services. Trends in Ecology and Evolution 18, 331-336.

Maiorano, L., Falcucci, A., Boitani, L., 2006. Gap analysis of terrestrial vertebrates in Italy: Priorities for conservation planning in a human dominated landscape. Biological Conservation 133, 455-473.

Maron, M., Fitzsimons, J., 2007. Agricultural intensification and loss of matrix habitat over 23 years in the west Wimmera, south-eastern Australia. Biological Conservation 135, 587-593.

Martí, R., Del Moral, J. (Eds.), 2003. Atlas de las aves reproductoras de España. Dirección General de Conservación de la Naturaleza-SEO/BirdLife, Madrid.

Martínez-Fernández, J., Esteve-Selma, M., Calvo-Sendín, J., 2000. Environmental and socioeconomic interactions in the evolution of traditional irrigated lands: a dynamic system model. Human Ecology 28, 279-299.

Marzluff, J., 2001. Worlwide urbanization and its effects on birds. In: Marzluff, J., Bowman, R., Donnelly, R. (Eds.), Avian Ecology and Conservation in an Urbanizing World. Kluwer Academic Publishers, Boston, pp. 19-48.

Marzluff, J., Bowman, R., Donnelly, R., 2001. A historical perspective on urban bird research: trends, terms, and approaches. In: Marzluff, J., Bowman, R., Donnelly, R. (Eds.), Avian ecology and conservation in an urbanizing world. Kluwer Academic Publishers, Boston.

Matson, P., Parton, W., Power, A., Swift, M., 1997. Agricultural intensification and ecosystem properties. Science 277, 504-509.

McDonnell, M., Pickett, S., 1990. Ecosystem structure and function along urban-rural gradients: an unexploited opportunity for ecology. Ecology 71, 1232-1237.

McKinney, M., 2002. Urbanization, Biodiversity, and Conservation. BioScience 52, 883-890.

Miller, J., 2005. Biodiversity conservation and the extinction of experience. Trends in Ecology and Evolution 20, 430-434.

Miller, J., Hobbs, R., 2002. Conservation where people live and work. Conservation Biology 16, 330-337.

MMARM, 2008. Anuario de estadística agroalimentaria 2007. In: Agroalimentarias, S.G.d.E. (Ed.). Ministerio de Medio Ambiente, Medio Rural y Marino.

Murcia, C., 1995. Edge effects in fragmented forests: implications for conservation. Trends in Ecology and Evolution 10, 58-62.

Murgui, E., 1996. Aproximación al conocimiento de la avifauna invernante de la Sierra Calderona. El Serenet 1, 2-18.

Murgui, E., 1998. Estructura de la comunidad de aves reproductoras de la Sierra Calderona. El Serenet 3, 1-20.

Murgui, E., 2001. Factors influencing the distribution of exotic bird species in Comunidad Valenciana (Spain). Ardeola, 149-160.

Murgui, E., 2005. València. In: Kelcey, J., Rheinwald, G. (Eds.), Birds in European Cities. GINSTER Verlag, St. Katherinen, pp. 335-358.

Murgui, E., 2006. Influencia de la estructura del paisaje a diferentes escalas espaciales sobre las comunidades y poblaciones de aves urbanas. University of Valencia, Ph.D. Thesis.

Murgui, E., 2007a. Effects of seasonality on the species–area relationship: a case study with birds in urban parks. Global Ecology and Biogeography 20, 12-18.

Murgui, E., 2007b. Factors influencing the bird community of urban wooded streets along an annual cycle. Ornis Fennica 84, 66-77.

Murgui, E., 2009. Influence of urban landscape structure on bird fauna: a case study across seasons in the city of Valencia (Spain). Urban Ecosystems 12, 249-263.

Murgui, E., Valentín, A., 2003. Relación entre las carácterísticas del paisaje urbano y la comunidad de aves introducidas en la ciudad de Valencia (Spain). Ardeola 50, 201-214.

Nakamura, T., Short, K., 2001. Land-use planning and distribution of threatened wildlife in a city of Japan. Landscape and Urban Planning 53, 1-15.

Oberndorfer, E., Lundhol, J., Bass, B., Coffman, R., Doshi, H., Dunnett, N., Gaffin, S., Köhler, M., Karen, K., Liu, Y., Rowe, B., 2007. Green Roofs as Urban ecosystems: ecological structures, functions, and services. BioScience 57, 823-833.

Ontiveros, D., Real, J., Balbontín, J., Carrete, M., Ferreiro, E., Ferrer, M., Mañosa, S., Pleguezuelos, J., Sánchez-Zapata, J., 2004. Biología de la conservación del Aguil-azor Perdicera Hieraaetus fasciatus en España: Investigación científica y gestión. Ardeola 51, 461-470.

OSE, 2007. Sostenibilidad en España 2006. Ministerio de Medio Ambiente Fundación Universidad de Alcalá.

Pacione, M., 2001. Urban Geography. A Global Perspective. Routledge, London.

Palomino, D., Carrascal, L.M., 2006. Urban influence of bird at a regional scale: A case study with the avifauna of northern Madrid Province. Landscape and Urban Planning 77, 276-290.

Pautasso, M., 2007. Scale dependence of the correlation between human population presence and vertebrate and plant species richness. Ecology Letters 10, 16-24.

Porter, E., Forschner, B., Blair, R., 2001. Woody vegetation and canopy fragmentation along a forest-to-urban gradient. Urban Ecosystems 5, 131-151.

Possingham, H., Andelman, S., Burgman, M., Medellín, R., Master, L., Keith, D., 2002. Limits to the use of threatened species. Trends in Ecology and Evolution 17, 503-507.

Pressey, R.L., Cabeza, M., Watts, M.E., Cowling, R.M., Wilson, K.A., 2007. Conservation planning in a changing world. Trends in Ecology and Evolution 22, 583-592.

Purroy, F., 2003. Petirrojo Erithacus rubecula. In: Martí, R., Del Moral, J. (Eds.), Atlas de las aves reproductoras de España. Dirección General de Conservación de la Naturaleza-Sociedad Española de Ornitología

Madrid, pp. 416-417.

Ramírez, A., Tellería, J., 2003. Efectos geográficos y ambientales sobre la distribución de las aves forestales ibéricas. Graellsia 59, 219-231.

Real, J., 2003. Aguila-azor Perdicera Hieraaetus fasciatus. In: Martí, R., Del Moral, J. (Eds.), Atlas de las aves reproductoras de España. Dirección General de Conservación de la Naturaleza-Sociedad Española de Ornitología, pp. 192-193.

Ribera, A., 2003. La fundación de Valencia y su impacto en el paisaje. In: Dauksîs, S., Taberner, F. (Eds.), Historia de la ciudad II. Territorio, sociedad y patrimonio. Colegio Territorial de Arquitectos de Valencia, Valencia.

Roberge, J., Svensson, S., 2003. How much time is required to survey land birds in forest-dominated atlas squares? Ornis Fennica 80, 137-147.

Rosenberg, K., Terrill, S., Rosenberg, G., 1987. Value of suburban habitats to desert riparian birds. Wilson Bulletin 99, 642-654.

Sala, O.E., Stuart-Chapin III, F., Armesto, J., Berlow, E., Bloomsfield, J., Dirzo, R., Huber, E., Huennke, L., Jackson, R., Kinzig, A., Leemans, R., Lodge, D., Mooney, H., Oesterheld, M., Poff, N., Skyes, M., Walker, B., Walker, M., Wall, D., 2000. Global biodiversity scenarios for the year 2100. Science 287, 1770-1774.

Saunders, D., Briggs, S., 2002. Nature grows in straight lines- or does she? Whata are the consequences of the mismatch between human-imposed linear boundaries and ecosystem boundaries? An Australian example. Landscape and Urban Planning 61, 71-82.

Saunders, D., Hobbs, R., Margules, C., 1991. Biological consequences of ecosystem fragmentation: a review. Conservation Biology 5, 18-32.

Savard, J., Clergeau, P., Mennechez, G., 2000. Biodiversity concepts and urban ecosystems. Landscape and Urban Planning 48, 131-142.

Shaw, L., Chamberlain, D., Evans, M., 2008. The House Sparrow Passer domesticus in urban areas: reviewing a possible link between post-decline distribution and human socioeconomic status. Journal of Ornithology 149, 293-299.

Shochat, E., Warren, P., Faeth, S., McIntyre, N., Hope, D., 2006. From patterns to emerging processes in mechanistic urban ecology. Trends in Ecology and Evolution 21, 186-191.

Stuart-Chapin III, F., Zavaleta, E., Eviner, V., Naylor, R., Vitousek, P., Reynolds, H., Hooper, D., Lavorel, S., Sala, O., Hobbie, S., Mack, M., Díaz, S., 2000. Consequences of changing biodiversity. Nature 405, 234-242.

Tellería, J., 2004. Migración de aves en el Paleártico Occidental: aspectos ecológicos y evolutivos. In: Tellería, J. (Ed.), La Ornitología hoy. Homenaje al profesor Francisco Bernis Madrazo. Editorial Complutense S.A., Madrid, pp. 109-126.

Tellería, J., Asensio, B., Díaz, M., 1999. Aves Ibéricas II. Paseriformes. J.M. Reyero Editor, Madrid.

Tucker, G., Heath, M., 1994. Birds in Europe: their conservation status. BirdLife International, Cambridge.

Turner, W., Nakamura, T., Dinetti, M., 2004. Global urbanization and the separation of humans from nature. BioScience 54, 585-590.

UN, 2005. World Population Prospects: The 2004 Revision Population Database. United Nations Population Division.

Urios, V., Escobar, J., Pardo, R., Gómez, J. (Eds.), 1991. Atlas de las aves nidificantes de la Comunidad Valenciana. Conselleria d´Agricultura i Pesca. Generalitat Valenciana, Valencia.

Vandermeer, J., Perfecto, I., 2007. The agricultural matrix and a future paradigm for conservation. Conservation Biology 21, 274-277.

Vitousek, P., Mooney, H., Lubchenco, J., Melillo, J., 1997. Human domination of Earth´s Ecosystem. Science 277, 494-499.

Whittaker, R., Willis, K., Field, R., 2001. Scale and species richness: towards a general, hierarchical theory of species diversity. Journal of Biogeography 28, 453-470.

Wiens, J., 1989. Spatial scaling in ecology. Functional Ecology 3, 385-397.

Zar, J., 1996. Biostatistical Analysis. Prentice Hall, New Jersey.

In: Biogeography
Editors: M. Gailis, S. Kalniņš, pp. 203-217

ISBN: 978-1-60741-494-0
© 2010 Nova Science Publishers, Inc.

Chapter 6

HISTORICAL BIOGEOGRAPHY OF CLAVATORACEAE (FOSSIL CHAROPHYTES)

Carles Martín-Closas[1] and Wang Qifei[2]***

[1]Departament d'Estratigrafia, Paleontologia i Geociències Marines, Facultat de Geologia, Universitat de Barcelona, 08028 Barcelona, Catalonia (Spain)
[2]Nanjing Institute of Geology and Palaeontology-Chinese Academy of Science, 39 East Beijing Road, Nanjing 210008, People's Republic of China

ABSTRACT

The extinct charophyte family Clavatoraceae was a significant component of the early Cretaceous lacustrine macrobenthos before the radiation of aquatic angiosperms. The most plesiomorphic and oldest representatives of this family occurred during the Late Jurassic and were scattered in a few localities in the central part of the Peri-Tethyan domain (Western Europe and Northern Africa) and on the margins of the North American Interior Seaway (Morrison Formation). In the beginning of the Early Cretaceous (Berriasian-Barremian), clavatoraceans were dominant in the Peri-Tethyan region and expanded eastwards, reaching the Chinese basins during the Valanginian and Hauterivian. During Barremian and Aptian times clavatoraceans reached their maximum diversity in the Peri-Tethyan region and their maximum palaeogeographic extension worldwide. Two species achieved a cosmopolitan distribution (Eurasia, North and South America and Northern and Eastern Africa), while four species were subcosmopolitan, mainly in the Northern Hemisphere. In contrast, the remaining clavatoracean species showed a marked endemic distribution in specific areas of the Tethyan region, which at that time was an archipelago of large islands enhancing allopatric speciation. During the Albian, the clavatoraceans began to decline. After a significant gap in the fossil record, Late Cretaceous clavatoraceans were either relict forms of previous cosmopolitan species in Northeast Asia or belonged to newly evolved species of endemic distribution in Southern Europe. The family Clavatoraceae became extinct near the Cretaceous-Tertiary boundary.

---

* Corresponding author: cmartinclosas@ub.edu
** Corresponding author: qfwang@nigpas.ac.cn)

A historical biogeographic analysis of the clavatoraceans shows that the Tethyan region was a main focus of speciation for the family. A few species that originated in the Peri-Tethyan archipelago migrated elsewhere, sometimes reaching a worldwide distribution contemporaneous with the palaeogeographic extension of wetlands. Certain adaptations, such as the conjoint disposition of gametangia or the capability to colonise new biotopes provide keys to understanding the worldwide migration of particular species. The dispersal of most Tethyan species required substantial time spans - usually several million years – to achieve a cosmopolitan biogeographic range. The animal vectors of clavatoracean propagules are unknown, but may include ancestral birds and dinosaurs. The decline of the family Clavatoraceae in the Albian was marked by the extinction of many endemic species and by significant biogeographic range restrictions for cosmopolitan species. In the Latest Cretaceous, only two completely isolated species remained prior to extinction.

Keywords: Charophytes, biogeography, Cretaceous, cosmopolitism, endemism

INTRODUCTION

Clavatoraceans were the dominant charophytes in the non-marine basins of the Central Tethyan Domain (Europe and Northern Africa) during the Early Cretaceous, 99.6 to 145.5 million years ago according to Gradstein et al. (2004). They have been reported in all continents except Australia and Antarctica. Some species, such as *Atopochara trivolvis,* are known to most palaeontologists and biostratigraphers working on the non-marine Lower Cretaceous and represented a central component of the lacustrine biotas at that time. The clavatoracean fossil record is excellent due to the biomineralization of their fructifications, which was activated enzymatically after fertilization. The geographic occurrence of clavatoracean fructifications is well documented due to their excellent preservation in the fossil record and to their use in the dating of non-marine sequences for stratigraphic correlation and oil exploration purposes. The enormous amount of palaeogeographic information available allows us to describe the historical biogeography of the entire family Clavatoraceae, which is the main objective of this study.

FOSSIL REMAINS OF CLAVATORACEANS

Clavatoraceans are known in the fossil record thanks to the biocalcification of their oosporangia. These reproductive structures were formed by a bottle-shaped oogonium coated with a vegetative structure called a utricle. The utricle allows us to distinguish the three clavatoracean subfamilies Atopocharoidae, Dictyoclavatoroidae and Clavatoroidae (Grambast, 1966; Martín-Closas, 1996). The subfamily Atopocharoideae had utricles with triradial symmetry; Dictyoclavatoroidae had almost asymmetrical utricles or had a relict bilateral structure. Clavatoroidae utricles were clearly bilateral in symmetry. In addition, oogonia did not calcify in Atopocharoidae and Dictyoclavatoroidae, which show the impressions of oogonial spiral cells on the inner wall of the utricle. Clavatoroidae is the only subfamily to calcify the oosporangium, forming a gyrogonite after fertilization. In this case, the spiral cells were calcified as empty tubes, which is a unique way of gyrogonite calcification in charophytes.

In contrast with the excellent fossil record of clavatoracean fructifications, vegetative remains are scarce and complete specimens are rare. Exceptionally well-preserved clavatoracean thalli were first described as silicified fossils from the English Purbeck by Harris (1939). Musacchio (1971) found a very different type of silicified thallus bearing clavatoracean fructifications in the volcanosedimentary deposits of the Barremian of the Neuquén Basin, Argentina. Thus far, well-preserved calcified assemblages of clavatoraceans have been described only in the Barremian of the Las Hoyas Lagestätte in Central Spain by Martín-Closas and Diéguez (1998). Although clavatoracean thalli probably included a large number of different types, one structural element appears to be unique to this family: the spine-cell rosettes in the internodes, coating the cortical cells. These structures, formed by radial cushions of calcified spines, were most likely adaptations against herbivory of aquatic invertebrates, probably arthropod larvae.

BIOGEOGRAPHY OF THE FIRST CLAVATORACEANS

The oldest representative of clavatoraceans in the fossil record is *Echinochara peckii* from the Upper Jurassic of North America and Europe. This species occurred in the Kimmeridgian (around 156 million years ago) of the Morrison Formation, USA and Northwestern Germany (Mädler 1952, Peck, 1957; Schudack et al., 1998), and perhaps earlier, in the Oxfordian of Switzerland (Mojon, 1989a). These data are consistent with a phylogenetic analysis of the family by Martín-Closas (1996), which shows that Atopocharoidae, including genus *Echinochara*, was the most plesiomorphic clade. Dictyoclavatoroidae, the second derived clade, first appeared in the Kimmeridgian of Portugal (Helmdach and Ramalho, 1976), whereas Clavatoroidae, the most derived clade, first appeared in the Tithonian of Lower Saxony, Germany (Schudack, 1993).

This information appears to indicate that from the very beginning clavatoraceans inhabited the so-called paratropical belt of the Late Jurassic and Early Cretaceous world, growing in alkaline and freshwater lakes and wetlands in a large area around the Central Tethys Sea (present-day Europe and Northern Africa) and around the margins of the North American Interior Seaway. The biogeographic origin of Clavatoraceae cannot be further clarified based on the present state of knowledge, particularly since the most plesiomorphic components of the family (subfamilies Atopocharoidae and Dictyoclavatoroidae) show clearly that there was a lack of calcification in the oogonia of the first representatives of the family. This may indicate that the clavatoracean history could be incompletely documented in the fossil record before the Kimmeridgian.

EARLY CRETACEOUS RADIATION IN THE PERI-TETHYS REGION

The family Clavatoraceae displayed an extensive radiation at the beginning of the Berriasian (around 145.5-140 million years ago), which lasted with different evolutionary ratios until the end of the Aptian (112 million years ago). During this period most of the species of this family appeared with some achieving a worldwide biogeographic distribution (Grambast, 1974; Feist et al., 2005). At the beginning of the Cretaceous (between 145.4 and 130 million years ago), including the Berriasian, Valanginian and Hauterivian, clavatoraceans were abundant in the macrobenthos of many freshwater and alkaline lakes of the paratropical Central Tethys. Thus, their fossil record is excellent in the lacustrine basins of the Iberian Peninsula (Martín-Closas, 2000), the English Purbeck (Feist et al., 1995) and the Swiss Jura (Mojon, 2000). In these areas clavatoraceans quickly become dominant and relegated the two other important charophyte families, (porocharaceans and early characeans) to environments that were marginal for charophytes, i.e. brackish the fluviatile facies. Porocharaceans formed almost monotypic associations in Neocomian brackish facies (Mojon, 1989b). The brackish habitat requires specific physiological adaptations for charophytes to survive (Winter and Kirst, 1990), and only a reduced number of species developed such adaptations. Early characeans formed abundant assemblages in fluviatile siliciclastic environments, which also require specific tolerances for charophytes to grow, such as a higher resistance to short-lived habitats and changes in trophic conditions. In contrast to the abundance of clavatoraceans in the alkaline lakes of the Tethyan paratropical belt, their presence in basins from higher latitudes was moderate: they represent only subsidiary elements of floras dominated by porocharaceans and early characeans (*Porochara, Feistiella, Latochara, Aclistochara*), even in freshwater facies. This also applies to the Tithonian-Berriasian basins in Northern Europe and the United States (Schudack, 1993).

The Barremian and Aptian (130 to 112 million years ago) represent the two stages of maximum diversity for clavatoraceans, especially in the Central Tethyan Realm. The non-marine basins of the Iberian Peninsula (Spain and Portugal), the Atlas Mountains (Morocco and Algeria) and the Pre-Dobrogean Depression (Romania and Georgia) were the most prominent clavatoracean areas during the Barremian and Aptian (Neagu and Georgescu-Donos, 1973; Feist et al., 1999; Martín-Closas, 2000). In these basins, clavatoraceans were dominant in alkaline and freshwater lakes, which continued to be their preferred habitats and which had reached a wide geographic extension by this period. This was produced by an increase in the tectonic rifting of the Peri-Tethyan basins (Salas et al., 2001). However, some species such as *Atopochara trivolvis,* also colonised brackish and fluviatile environments (Martín-Closas and Wang, 2008). They are found in such environments along with representatives of the other charophyte families, i.e. with *Porochara* and other porocharaceans in brackish environments and *Mesochara, Aclistochara* and *Sphaerochara* and other early characeans in fluviatile, siliciclastic facies.

The Early Cretaceous radiation of clavatoraceans in the Peri-Tethyan basins ended in the Latest Aptian, when a decrease in the number of species and especially in their biogeographic range occurred. The clavatoracean extinctions during the Latest Aptian and Albian have been attributed to the rise and diversification of aquatic angiosperms and to the selective pressure they exerted in the colonization of biotopes and life-habits that were unexplored during charophyte evolution, such as the floating habit on the water surface and mesotrophic and eutrophic habitats (Martín-Closas and Serra-Kiel, 1991; Martín-Closas, 2003). In addition, in the Central Tethys, the Albian represents a change in the hydrology and chemistry of lacustrine bodies, which evolved from alkaline lakes and swamps related to carbonate platforms into acidic ponds related to peat swamps in deltaic environments (Salas et al., 2001).

CLAVATORACEAN COSMOPOLITISM AND ENDEMISM

During the Early Cretaceous, the clavatoraceans displayed a dual palaeobiogeographic pattern: most species were limited in their biogeographic range to the Central Tethys, even to specific islands of the Central Tethyan Archipelago, while only 6 out of 23 species were able to extend to larger territories or achieve a cosmopolitan distribution. Two species illustrate this particular biogeographic pattern, *Globator maillardii* and *Atopochara trivolvis*. They represent two closely-related atopocharoideans according to the phylogenetic hypotheses proposed by Grambast (1974) and Martín-Closas (1996).

Genus *Globator* contains a single evolutionary species, *Globator maillardii* with many anagenetic morphotypes described as varieties (*Globator maillardii* var. *praecursor, maillardii, incrassatus, nurrensis, steinhauseri, mutabilis,* and *trochiliscoides*) evolving gradually over time. *Globator* represents one of the best biochronostratigraphic markers of the non-marine Early Cretaceous of the Tethyan Realm. As such, it has been widely used for dating non-marine sequences in the European basins (Grambast, 1974; Riveline et al., 1998; Mojon, 2002) and there is an excellent record of its biogeographic distribution. This species is known exclusively from the Peri-Tethyan basins in Europe and Northern Africa, during the Tithonian to the Early Aptian, when it became extinct. The absence of records elsewhere represents a real lack rather than a palaeontological bias, especially if we take into account the usefulness of this species to applied geology, which is why most fossil charophytes are sought after and studied. This species is a prime of a Central Tethyan endemism (Figure 1).

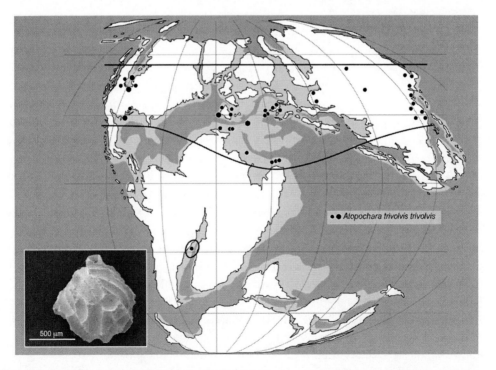

Figure 1. Distribution map of *Globator maillardii* showing the total range during the Early Cretaceous (endemic to Peri-Tethyan non-marine basins). Paleogeographic map of the world at -120 MA modified from Blakey (2006). Picture of *Globator maillardii trochiliscoides* from the Early Barremian of Pas du Frou, Subalpine Chains (France).

The evolutionary species *Atopochara trivolvis*, on the other hand, is probably the best known case of cosmopolitan distribution in the charophyte fossil record. Palamarev (1971), when many clavatoraceans were still unknown in large parts of the planet, identified the world-wide distribution of *Atopochara trivolvis* during the Aptian. *Atopochara trivolvis* is also an evolutionary species formed by a succession of gradualistic morphotypes or "chronospecies" considered to represent only the anagenetic changes within the same lineage over time (Martín-Closas and Schudack, 1997). The palaeogeographic distribution of these successive morphotypes, along with other chronostratigraphic evidence, clearly describes the historical biogeography of this species (Martín-Closas and Wang, 2008). The first representative, *Atopochara trivolvis* var. *horrida*, is recorded in the western part of the Central Tethyan Archipelago as early as the Early Berriasian, 145.5 million years ago. This original range lasted for several million years. The next morphotype (*A. trivolvis* var. *micrandra*), which first occurred during the Valanginian (140.2 million years ago), shows only a limited expansion to the eastern coast of the Central Tethys (Pre-Dobrogean Basin, Ukraine), in addition to Western Europe. However, once the species established a foothold on the Western Asiatic mainland, it expanded eastwards rapidly enough to cross the whole continent, reaching Southern China (Hengyang basin in Hunan Province) four million years later, during the Hauterivian (136.4 million years ago), where it took the morphotype *ancora* (Hu and Zeng, 1981; Wang and Lu, 1982).

During the next stage, the Barremian (130 to 125 million years ago according to Gradstein et al., 2004), the morphotype *triquetra* extended throughout the entire Eurasiatic continent, where it occurs in many basins of Northern and Southern China (Wang and Lu, 1982), Central Asia and even in Southeast Asia, beyond the paratropical latitudes that constituted its initial range. In the Barremian of Europe, *Atopochara trivolvis* var. *triquetra* was extremely abundant in almost all non-marine basins and quickly reached the northern coast of the African mainland (Andreu et al., 1988). It is probable that from that starting point, the species crossed the Guinea Corridor, which at that time held together Africa and South America, and reached the Argentinean and Brazilian basins during the Early Barremian (Musacchio, 2000). Crossing of Central Africa to reach South America is a significant milezone in the worldwide expansion of *Atopochara trivolvis* since an extremely arid climate, i.e. with few available habitats for charophytes to grow, is attributed to this region during the Barremian (Scotese, 2000)

By the Aptian, 125 million years ago, *Atopochara trivolvis* (morphotype *trivolvis*) is also recorded in the North American Interior Seaway, from Texas to the Rocky Mountains (Peck, 1957; Soulié-Märsche, 1994), in addition to all of the previously colonised continents. The colonization of North America represented the maximum extension of the species, which was truly cosmopolitan in the Northern Hemisphere and extended into South America as well (Fig. 2). Given our present state of knowledge, the late colonization of North America is difficult to understand and may be biased by a poor characterisation of the North American Barremian sequences (Sames et al. 2008). Whatever the actual time of arrival in North America, it is clear that during the Barremian-Aptian interval, over a period of about 18 million years, the species occurred in most of the Northern Hemisphere.

The dispersal of *Atopochara trivolvis* across the continents and seaways raises a number of questions. First, we should attempt to identify which features enabled the species to expand worldwide, while its close relative, *Globator maillardii* remained confined to its original biogeographic range in the Central Tethys. Second, it would be useful to determine which animal vector allowed Early Cretaceous charophytes to migrate over long distances. Comparison with extant charophytes may be useful in the exploration of these difficult questions.

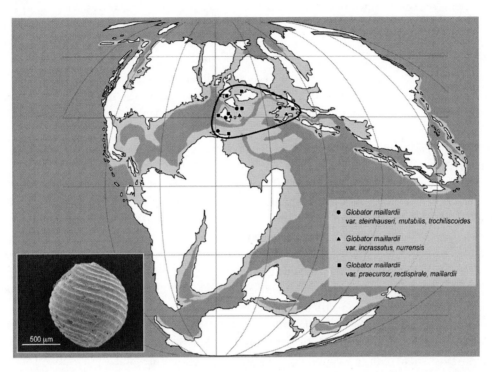

Figure 2. Distribution map of *Atopochara trivolvis* showing the maximum range, during the Aptian, when it displayed a cosmopolitan distribution. Paleogeographic map of the world at -120 MA modified from Blakey (2006). Picture of *Atopochara trivolvis triquetra* from the Upper Barremian of Las Hoyas, Iberian Chain (Spain).

The dispersal of extant charophytes is contingent to the transport of gyrogonites and oospores by migratory birds, especially waterfowl. The study of stomach contents of ducks of the Ebro Delta (Catalonia, Spain) resulted in the discovery of abundant charophyte remains, including oospores (Llorente Cabrera, 1984). Some species of ducks, such as the red-crested pochard (*Netta rufina*) are massive consumers of charophytes (Schmieder et al., 2006). Biogeographic studies involving species of the extant genus *Chara* showed that the extant cosmopolitan species are monoecious, i.e. they display male and female organs within the same plant. On the other hand, dioecious species, which develop into separate male and female plants, display biogeographic distributions that comprise, at most, large areas within one particular continent or that have no significant geographic barriers (Proctor, 1980). The best example of this biogeographic pattern is shown by species growing in oceanic islands of the Central Pacific; all of these species are monoecious. Clearly, monoecious species are able to colonise remote areas more easily than species with separate male and female plants (i.e. dioecious species). After a successful dispersal event, monoecious species will quickly develop fertile populations ready to supply new propagules for the next dispersal. On the other hand, dioecious populations are usually unisexual. Fertile oogonia are rare in these populations: and even in the case waterfowl-mediated dispersal, the new population will continue to be unisexual, thus complicating further dispersal.

Applying the data based on living charophytes to the Cretaceous species *Atopochara trivolvis* makes sense, since the fossil record of the species' fructifications clearly shows that the antheridia were borne upon the oogonium, which means that this species was monoecious. Antheridial marks are clearly visible in the calcified utricle around the oogonium of the earliest morphotypes (*Atopochara trivolvis* var. *horrida, maillardii, ancora, vidua, triquetra* and *trivolvis*). Later morphotypes do not show antheridial marks although this was due to the calcification process, as explained by Martín-Closas and Wang (2008). In conclusion, the monoecious character of *Atopochara trivolvis* would have enabled its rapid dispersal over long distances.

The comparison between Cretaceous clavatoraceans and extant representatives of genus *Chara* is less conclusive as regards as the animal vectors of dispersal. To the best of our knowledge, the capability of long-range flight by birds in the Barremian (Enantiornithes) was reduced. In recent years, however, aquatic birds belonging to the modern Ornithurae clade were described in the Aptian of Northern China (You et al., 2006) and show some adaptations to long-range flight, such as a prominent sternum. Unfortunately, the skull of *Gansus yumenensis*, is still unknown, which creates uncertainty as to whether this bird was herbivorous. Other possible dispersal vectors were certain herbivorous dinosaurs living in Barremian and Aptian wetlands, such as iguanodontids, which migrated in herds and could easily disperse clavatoracean utricles. This possibility has yet to be documented: thus far, only a small quantity of dinosaur digestive tract contents and coprolites have been analysed. In addition, unlike bird-mediated dispersal, the successful dispersal of *Atopochara trivolvis* by dinosaurs would require a continuous landscape of non-marine facies, preferably wetlands, across continents and even across different palaeoclimatic belts. This possibility has only been documented for certain parts of Europe and China during the Barremian and Aptian.

BIOGEOGRAPHIC PATTERNS DURING DECLINE TO EXTINCTION

The decline of clavatoraceans began in the Late Aptian and Albian (112 to 99.6 million years ago according to Gradstein et al., 2004), when a number of species disappeared even in environments where the family was more prolific, such as the lacustrine basins of the Central Tethys. As explained above, this decline paralleled the rise of aquatic angiosperms in the region along with the replacement of alkaline lakes by siliciclastic swamps (Martín-Closas and Serra-Kiel, 1991; Martín-Closas, 2003). Cosmopolitan clavatoraceans, such as *Atopochara trivolvis*, appear more resistant than endemic species to this initial decline. Thus, the Albian records of *Atopochara trivolvis* still cover the entire paratropical belt of the Northern Hemisphere (China, Europe and North America) but the number of occurrences is extremely limited compared previous stages. In China, Albian *Atopochara trivolvis* also shows some local morphological variation, suggesting that the population flows between different localities were not as intense as they had been in the past.

After the initial decline of clavatoraceans during the late Early Cretaceous, the charophyte fossil record contains a significant gap, lasting for at least for 15 million years, from the Cenomanian (99.6 to 93.5 million years ago) and to the end of the Santonian (83.5

million years ago). Thus, a few Cenomanian outcrops in Spain, China and the United States provide the only data on charophyte floras during the early Late Cretaceous (Peck, 1957; Feuillée and Grambast, 1961; Van Itterbeeck et al., 2005). During the next stage, the Turonian (93.5 to 89.3 million years ago), only one locality in the world has been reported with charophytes (Feist, 1981). In these early Late Cretaceous localities, clavatoraceans continue to be dominant elements of the charophyte floras. During 5.8 million years (Coniacean and Santonian stages) the charophyte fossil record is unknown and does not recover until the Campanian (83.5 million years ago). At that time, charophytes floras displayed a completely different physiognomy, with characeans instead of clavatoraceans dominating worldwide (Grambast, 1974). The complete absence of a charophyte fossil record during several stages of the Late Cretaceous has been attributed to a general reduction of non-marine basins due to the maximum sea-level during the Turonian (Martín-Closas and Serra-Kiel, 1991): this absence hinders the accumulation of evidence about the clavatoracean extinction patterns.

In the Latest Cretaceous (from 70.6 to 65.5 million years ago), two clades of clavatoraceans survived but were relegated to subordinate ecological and floristic positions in charophyte assemblages. On the one hand, clavatoroideans were represented in brackish facies from Southern Europe by a few species of genus *Heptorella* (Grambast, 1971; Feist et al., 2005). The nearest known ancestors of this species were Early Cretaceous charophytes endemic to the eastern part of the Central Tethys, in Lebannon (Martín-Closas, 1996). On the other hand, atopocharoideans were represented by a local morphotype of the cosmopolitan *Atopochara trivolvis*, known as *Atopochara trivolvis ulanensis*, which occurs in a few localities of Northern China and Mongolia in fluvial environments (Wang and Lu, 1982; Van Itterbeeck, 2005). This late representative of *Atopochara trivolvis* representative displays a plesiomorphic morphology in comparison to previous representatives and recalls certain Early Cretaceous morphotypes of the species. This leads to the suggestion that they may represent isolated and relict populations (Martín-Closas and Wang, 2008). Both *Heptorella* and the latest representatives of *Atopochara* disappeared around the Cretaceous-Tertiary boundary which marks the final extinction of the family Clavatoraceae.

Based on these incomplete data, it is difficult to analyse the historical biogeography of the Late Cretaceous clavatoraceans. It appears that in comparison to most of the Early Cretaceous, the Late Cretaceous was a time of biogeographic range reduction, barriers to population flows and relict or endemic species development.

CONCLUSION

The excellent fossil record of the charophyte family Clavatoraceae allows us to assess the historical biogeography of a fossil plant group, throughout the Cretaceous period. It also provides a broad time perspective for significant biogeographic patterns such as initial range, endemism, dispersal, cosmopolitism and biogeography during decline and extinction.

Beginning with their origin in the Late Jurassic, clavatoraceans were represented in large regions of the so-called paratropical climate belt of the Northern Hemisphere. The first records include a number of localities from the non-marine basins around the Central Tethyan region and the North American Interior Seaway.

The radiation of the family during the Early Cretaceous occurred according to a dual biogeographic scenario. Most of the clavatoracean species were endemic to the Central Tethyan region, which had an archipelago-like palaeogeography at that time, probably enhancing allopatric speciation. A well-known example of Tethyan endemism is the evolutionary species *Globator maillardii*, which survived for more than 30 million years. Only a few species, primarily in the Northern Hemisphere, expanded to larger regions, achieving cosmopolitan distribution between the Barremian and Aptian (from 130 to 112 million years ago). This was the case for the evolutionary species *Atopochara trivolvis*, another long-lasting lineage. Expanding to a cosmopolitan distribution was not a rapid process, but required several million years of progressive colonization. Cosmopolitism was probably enhanced by the conjoint arrangement of gametangia, as in extant *Chara*. Adaptability to new environments also appears to be a significant factor in the colonization of larger territories. The animal vectors of clavatoracean dispersal are poorly understood. The first evidence of waterfowl with long-range flight capability (the main dispersers of extant charophytes), is contemporaneous with the period during which some clavatoraceans became cosmopolitan. Other possible candidates for charophyte dispersal over large distances are the dinosaurs, but this possibility needs further study to be confirmed.

The historical biogeography of clavatoraceans during their decline to extinction can be divided into two stages. In the late Early Cretaceous, during the Late Aptian and Albian, a number of Tethyan endemic species became extinct. Cosmopolitan species survived for longer periods but suffered from biogeographic range reduction and limited exchange between isolated populations. During the Latest Cretaceous, after a significant gap in the fossil record, the last clavatoraceans were relegated to subordinate ecological and floristic roles. Their biogeographic range was extremely limited and isolated before their complete extinction near the Cretaceous-Tertiary boundary.

ACKNOWLEDGEMENT

This study is a contribution to project CGL2008-00809/BTE of the Spanish Ministry of Science and Technology and to projects 40572009 and 40632010 of the Chinese National Nature Science Foundation. The English text was corrected by Robin Rycroft (Universitat de Barcelona).

REFERENCES

Andreu, B., Canerot, J., Charriere, A. & Feist, M. (1988). Mise en évidence du Wealdien (Barrémien) dans le Moyen-Atlas (région de Boulmane, Maroc). *Comptes Rendus de l'Académie des Sciences* Paris, *307*, sér. II, 2069-2075.

Blakey (R) 2006. Palaeogeographic Maps of the World. 2006/06/09. Available from http://geology.about.com/od/paleomaps/Global_Paleogeographic_ Maps.htm

Feist, M. (1981). Charophytes du Crétacé Moyen et données nouvelles sur l'évolution des Clavatoracées. *Cretaceous Research, 2*, 319-330.

Feist, M., Lake, R. D. & Wood, C. J. (1995). Charophyte biostratigraphy of the Purbeck and Wealden of Southern England. *Palaeontology, 38*, 407-442.

Feist, M., Charrière, A. & Haddoumi, H. (1999). Découverte de charophytes aptiennes dans les couches rouges continentales du Haut-Atlas oriental (Maroc). *Bulletin de la Société Géologique de France, 170*, 611-618.

Feist, M., Grambast-Fessard, N., Guerlesquin, M., Karol, K., Lu, H., Mccourt, R. M., Wang, Q. & Shenzen, Z. (2005). *Treatise on Invertebrate Paleontology. Part B., Protoctista 1. Volume 1: Charophyta*. Boulder, USA: The Geological Society of America.

Feuillée, P. & Grambast, L. (1961). Présence d'*Atopochara multivolvis* Peck dans le Cénomanien d'Oña (prov. de Burgos, Espagne). *C.R. Sommaire des Séances de la Société Géologique de France, Séance du 19 juin 1961*, 202-203.

Gradstein, J. G., Ogg, A. G. & Smith, A. G. (2004). A Geologic Time Scale. New York, USA: Cambridge University Press.

Grambast, L. (1966). Structure de l'utricule et phylogénie chez les Clavatoracées. *Comptes Rendus des Séances de l'Académie des Sciences Paris, 262*, 2207-2210

Grambast, L. (1971). Remarques phylogénétiques et biochronologiques sur les *Septorella* du Crétacé terminal de Provence et les Charophytes associées. *Paléobiologie Continentale, 2*, 1-38.

Grambast, L. (1974). Phylogeny of the Charophyta. *Taxon, 23*, 463-481.

Harris, T. M. (1939). *British Purbeck Charophyta*. London, UK: British Museum (Natural History).

Helmdach, F. F. & Ramalho M. M. (1976). *Bisulcocypris algarbiensis* n. sp., un nouvel ostracode du Malm portugais, *Revue de Micropaléontologie, 19*, 156-161.

Hu, J. & Zeng, D. (1985). Charophyta assemblages of the Cretaceous to Paleogene in Hunan Province. *Oil & Gas Geology, 4*, 409-418 (in Chinese).

Llorente-Cabrera, G.A. (1984). *Contribución al conocimiento de la Biología y Ecología de cuatro especies de anátidas en el Delta del Ebro, Abstracts of PhD Thesis*. Barcelona, SPAIN: Publicacions de la Universitat de Barcelona.

Mädler, K. (1952). Charophyten aus dem Nordwestdeutschen Kimmeridge. *Geologisches Jahrbuch, 67*, 1-46

Martín-Closas, C. (1996). A phylogenetic system of Clavatoraceae (Charophyta). *Review of Palaeobotany and Palynology, 94*, 259-293.

Martín-Closas, C. (2000). Els caròfits del Juràssic superior i Cretaci inferior de la Península Ibèrica. *Arxius de la Secció de Ciències de l' Institut d'Estudis Catalans, 125*, 1-304.

Martín-Closas, C. (2003). The fossil record and evolution of freshwater plants. A review. *Geologica Acta, 1*, 315-338.

Martín-Closas, C. & Diéguez, C. (1998). Charophytes from the Lower Cretaceous of the Iberian Ranges (Spain), *Palaeontology, 41*, 1133-1152.

Martín-Closas, C. & Serra-Kiel, J. (1991). Evolutionary patterns of Clavatoraceae (Charophyta) analysed according to environmental change during Malm and Lower Cretaceous. *Historical Biology, 5*, 291-307.

Martín-Closas, C. & Schudack, M.E. (1997). On the concept of species in fossil Charophyta. A reply to Feist & Wang. *Taxon, 46*, 521-525.

Martín-Closas C. & Wang, Q. (2008). Historical biogeography of the lineage *Atopochara trivolvis* PECK 1941 (Cretaceous Charophyta). *Palaeogeography, Palaeoclimatology, Palaeoecology, 260*, 435-451.

Mojon, P. O. (1989 a). Charophytes et ostracodes laguno-lacustres du Jurassique de la Bourgogne (Bathonien) et du Jura Septentrional Franco-Suisse (Oxfordien). Remarques sur les discontinuités emersives du Kimmeridgien du Jura, *Revue de Paléobiologie, volume spécial 3,* 1-18.

Mojon, P. O. (1989 b). Polymorphisme ecophenotypique et paléoécologique des Porocharcées (Charophytes) du Crétacé Basal (Berriasien) du Jura Franco Suisse, *Revue de Paléobiologie, 2,* 505-524.

Mojon, P. O. (2002). Les formations mésozoiques à charophytes (Jurassique moyen-Crétacé inférieur) de la marge téthysienne nord-occidentale (Sud-est de la France, Suisse Occidentale, Nord-est de l'Espagne). Sédimentologie, Micropaléontologie, Biostratigraphie. *Géologie Alpine, Mémoire Hors Série, 41,* 1-386.

Musacchio, E. A. (1971). Charophytas de la formación La Amarga (Cretácico inferior), Provincia de Neuquén, Argentina. *Revista del Museo de la Plata, 4,* 19-38.

Musacchio, E. A. (2000). Biostratigraphy and Biogeography of Cretaceous Charophytes from South America. *Cretaceous Research, 21,* 211-220.

Neagu, T. & Georgescu-Donos, M. O. 1973. Characeae Eocretacice die Dobrogea de Sud (Valea Akargea-Pestera). *Studi si cercetari de Geologie, Geofisica, Geografie, Serie Geologie, 18,* 171-185.

Palamarev, E. (1971). Fossile Charophyten aus der Unterkreide Nordbulgariens. *Mitteilungen des Botanischen Instituts , 21,* 145-159.

Peck, R. E. (1957). North American Charophyta. *Geological Survey Professional Paper, 294 A,* 1-44.

Salas, R., Guimerà, J., Mas, R., Martín-Closas, C., Meléndez, A. & Alonso, A. (2001). Evolution of the Mesozoic Central Iberian Rift System and its Cainozoic inversion (Iberian Chain). *Mémoires du Muséum national d'Histoire naturelle, 186,* 145-185.

Proctor, V. W. (1980). Historical biogeography of *Chara* (Charophyta) an appraisal of the Braun-Wood classification plus a falsifiable alternative for future consideration. *Journal of Phycology, 16,* 218-233.

Riveline, J., Berger J. P., Bilan W., Feist, M., Martín-Closas, C., Schudack, M. & Soulié-Märsche, I. (1996). European Mesozoic-Cenozoic Charophyte Biozonation, *Bulletin de la Société Géologique de France, 167,* 453-468.

Sames, B, Schudack, M. E. & Cifelli, R. L. (2008). Western Interior Early Cretaceous hiatus likely to be much shorter than previously reported—new biostratigraphic results derived from nonmarine ostracod correlations. *Roger Kaesler Memorial Meeting,* Abstract Session T43.

Schmieder, K., Werner, S. & Bauer, H. G. (2006). Submersed macrophytes as a food source for wintering waterbirds at Lake Constance. *Aquatic Botany, 84,* 245-250.

Scotese (C.R.). PALEOMAP Project . 2000/05/03. Available from: http://www.scotese.com

Schudack, M. E. (1993). Die Charophyten im Oberjura und Unterkreide Westeuropas. Mit einer phylogenetischer Analyse der Gesamtgruppe, *Berliner geowissenschaftliche Abhandlungen, Reihe A, 8,* 1-209.

Schudack, M. E., Turner C. E. & Peterson F. (1998). Biostratigraphy, paleoecology and biogeography of charophytes and ostracodes from the Upper Jurassic Morrison Formation, Western interior, USA, *Modern Geology, 22,* 379-414.

Van Itterbeeck, J., Horne, D. J., Bultynck, P. & Vandenberghe, N. (2005). Stratigraphy and palaeoenvironment of the dinosaur-bearing Upper Cretaceous Iren Dabasu Formation, Inner Mongolia, People's Republic of China. *Cretaceous Research, 26*, 699-725.

Wang, Z. & Lu, H. N. (1982). Classification and evolution of Clavatoraceae with notes on its distribution in China. *Bulletin Nanjing Institute of Geology and Palaeontology, Academia Sinica, 4*, 77-104 (in Chinese).

Winter, U. & Kirst, G. O. (1990). Salinity response of a freshwater charophyte, *Chara vulgaris. Plant, Cell end Environment, 13*, 123-134.

You, H. L., Lamanna, M. C., Harris, J. D., Chiappe, L. M., O'Connor, J., Ji, S. A., Lü, J. C., Yuan, C. X., Li, D. Q., Zhang, X., Lacovara, K. J., Dodson, P. & Ji, Q. (2006). A Nearly Modern Amphibious Bird from the Early Cretaceous of Northwestern China. *Science, 312*, 1640-1643.

In: Biogeography
Editors: M. Gailis, S. Kalninš, pp. 217-229

ISBN: 978-1-60741-494-0
© 2010 Nova Science Publishers, Inc.

Chapter 7

BIOGEOGRAPHY AND CITIZEN-SCIENCE

Vincent Devictor[1] and Coralie Beltrame[2]
[1]Edward Grey Institute, Department of Zoology, University of Oxford,
Oxford OX1 3PS, UK
[2]Sation Biologique de la Tour du Valat, le Sambuc, F-13200 Arles, France

ABSTRACT

Investigating biogeographic patterns and processes requires considerable amounts of data collected over large spatial and/or temporal scales. Availability of such large datasets has recently increased thanks to the rapid developments of geographical information systems, satellite images, and data accessibility through the Internet. But beyond these technical advances, many biogeographic studies are now based on data collected by volunteers from the general public, so-called citizen scientists. The shared principles of these programs most likely to improve large-scale investigations have hardly been highlighted. In this chapter, we first browse existing citizen-science monitoring programs particularly useful for biogeography. We then highlight whether and how these data are valuable to address current challenges in biogeography and large-scale conservation targets. Using concrete examples, we further explain why these data should be particularly efficient to develop the preventive and educational component of conservation biogeography.

INTRODUCTION

Biogeography is by definition highly demanding in large-scale datasets. Traditional descriptive approaches are increasingly coupled with the study of large-scale ecological processes shaping species distribution (so-called macroecology). Global land-use and climate changes have also led scientists to address conservation issues at large-scales. Biogeography is therefore now a growing field of ecology addressing most timely conservation and ecological issues (Lomolino & Heaney 2004). The recent increase in biogeographic studies is also both the stimulus and the response to the increase in large-scale datasets availability (Whittaker et al. 2005). Species-occurrence data at large scale have indeed exploded during the last decade, together with environmental electronic coverages, GIS technology, spatial statistics and large-scale niche-modelling techniques (e.g., now $> 10^8$ records of species distribution are freely available through the http://www.gbif.org portal, or from http://www.natureserve.org; while the electronic layer of several abiotic variables for the entire planet can be downloaded from http://www.worldclim.org).

However, these datasets are highly heterogeneous in their coverage and quality, as well as in their ability to address biogeographical and large-scale conservation issues. Surprisingly, although major methodological advances are recurrently highlighted in biogeography (e.g., capture-recapture algorithms, Eraud et al. 2007; spatial statistics, Diniz-Filho et al. 2008; handling spatial autocorrelation, Kissling & Carl 2008; predictive habitat techniques, Crossman & Bass 2008), drivers of progress in data collection have generally been ignored.

In this respect, the value of employing volunteers from the general public to collect data (i.e., citizen scientists) has been recognized for a long time (see Maltby 2003 for a review), and there has been a recent surge in biogeographic studies based on citizen science (McCaffrey 2005). However, the key strengths and limits of these datasets for biogeography and conservation have hardly been highlighted (but see Evans et al. 2005). Moreover, beyond academic scientific advances, citizen-science projects should also be a valuable framework to develop the preventive and educational components of conservation biogeography (Cooper et al. 2007). Indeed, the aim of these data-collection programs is also to promote public engagement with research and conservation programs while value judgments from the general public are now recognized as essential instruments to improve design and communication of biodiversity policies (Miller 2006, Evans et al. 2007, Fischer & Young 2007, D'elia et al. 2008).

In this chapter, we first review current citizen-science programs providing useful datasets for biogeography. We further highlight the key statistical strengths of these data and how they can be used to study global change impacts on biodiversity. We finally show why and how citizen-science monitoring programs are also of considerable interest to implement large-scale programs of environmental education.

I. THE ROLE OF CITIZEN-SCIENCE IN BIOGEOGRAPHY: SETTING THE SCENE

The principal goal of citizen-science programs is to monitor large spatial and temporal-trends of specific species. Interestingly, these programs are deeply anchored in free and easy access principles: they all have a specific website, which can be easily found on the Internet using their name (thereafter given in italic). For instance, the long-term temporal trend of many bird species can be derived from the *Christmas Bird Census,* which was launched in 1900. More than 50 other large-scale citizen science programs are also based on bird census, among which twelve programs are currently running in the USA (reviewed in The Duluth Audubon Society Website, http://www.duluthaudubon.org/citizen_science-birding.htm). A *Breeding Bird Survey,* largely based on volunteers is also now monitored in most European countries (*PECBM,* Gregory et al. 2005).

In fact, birding citizen-science programs can be considered as a showcase of the happy wedding between science citizens and conservation (Grenwood 2007). The reasons for this success are that birds are easy to census (Pereira & Cooper 2005) and that many volunteers are willing to contribute because this taxonomic group is attractive (Elzinga et al. 2001). However, several projects are increasingly being designed for other taxonomic groups such as amphibians (*FrogWatch*), spiders (*Spider WebWatch*), worms (*WormWatch*), mammals (*RoadWatch*), plants (*BudBirst*), mushrooms (*Mushroom observer*), fire flies (*FireFly*), moths (*National moths night*), butterflies (*Observatoire des Papillons des Jardins*).

Although several citizen-science programs are dedicated to focus on particular regional areas, most of these programs are covering national or continental-wide areas relevant for large-scale investigations (France, *Observatoire des Papillons des Jardins*; Europe, *EBCC*; World, *eBird*; India, *MigrantWatch*) and involve a very large number of people (e.g. the *CBC* involves more than 50,000 people). Therefore, from backyards and city streets to forests and farmlands, citizen-scientists represent, somehow, the world's largest research team (Irwin 1995).

The large-scale datasets from citizen science have contributed to study population structures and dynamics, species distributions and behaviors, but also to assist with the conservation of various organisms (Delaney et al. 2008, Greenwood 2007). Yet, more recent programs have focused on specific targets (e.g., dates of the phenophases of trees, shrubs, or flowers, *BudBurst*; effects of forest fragmentation on North American birds, *BFL* project; the dispersal of invasive species, *The National Institute of Invasive Species Science*; the spread of an infectious disease in a wildlife population, *The House Finch Disease Survey*). These data are also widely used at a national and international level to build indicators of sustainable development (Gregory et al. 2005).

However, scientific results based on citizen-science are often questioned because they rely on correlative results involving non-professionals, instead of being based on traditional hypothetico-deductive approaches carried by professionals. Yet, major advances of theoretical macroecology and biogeography are based on large-scale datasets of coarse resolution (Blackburn & Gaston, 2003). It is therefore crucial to highlight the strengths and the weaknesses of citizen-science datasets to gain confidence in biogegraphic studies based on these data.

II. STRENGTHS AND CAVEATS OF CITIZEN-SCIENCE DATASETS FOR BIOGEOGRAPHY

A. Statistical Properties and Relevance of Citizen-Science Datasets

1. Studying large temporal and spatial trends and testing explicit predictions

For a given sampling effort, there is an inherent trade-off between monitoring biodiversity in a few plots continuously, versus monitoring many plots sporadically. The first approach gives very detailed information on what is happening at a few points over space and/or time. The second, which is generally what citizen-programs are collecting, provides a way of extrapolating very local results to a broader scale. When datasets are based on randomization and replication of many surveyed plots, it is then reasonable to draw broad conclusions, although the raw data of a particular plot only capture a small amount of information. Moreover, citizen-science programs are highly valuable to set perennial data collection at large-scale: the program still runs despite the turnover in citizen-scientists.

Specific methods and statistics can then be developed to handle citizen-science datasets a posteriori when needed (Link et al. 2006).

But the relevance of using citizen-science data in biogeography largely depends on the question being asked. For instance, citizen-science datasets are highly useful to estimate large-temporal and spatial trends (e.g., the decline of given species in the last 20 years, or the range expansion of an invasive species at a continental scale). These trends can be highly informative *per se*, and are unlikely resulting from the process of data collection, but clearly reflect large-scale processes. In other words, when using large-scale datasets from citizen-science, the quality and statistical properties of datasets matter, but so do ecology. In this respect, large-scale datasets from monitoring programs are highly valuable to test explicit predictions stated *a priori* rather than to draw hazardous correlations (Yoccoz et al. 2001). Using this approach, monitoring data were shown to be useful to highlight large-scale processes shaping community structure and composition (Julliard et al. 2006), to test general ecological predictions about species responses to land use-changes (Devictor et al. 2008), global warming (Julliard *et al.* 2004), or impacts of acid rain (Hames et al. 2007). In fact, data from citizen-science can generally best be viewed as valuable in terms of relative rather than absolute numbers. Indeed, large-scale monitoring programs are often impaired by intrinsic biases such as heterogeneity in detectability.

2. The problem of imperfect detectability

In practice, an important methodological source of bias has been emphasized concerning animal or plant surveys, the so-called heterogeneity of species detection. This bias is inherent to the fact that counts of individuals or species are the result of two processes: the true presence (or absence) of a species or individual, and the ability of the observer to detect an individual. Parameter estimations from citizen-science can be biased either at population (Royle et al. 2005), or at community level (Boulinier et al. 1998). Indeed, if variation in detectability among species and/or individuals is not accounted for, an unknown part of the variation in presence or abundance of a given species will result from variation in detectability, regardless of its true variation (Bas et al. 2008). Therefore, using counts (per unit effort) as an index of abundance was early considered to be neither scientifically sound nor reliable (Burnham 1981).

Hopefully, for established protocol of most citizen-science monitoring programs (such as point counts), there is a body of scientific literature (e.g., so called capture-recapture algorithms) allowing conversion of raw detections to actual population estimates (see Williams et al. 2002 for a review). Moreover, most programs rely on only a selected list of the most common species. These chosen species are easier to identify and more abundant. They thus provide less biased data as they carry less false (and non-detection) events.

3. The statistical power and robustness of citizen-science datasets

Most monitoring programs are designed to maintain a major source of variation consistent over time (observer ability, route length, duration, time of day). Impacts of biases on trends produced by these data are thus minimized because they are held constant. This basic standardization greatly enhances the statistical power of the data when testing change in relative abundance in space and/or time. But the major strength of datasets from citizen-science is their large sample size (Greenwood 2007) which ensures a great statistical power (e.g., the probability of detecting a trend of interest using a regression) and high robustness (e.g., the stability of the trend to change in datasets). In fact, as soon as an adequate quantity of data is available, there are a lot of techniques for extracting useful information from imperfect datasets (Nakagawa & Cuthill 2007). Besides, although based on time saving techniques (without specialist training), data from citizen-scientists and from specialists often yield similar results (Newman et al. 2003, Delaney et al. 2008).

Interestingly the robustness of particular results derived from citizen-science can be explicitly assessed. For instance, one can check whether population trends are consistent among different citizen-science surveys (Lepage & Francis 2002) or using cross-validation (Henry et al. 2008). The former method would confirm that results are not dependent on the particular observers or protocol used (e.g., population trends from the Dutch and British common breeding bird surveys compare well with the French breeding bird survey, Julliard et al. 2004). In cross-validation, part of the data is used for building a statistical model, and part of the data is used to assess whether and how this model is affected by change in the data used. More generally, integrating biodiversity information across monitoring schemes was shown to be highly informative to build pan-European population trends (Gregory et al. 2005) and building composite trends from different citizen-science program should increase in the future by the development of appropriate methods (Henry et al. 2008).

B. A Tool-Box for Investigating Global Changes impacts on biodiversity

The best way to study consequences of global changes on biodiversity is to use data from large multi-site/multi-species monitoring programs, best able to provide considerable amounts of standardized data across taxa (Balmford 2005). The trends of common species estimated from citizen-science provide biodiversity indicators both scientifically sound and useful for decision makers (Gregory et al. 2005) and can also be used for protected area assessment (Devictor et al. 2007). Large-scale data on common species can also serve as proxy for measuring complex patterns or processes. For instance, spatial patterns of species richness as well as turnover in community composition are often better described by recording distributions of common rather than rare species (Lennon et al. 2004; Gaston et al. 2007).

But apart from counting species and individuals, citizen-science could also be useful to set large-scale experiment valuable for biogeography. For instance, *Monarch Watch* is a citizen-science project involving volunteers across the United States and Canada who tag individual butterflies. The tagging program helps answer questions about the geographic origins of species, the timing and pace of the migration, mortality during migration, and changes in geographic distribution.

III. Beyond Data Collection: The Role of Citizen-Science for Conservation Biogeography

A. Reconnecting People to Nature from Citizen-Science

By 2050, as many as three quarters of the world's human population will live in cities and suburbs (Cohn 2005). Humans may thus have progressively fewer opportunities for first-hand experience of wild species and progressively get disconnected from nature. Biology and ecology are only taught at school and naturalist books are often quite technical for beginners. In fact, it is likely that although most urban people appreciate nature (at least for outdoor recreation), they don't know how to deal with it: how to behave, what to watch, what can be a source of interest and wonder. Getting people interested in nature conservation in their everyday life is thus clearly not an easy task. Yet, real contacts with wild species may be vital in stimulating an appreciation of the natural world and the desire to conserve it (Miller 2005).

Most of the programs of citizen science explicitly deal with this difficulty in proposing practical tools for neophytes to recognize the species under study (e.g., in *FeederWatch* tips for distinguishing between similar looking species as well as specific guides can be downloaded from the website of the program). They often give just the good piece of information on everyday species to help beginners and step by step procedures (for instance, the colours of the light to differentiate the three genuses of fireflies in the program *Firefly Watch*). Using citizen-science could therefore be a promising way for reconnecting people to nature (Miller 2006).

Moreover, the strength of citizen-science programs also relies on the curiosity and pleasure of the volunteers to learn and observe things they have never noticed in their familiar place. Obviously, everybody would not find it interesting. But many citizen-scientists are astonished by their own, unsuspected capacity of observation as well as by the diversity of species (as well as the ecology and behaviours of the species) they observe in their usual surroundings (Cooper et al. 2007). Citizen-science is thus a way to realize how common species are related to "the scale of human experience" (Horwitz et al. 2001, Miller 2005). Indeed, in citizen science, everyone can be involved in a simple field-based science which is explicitly dedicated to tackle issues of importance where people live and work.

B. Reconnecting People to Science from Citizen-Science

Occidental societies consider scientific proofs and expert certifications as marks of value and safety. Citizens are curious about science, as proved by the numerous scientific magazines, emissions and expositions. But, at the same time, people are also more and more suspicious about scientific progress, often because it is followed by technological or biotechnological risks. Moreover, scientists are often hyper-specialized persons, not prone to vulgarise their results (they are often not expected and/or reluctant to spend time in explaining their researches). By contrast, citizen science programs directly involve people and their children into real scientific programs and are great opportunities to demystify scientists and their approaches. The aim is not, as in a pedagogic action, to have them "playing the scientist" (which would be less successful, especially with adults) but to let people think scientifically by themselves in being a scientist coordinated by a recognized institution (Trumbull et al. 2000).

The success of the program relies on everyone knowing that data are not collected only for fun: important questions are being addressed and results are to be published. This concrete application of their action is also a guarantee for data quality as it gives a sense of responsibility to the participants. As you trust in people's capacity and willingness, most of them will do their best. Interestingly, while some programs require specific skills (e.g. to recognize bird songs in Breeding Bird Survey), others explicitly rely on people with no particular skills. Again, this flexibility fills the gap between people who "know" nature and others who only like it.

Interestingly, getting volunteers involved in a real scientific work also have important implications for professional scientists. First, they have to explain clearly the global objectives of the program: what we want to study and why. In this respect, placing the program in a larger context can be very motivating is a good opportunity to highlight the key strength and reasons for such investigations. Second, the protocol (and any aspect of the particular methodology chosen) has to be clearly set. Making sure that the methodology is meaningful for observers will help people to understand the scientific approach. For instance, in the *Observatoire des Papillons des Jardins* program (a citizen-science monitoring program of French butterflies), people have to note only the greatest number of butterflies they see *at once* in their garden during one month. In this program, focusing on the maximum individuals recorded during a single observation session allows avoiding double counting of an individual. Third, scientists must inform participants about the program (e.g., number of participants, data collected and species seen) and communicate regularly their results. For these reasons, most programs propose regular feedback to observers, including visualization of data online or providing specific newsletters focusing on the most important results.

To reinforce their scientific education component, some programs also provide specific materials to be used by children with their parents or with their teachers at schools (see *Observatoire des Papillons des Jardins* for instance). These additional materials can explain how to perform a mini-statistical analysis in order to explain the scientific approach or to initiate experiences related to the species under study (caterpillar breeding and metamorphosis observation...). Even students may be involved (see *BudBurst* for instance). But we believe that citizen-scientists can not only help to set surveys, they can also become citizen-conservationists in being involved in specific large-scale conservation targets.

C. Using Citizen-Science to Spread Large-Scale Conservation Guidelines

Many citizen-science programs deal with conservation issues and the fate of particular species facing global changes. The threats that weigh on the group monitored is clearly explained to citizen-scientists. Citizen science programs can therefore go one step further in showing empirically what anyone can do to avoid the loss of biodiversity. For volunteers involved in citizen-science, the personal link created with the group of species monitored, as well as the awakening of a naturalist perception of their direct surroundings can be easily tuned to ecocitizen messages. This might be particularly relevant with monitoring programs based on people backyards (*Observatoire des Papillons des Jardins*, *FeederWatch*). Indeed, one feels free to do more or less what he wants in his private space. Simple acts such as stopping the input of chemical fertilizers and insecticides as well as the changing of lawns to meadow (e.g., at least in a part of the backyard), can provide very rapid results on insect conservation.

Citizen-scientists may thus be the best allies of managers and stake-holders to set-up large-scale conservation targets. Involving citizen participants directly in monitoring and active management of residential lands can generate very powerful management efforts, leading to positive, cumulative, and measurable impacts on biodiversity (Cooper et al. 2007). For instance, recreational fishers can be successfully involved to reduce drivers of fishery declines (Granek et al. 2008). More adaptive conception of ecological restoration, informed by local knowledge and citizen observations may be a better option than any top-down conservation restrictions (Evans et al. 2007).

CONCLUSION

The relevance of using any datasets in biogeography is obviously dependent on whether the question being addressed matches the quality of the data. In being geographically explicit, standardized, and by covering large spatial and/or temporal scales, citizen-science programs share specific characteristics generally needed for large-scale investigations. Given the success of these programs (e.g. more than 200 scientific publications investigating large-scale patterns and processes, including papers published in the best scientific journals, are based on citizen-science), we believe that they should increase and be encouraged in the future.

But citizen science has also helped to democratize large-scale conservation issues as they often celebrate some of the most common species that live with us in cities, towns and the countryside. In this respect, "familiar species", "wider countryside", "ordinary nature", and "everyday nature", are terms now frequently used in conservation biology and land-use policy. Nature protection is no longer solely considered as set apart from human activities and restricted to emblematic or rare species, nor the subject of pure academic science. The protection of a social nature encompassing a variety of environments and cultural contexts has gained credence (Kaplan et al. 1999). Cultural evolution is required, in both the scientific community and the public at large, to go beyond the measurement of biodiversity loss. We believe that people's actions can help to develop a more citizen-based biogeography and change the fate of declining common species just as they can help scientists to quantify global change impacts. The role played by biogeographical science in the emergence of conservation guidance is now acknowledged (Whittaker et al. 2005). We believe that developing conservation biogeography towards common and familiar species using citizen-science should provide a good opportunity to go beyond the data, to look at the values and visions that people hold for their own landscape.

REFERENCES

Balmford, A., Crane, P., Dobson, A. P., Green, R. E. & Mace, G. M. (2005). The 2010 challenge: data availability, information needs, and extraterrestrial insights. *Philosophical Transactions of the Royal Society B, 360,* 221-228.

Bas, Y., Devictor, V., Moussus, J. P. & Jiguet, F. (2008). Accounting for weather and time-of-day parameters when analysing count data from monitoring programs. *Biodiversity and Conservation, 17,* 3403-3416.

Blackburn, T. M. & Gaston, K. J. (2003). *Macroecology: concepts and consequences.* Blackwell Publishing, Oxford.

Boulinier, T., Nichols, J. D., Sauer, J. R., Hines, J. E. & Pollock, K. H. (1998). Estimating species richness: the importance of heterogeneity in species detectability. *Ecology, 79,* 1018–1028.

Burnham, K. P. (1981) Summarizing remarks: environmental influences. In: Ralph CJ, Scott JM (eds) Estimating numbers of terrestrial birds. *Studies in Avian Biology, 6,* 324–325.

Cohn, J. P. (2005). Urban wildlife. *BioScience, 55,* 201–205.

Cooper, C. B., Dickinson, J., Phillips, T. & Bonney, R. (2007). Citizen science as a tool for conservation in residential ecosystems. *Ecology and Society, 12,* 11. [online] URL: http://www.ecologyandsociety.org/vol12/iss2/art11/.

Crossman, N. D. & Bass, D. A. (2008). Application of common predictive habitat techniques for post-border weed risk management. *Diversity and Distribution, 14,* 213-224.

Diniz-Filho, J. A. F., Fernando, T., Rangel, L. V. B. & Bini, L. M. (2008). Model selection and information theory in geographical ecology. *Global Ecology and Biogeography, 17,* 479–488.

Delaney, D. G., Sperling, C. D., Adams, C. S. & Leung, B. (2008). Marine invasive species: validation of citizen science and implications for national monitoring networks. *Biological Invasions, 10,* 117–128.

D'elia, J., Zwartjes, M. & McCarthy, S. (2008). Considering legal viability and societal values when deciding what to conserve under the U.S. Endangered Species Act. *Conservation Biology, 22,* 1072-1074.

Devictor, V., Godet, L., Julliard, R., Couvet, D. & Jiguet, F. (2007). Can common species benefit from protected areas? *Biological Conservation, 139,* 29-36.

Devictor, V., Julliard, R., Clavel, J., Jiguet, F., Lee, A. & Couvet, D. (2008). Functional biotic homogenization of bird communities in disturbed landscapes. *Global Ecology and Biogeography, 17,* 252–261.

Elzinga, C. L., Salzer, D. W., Willoughby, J. W. & Gibbs, J. P. (2001). *Monitoring Plant and Animal Populations,* Blackwell Science.

Eraud, C., Boutin, J. M., Roux, D. & Faivre, B. (2007). Spatial dynamics of an invasive bird species assessed using robust design occupancy analysis: the case of the Eurasian collared dove (*Streptopelia decaocto*) in France. *Journal of Biogeography, 34,* 1077–1086.

Evans, C., Abrams, E., Reitsma, R., Roux, K., Salmonsen, L. & Marra, P. P. (2005). The Neighborhood Nestwatch Program: participant outcomes of a citizen-science ecological research project. *Conservation Biology, 19,* 589–594.

Evans, J. M., Wilkie, A. C., Burkhardt, J. & Haynes, R. P. (2007). Rethinking Exotic Plants: Using Citizen Observations in a Restoration Proposal for Kings Bay, Florida. *Ecological Restoration, 25,* 199-210.

Fischer, A. & Young, J. C. (2007). Understanding mental constructs of biodiversity: Implications for biodiversity management and conservation. *Biological Conservation, 136,* 271-282.

Gaston, K. J., Davies, R. G., Orme, C. D. L., Olson, V., Thomas, G. H., Bennett, P. M., Owens, I. P. F. & Blackburn, T. M. (2007). Spatial turnover in the global avifauna. *Proceedings of the Royal Society of London, Series B, 274,* 1567-1574.

Granek, E. F., Madin, E. M. P., Brown, M. A., Figueira, W., Cameron, D. S., Hogan, Z., Kristianson, G., de Villiers, P., Williams, J. E., Post, J., Zahn, S. & Arlinghaus, R. (2007). Engaging recreational fishers in management and conservation: global case studies. *Conservation Biology, 22,* 1125-1134.

Greenwood, J. J. D. (2007). Citizens, science and bird conservation. *Journal of Ornithology, 148,* 77-124.

Gregory, R. D., van Strien, A., Vorisek, P., Meyling, A. W. G., Noble, D. G., Foppen, R. P. B. & Gibbons, D. W. (2005). Developing indicators for European birds. *Philosophical Transactions of the Royal Society B, 360,* 269–288.

Hames, R. S., Rosenberg, K. V., Lowe, J. D., Barker, S. E. & Dhondt, A. A. (2002). Adverse effects of acid rain on the distribution of the Wood Thrush Hylocichla mustelina in North America. *Proceedings of the National Academy of Sciences, 99,* 11235-11240.

Henry, P. H., Szabolcs Lengyel, S., Nowicki, P., Julliard, R., Clobert, J., Čelik, Y., Bernd Gruber, B., Schmeller, D. S., Babij, V. & Henle, K. (2008). Integrating ongoing biodiversity monitoring: potential benefits and methods. *Biodiversity and Conservation,* 17, 3357-3382.

Horwitz, P., Lindsay, M. & O'Connor, M. (2001). Biodiversity, endemism, sense of place, and public health: inter-relationships for Australian inland aquatic systems. *Ecosyst Health 7,* 253-265.

Irwin, A. (1995). *Citizen science : a study of people, expertise, and sustainable development.* London ; New York: Routledge.

Julliard, R., Jiguet, F. & Couvet, D. (2004) Common birds facing global changes: what makes a species at risk? *Global Change Biology, 10,* 148-154.

Julliard, R., Clavel, J., Devictor, V., Jiguet, F. & Couvet, D. (2006). Spatial segregation of specialists and generalists in bird communities. *Ecology Letters, 9,* 1237–1244.

Kaplan, R., Ryan, R. L. & Kaplan, S. (1999). *With people in mind: design and management for everyday nature.* Island Press. 239 p.

Kissling, W. D. & Carl, G. (2008). Spatial autocorrelation and the selection of simultaneous autoregressive models. *Global Ecology and Biogeography, 17,* 59–71.

Lennon, J. J., Koleff, P., Greenwood, J. J. D. & Gaston, K. J. (2004). Contribution of rarity and commonness to patterns of species richness. *Ecology Letters, 7,* 81-87.

LePage, D. & Francis, C. M. (2002). Do feeder counts reliably indicate bird population changes? 21 years of winter bird counts in Ontario, Canada. *Condor, 104,* 255-270.

Link, W. A., Sauer, J. R. & Niven, D. K. (2006). Hierarchical model for regional analysis of population change using Christmas bird count data, with application to the American black duck. *Condor, 108,* 13-24.

Lomolino, M. V. & Heaney, L. R. (eds) (2004) Frontiers of Biogeography: new directions in the geography of Nature, Sinauer Press, Inc. Sunderland, MA.

Maltby, J. W. (2003). A Brief History of Science for the Citizen. Halsgrove. 208 p.

McCaffrey, R. E. (2005). Using Citizen Science in Urban Bird Studies. *Urban habitats, 3,* 70-86.

Miller, J. R. (2005). Biodiversity conservation and the extinction of experience. *Trends in Ecology & Evolution, 20,* 430-434.

Miller, J. R. (2006). Restoration, reconciliation, and reconnecting with nature nearby. *Biological Conservation, 127,* 356-361.

Nakagawa, S. & Cuthill, I. C. (2007). Effect size, confidence interval and statistical significance: a practical guide for biologists. *Biological Reviews, 82,* 591 – 605.

Newman, C., Buesching, C. D. & Macdonald, D. W. (2003). Validating mammal monitoring methods and assessing the performance of volunteers in wildlife conservation—"Sed quis custodiet ipsos custodies ?" *Biological Conservation, 113,* 189–197.

Pereira, H. & Cooper, H. D. (2005). Towards the global monitoring of biodiversity change. *Trends in Ecology & Evolution, 21,* 123-129.

Royle, J. A., Nichols, J. D. & Kéry, M. (2005). Modelling occurrence and abundance of species when detection is imperfect. *Oikos, 110,* 353–359.

Trumbull, D. J., Bonney, R., Bascom, D. & Cabral, A. (2000). Thinking scientifically during participation in a citizen-science project. *Science Education, 84,* 265-275.

Whittaker, R. J., Araújo, M. B., Jepson, P., Ladle, R. J., Watson, J. E. M. & Willis, K. J. (2005). Conservation biogeography: assessment and prospect. *Diversity and Distributions, 11,* 3-23.

Williams, B. K., Nichols, J. D. & Conroy, M. J. (2002). Analysis and Management of Animal Populations. Academic Press, San Diego, California.

Yoccoz, N. G., Nichols, J. D. & Boulinier, T. (2001). Monitoring of biological diversity in space and time. *Trends in Ecology & Evolution, 16,* 446–453.

In: Biogeography
Editors: M. Gailis, S. Kalninš, pp. 229-240

ISBN: 978-1-60741-494-0

Chapter 8

CONSERVATION BIOGEOGRAPHY: A VIEWPOINT FROM EVOLUTIONARY BIOGEOGRAPHY

Isolda Luna-Vega[1], Juan J. Morrone[2] and Tania Escalante[2]

[1]Departamento de Biología Evolutiva, Facultad de Ciencias, Universidad Nacional Autónoma de México (UNAM), Apdo. Postal 70-399, 04510 Mexico, D.F., Mexico.
[2]Museo de Zoología "Alfonso L. Herrera", Departamento de Biología Evolutiva, Facultad de Ciencias, Universidad Nacional Autónoma de México (UNAM), Apdo. Postal 70-399, 04510 Mexico, D.F., Mexico.

ABSTRACT

The relevance of biogeographical evolutionary analyses for conservation science is discussed and highlighted. The general methods that can be applied to prioritize areas for protection at regional global scales are briefly commented and exemplified with Mexico as a case study. We conclude that a biogeographical atlas, representing a synthesis of distributional patterns of taxa from a country or area, may help determine priorities for the selection of areas for conservation.

INTRODUCTION

The main goals of biogeography are to analyze the distributional patterns of taxa and hypothesize on the processes that may have shaped them. Biodiversity has an important spatial component, because different areas have different levels of taxonomic representation, so it is essential to document patterns of biodiversity at different spatial scales (Morrone 1999a). Whittaker et al. (2005) recently examined the role of biogeography in biodiversity conservation, recognizing "conservation biogeography" as a sub-discipline of both biogeography and conservation biology. Conservation biogeography consists in the application of biogeographic methods to problems concerning biodiversity conservation. Unfortunately, these authors concentrated in ecological biogeography, not including any evolutionary biogeographic approach. We consider that evolutionary biogeography (Morrone

2007) can also provide explicit methods to help prioritize putative or existing protected areas (Morrone and Crisci 1992; Espinosa and Morrone 1998; Morrone 1999ab, 2000). The objective of this contribution is to comment on these evolutionary biogeographic methods, and discuss how they can contribute to conservation biogeography.

The identification of priority areas for conservation is a basic task of conservation biogeography. But what should we choose to protect? Areas with higher species richness or with higher concentration of endemic species? Prance (1994), among other authors, argued that the identification of areas with species richness was essential for conservation. In some cases, however, many of the important components of biodiversity may not be present in areas with higher species richness, because some of them can have wide distributions, not being immediately endangered. Thirgood and Heath (1994) considered that any strategy intending to preserve biodiversity should choose sites with a high concentration of endemic species, mainly those with very restricted distributions, since the loss of these areas may imply the extinction of unique lineages.

Biodiversity conservation represents a major environmental concern, so it is essential to identify priority areas and rank them into a network of reserves (Margules and Pressey, 2000; Sarkar *et al.*, 2002; Halffter, 2005). During the last three decades, panbiogeographic and cladistic approaches have been used in biodiversity conservation (e.g. Grehan 1989, 1993, 1995; Vane-Wright et al. 1991; Morrone and Crisci 1992, 1993; Morrone 1999ab, 2000; Espinosa and Morrone 1998; Morrone and Espinosa 1998; Luna-Vega et al. 2000; Morrone and Márquez 2003, among others).

Following Espinosa and Morrone (1998), panbiogeographic and cladistic biogeographic methods can be used to select and rank priority areas for biodiversity conservation. Panbiogeography (Croizat 1958, 1964) is used first to recognize primary biogeographic homology and order the taxa studied in different biotic components (Morrone 2007). Additionally, nodes are recognized in the areas where different biotic components intersect, these nodes deserving a especial conservation status based on their complex biotic composition (Grehan 1993). Cladistic biogeography (Nelson and Platnick 1981) may be then applied, when phylogenetic analyses are available. In order to evaluate the relevance of the species implied, we can also use phylogenetic indices (i.e. Vane-Wright et al. 1991; see Salinas 2003). The result of these analyses can contribute to the development of a biogeographical atlas.

The challenge for conservation biology is to document the evolutionary structure of biodiversity through a natural classification system. In this context, the idea of "natural" is the same applied to systematics, implying the monophyly of the studied unities. It is important to incorporate the concept of "naturalness" to give a scientific basis to establish priorities for conservation. In the biogeographical atlases, panbiogeographic and cladistic biogeographic methods allow to apply the concept of spatial homology, by means of delimiting "natural" biogeographical areas supported by shared endemic species (Morrone and Espinosa 1998). Natural biotic classifications require a concept of biogeographic homology that groups areas according to their history. In the past, areas were related by their global similarity, resulting in some cases to be artifacts. To develop a biogeographic atlas, it is necessary to use different approaches of evolutionary biogeography, e.g. panbiogeography, methods for recognizing areas of endemism, and cladistic biogeography.

Selection of Taxa

The analysis begins with the selection of taxa inhabiting the areas of interest. Information is taken from monographs and databases of biological collections, preferably for those taxa with phylogenetic information available. Databases should be considered with caution, because in some cases the identification of taxa is doubtful (Contreras-Medina and Luna, 2007). However, it is possible to apply quality controls to the databases and improve their quality (Graham 2004; Martínez-Meyer and Sánchez-Cordero 2006; Rodríguez-Tapia and Escalante 2006).

Using more taxa sampled, we can have more confidence in the selection and ranking of areas. An adequate selection of taxa will emphasize those with restricted distribution and ecologically rare, because if we protect these species, we can warrant the protection of the common ones (Sarkar and Margules 2002, Sarkar et al. 2002, Sarkar et al. 2006).

Identification of Generalized Tracks and Nodes

Panbiogeography or track analysis is a biogeographical approach that attempts to reintroduce and re-emphasize the importance of the spatial or geographical dimension of biodiversity for the understanding of evolutionary patterns and processes (Craw et al. 1999). It was originally developed by Croizat (1958, 1964), who established an objective method to represent the spatial geometry of biodiversity. Some authors have proposed the application of panbiogeographic or track methods to identify priority areas for biodiversity conservation (Morrone and Crisci 1992; Grehan 1993; Morrone and Espinosa 1998; Luna-Vega et al. 2000; Contreras-Medina et al. 2001; Álvarez-Mondragón and Morrone 2004; García-Marmolejo et al. 2008).

A panbiogeographic analysis (Figure 1) requires mapping localities of different taxa and connecting them with individual tracks, according to their minimal geographical proximity. Tracks resulting from the coincidence of different individual tracks are considered generalized tracks, which indicate the pre-existence of ancestral biotas that were fragmented in the past due to tectonic and/or climate change (Craw et al. 1999). Areas where two or more generalized tracks intersect are considered as nodes, which represent the spatial and temporal overlap of different biotic and geological components (Morrone 2007). Nodes are particularly important for conservation because they contain biotic elements from different origins, and can qualify as 'hotspots'. Thus they allow to take into account not only the number of species, but the degree of difference between the biotas overlapping there (Morrone and Crisci 1992; Craw et al. 1999). The identification of nodes may be complemented with the calculation of phylogenetic indices to such taxa with available phylogenetic information (Morrone 1999b).

Identification of Areas of Endemism

Within each of the generalized tracks obtained, we may delineate areas of endemism. An area of endemism is defined by the superposition of the distributional areas of two or more different restricted taxa (Morrone 1994). The distributional area of a single taxon is the geographic space occupied by it. Data used to recognize the distributional area of each taxon are the localities where it has been recorded. To detect areas of endemism, we can apply different methods: parsimony analysis of endemicity (Rosen 1988; Morrone 1994), endemicity analysis (Szumik et al. 2002; Szumik and Goloboff 2004), nested clade analysis (Deo and DeSalle 2006) and network analysis (Dos Santos et al. 2008), among others. Parsimony Analysis of Endemicity or PAE (Figure 2) is the most widely used (Nihei 2006). It groups areas (analogous to taxa) by their shared taxa (analogous to characters) according to the most parsimonious cladogram. PAE data consist of area x taxa matrices, where taxa are coded for their absence (0) or presence (1) in each area in the data matrix, and the resulting cladograms represent nested sets of areas (Morrone 2007).

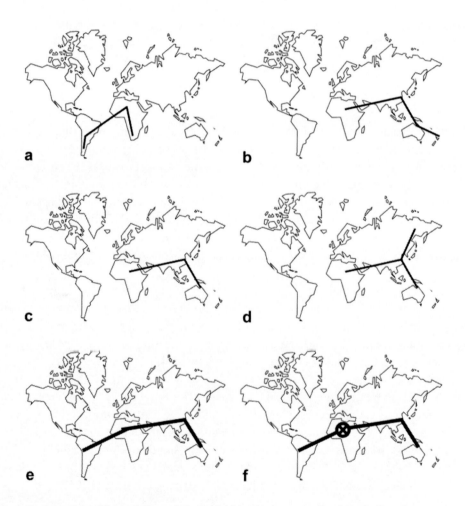

Figure 1. Panbiogeographic analysis. a-d, individual tracks; e, generalized track; f, node.

PAE allows to find those species that are related according to the localities where they are simultaneously present (Espinosa and Morrone 1998). Once these species are detected, they can be mapped using a Geographical Information System, in such a way that the naturalness of the areas of endemism is confirmed and defined by the maximum superposition area of the species involved.

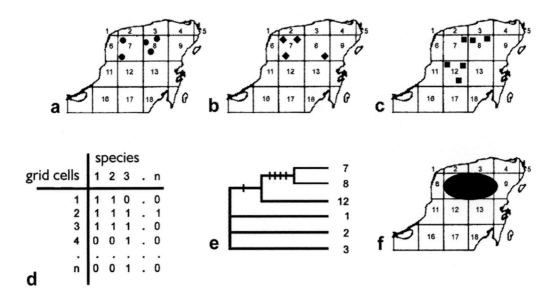

Figure 2. Parsimony analysis of endemicity. a, map with grid-cells; b, data matrix; c, cladogram; d, area of endemism.

Determining Relationships between Areas of Endemism

Cladistic biogeography searches for patterns of relationships among areas of endemism based on the phylogenetic relationships of the taxa inhabiting them (Humphries and Parenti 1999; Morrone 2007). It was originally developed by Nelson, Rosen and Platnick (Rosen 1976; Nelson and Platnick 1980, 1981). Interpretation of cladistic biogeographic results usually focuses on vicariance rather than on dispersal events, because vicariance affects different groups of organisms simultaneously (Nelson and Platnick 1981; Morrone 2007). A cladistic biogeographic analysis (Figure 3) implies the construction of area cladograms from at least two different taxonomic area cladograms and the derivation of general area cladogram(s).

Area cladograms are constructed replacing the names of the terminal taxa in a taxonomic cladogram with the names of the areas where they occur. This can be a trivial task, if every taxon is endemic to a unique area and every area includes a single taxon. Problems arise when there are widespread taxa (terminal taxon present in more than one area), missing areas (areas absent in the cladogram) and redundant distributions (areas containing more than one taxon). These problems complicate the analysis and taxon-area cladograms are treated under assumptions 0, 1 and 2, in order to obtain resolved area cladograms. Then, based on the information of the resolved area cladograms, a general area cladogram is derived. Methods available include component analysis (Humphries and Parenti 1999), Brooks Parsimony Analysis or BPA (Wiley 1987; Kluge 1988; Brooks 1990), three area statements (Nelson and Ladiges 1991, 1992, 1993) and paralogy-free subtrees (Nelson and Ladiges 1995; Ladiges et al. 1997, 2005; Contreras-Medina et al. 2007), among others.

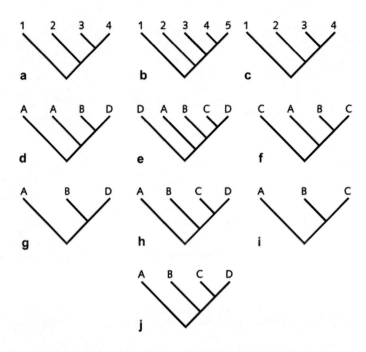

Figure 3. Cladistic biogeography. a-c, taxonomic cladograms; d-f, taxon area cladograms, g-I, resolved area cladograms; j, general area cladogram.

Ranking Areas

At a regional level, phylogenetic indices (Vane-Wright et al. 1991, Faith et al. 2004) can be used to rank the species, and with this, the particular value of each area can be calculated, summing the values of all the species. Also, the complementarity criterion may be applied to increase the efficiency of the phylogenetic index (Espinosa and Morrone 1998). The conservation coefficient (Dony and Denholm 1981) can be used to compare areas and to decide which one may be considered prioritary.

In order to rank the areas for conservation, we should consider that nodes should have the first priority, because they contain members of different biotas. Within each generalized track, the areas of endemism should be ranked according to their phylogenetic value. Finally, within each area of endemism, patches may be ranked according to their conservation coefficient.

Biogeographic Atlases

A biogeographic atlas allows to document efficiently patterns of biological diversity, with respect to conservation and sustainable use (Morrone and Espinosa 1998). Biogeographic atlases represent a synthesis of distributional patterns of taxa from a country or biogeographic area, represented by tracks, nodes, areas of endemism and area cladograms. Atlases provide information related to the identification of diversity centers and with their relative relevance that should be taken into consideration to determine priorities in the selection of areas for conservation. Additionally, atlases may help identify areas and taxonomic groups that deserve to be studied in detail, maximizing the scientific potential of their future study, and let incorporate other type of data (ecological, geological, urban planning, among others).

A biogeographic atlas includes detailed maps of individual tracks from different taxa inhabiting certain biogeographic area or country; generalized tracks and nodes indicating the different biotic components involved, as well as their biological relevance and geographic characteristics; areas of endemism at different levels according to the different taxa analyzed, listing the endemic and shared elements, as well as the geologic and ecologic characteristics of these areas; and general area cladograms detailing the relationships among the different areas of endemism. Based on the patterns detailed in a biogeographic atlas, those areas suitable to be conserved can be identified.

Mexico as a Case Study

Panbiogeographic and cladistic biogeographic analyses have been already published for several Mexican taxa. Moreover, some areas of endemism have been described for animals and plants. These can be included in Mexican biogeographic atlas. Morrone and Llorente (2006) compiled a biogeographic atlas of Mexican insects, which contains generalized tracks and nodes for several families of Coleoptera, Siphonaptera, Lepidoptera and Hymenoptera.

The generalized tracks obtained by Morrone (2001), Contreras-Medina and Eliosa (2001) and Morrone and Márquez (2003) for the country can be used as a base. They can be complemented with other regional works based on different taxa, v. gr. Luna-Vega et al. (1999, 2000, 2001), Morrone et al. (1999), Espinosa et al. (2000, 2001), Dávila-Aranda et al. (2002), Luna-Vega and Alcántara (2002), Aguilar-Aguilar et al. (2003, 2005), Espadas-Manrique et al. (2003), Álvarez and Morrone (2004), Escalante et al. (2004, 2005), Corona and Morrone (2005), Méndez-Larios et al. (2005), Morrone and Gutiérrez (2005), Andrés et al. (2006), Espinosa et al. (2006), Huidobro et al. (2006), Morrone (2006), Torres and Luna-Vega (2006), Contreras-Medina et al. (2007), and Herrera et al. (2008).

The generalized tracks, nodes and areas of endemism can be used as priority places due to their high diversity. Then, algorithms for richness and complementarity can use tracks or nodes as targets of conservation (see Margules and Pressey 2000; Margules and Sarkar 2007). As endemic taxa can have small distributional areas, as well as rare species, both can be used as targets in some software for systematic conservation planning. Richness, complementarity and rarity rules have been implemented in some softwares, i.e. ResNet and ConsNet have been used to design conservation area networks (http://www.consnet.org/).

CONCLUSIONS

As Whittaker et al. (2005) recently argued, biogeographical analyses are fundamental for conservation science. Despite this, in their broad definition of conservation biogeography, the evolutionary approach was not taken into account. Herein we discuss the relevance of evolutionary biogeography for taking decisions to prioritize areas for conservation. We conclude that a biogeographical atlas, representing a synthesis of distributional patterns of taxa from a country or area, may help determine priorities in the selection of areas of conservation. As Whittaker et al. (2005) wisely expressed, we need to take conservation decisions *now* to save as much as we can, and a biogeographic atlas will let us guide reserve network design at local and more global scales, with the intention to collaborate with the development of a "grand synthesis" of approaches and organismical elements that can serve to generate general assumptions about protected area planning. The idea is that evolutionary conservation biogeography will have also a key role in conservation in the mediate future.

ACKNOWLEDGMENTS

Frank Columbus invited us to contribute with this manuscript. Adolfo Navarro and Raúl Contreras made useful suggestions to a first draft of this manuscript.

REFERENCES

Aguilar-Aguilar, R., Contreras-Medina, R. & Salgado-Maldonado, Y. G. (2003). Parsimony analysis of endemicity (PAE) of Mexican hydrological basins based on helminth parasites of freshwater fishes. *J. Biogeogr., 30,* 1861-1872.

Aguilar-Aguilar, R., Contreras-Medina, R., Martínez-Aquino, A., Salgado-Maldonado, G. & González-Zamora, A. (2005). Aplicación del análisis de parsimonia de endemismos (PAE) en los sistemas hidrológicos de México: Un ejemplo con helmintos parásitos de peces dulceacuícolas. *In* Llorente, J. & Morrone, J. J. (eds.). Regionalización biogeográfica en Iberoamérica y tópicos afines: primeras Jornadas Biogeográficas de la Red Iberoamericana de Biogeografía y Entomología Sistemática (RIBES XII.I-CYTED), Las Prensas de Ciencias, Universidad Nacional Autónoma de México, Mexico, D. F. pp. 227-239.

Álvarez-Mondragón, F. & Morrone, J. J. (2004). Propuesta de áreas para la conservación de aves de México, empleando herramientas panbiogeográficas e índices de complementariedad. *Interciencia, 29,* 112–120.

Andrés, R., Morrone, J. Terrazas, T. & López-Mata, L. (2006). Análisis de trazos de las especies mexicanas de *Rhus* subgénero *Lobadium* (Angiospermae, Anacardiaceae). *Interciencia, 31,* 900-904.

Brooks, D. R. (1990). Parsimony analysis in historical biogeography and coevolution: methodological and theoretical update. *Syst. Zool., 39,* 14-30.

Contreras-Medina R. & Eliosa-León, H. (2001). Una visión panbiogeográfica preliminar de México. *In* Llorente J. & Morrone, J. J. (eds.). Introducción a la biogeografía en Latinoamérica: teorías, conceptos, métodos y aplicaciones. Las Prensas de Ciencias. Universidad Nacional Autónoma de México. Mexico, D.F. pp. 137-148.

Contreras-Medina, R., Luna-Vega, I. & Morrone, J. J. (2007). Application of parsimony analysis of endemicity (PAE) to Mexican gymnosperm distributions: grid-cells, biogeographic provinces and track analysis. *Biol. J. Linn. Soc., 92,* 405-417.

Corona, A. & Morrone, J. J. (2005). Track analysis of the species of *Lampetis* (*Spinthoptera*) Casey, 1909 (Coleoptera: Buprestidae) in North America, Central America and the West Indies. *Caribb. J. Sci., 41*, 37-41.

Craw, R. C., Grehan, J. R. & Heads, M. J. (1999). *Panbiogeography: Tracking the history of life.* Oxford Biogeography ser. 11. New York. 229 pp.

Croizat, L. (1958). *Panbiogeography.* Vols. 1 and 2. Published by the author, Caracas.

Croizat, L. (1964). *Space, time, form: The biological synthesis.* Published by the author. Caracas.

Dávila-Aranda, P., Arias-Montes, S., Lira-Saade, R., Villaseñor, J. L. & Valiente-Banuet, A. (2002). Phytogeography of the columnar cacti (tribe Pachycereeae) in Mexico: a cladistic approach. *In* Fleming, T. H. & Valiente-Banuet, A. (eds.).Columnar cacti and their mutualists: evolution, ecology and conservation, University of Arizona Press, Tucson. pp. 25-41.

Deo, A. J. & DeSalle, R. D. (2006). Nested areas of endemism analysis. *J. Biogeogr., 33*, 1511-1526.

Dony, J. M. & Denholm, I. (1981). Some quantitative methods of assessment of the conservation value of ecologically similar sites. *J. Appl. Ecol., 22*, 229-238.

Dos Santos, D. A., Fernández, H. R., Cuezzo, M. G. & Domínguez, E. (2008). Sympatry inference and network analysis in biogeography. *Syst. Zool., 57,* 432-448.

Escalante, T., Rodríguez, G. & Morrone, J. J. (2004). The diversification of the Nearctic mammals in the Mexican Transition Zone: a track analysis. *Biol. J. Linn. Soc., 83*, 327–339.

Escalante, T., Rodríguez, G. & Morrone, J. J. (2005). Las provincias biogeográficas del componente mexicano de montaña desde la perspectiva de los mamíferos continentales. *Rev. Mex. Biodiv., 76*, 199-205.

Espadas-Manrique, C., Durán, R. & Argáez, J. (2003). Phytogeographic analysis of taxa endemic to the Yucatán Peninsula using geographic information systems, the domain heuristic method and parsimony analysis of endemicity. *Divers. Distrib., 9*, 313-330.

Espinosa, D., Aguilar, C. & Escalante, T. (2001). Endemismo, áreas de endemismo y regionalización biogeográfica. *In* Llorente J. & Morrone, J. J. (eds.). Introducción a la biogeografía en Latinoamérica: teorías, métodos y aplicaciones. Las Prensas de Ciencias, Universidad Nacional Autónoma de México, Mexico, D.F. pp. 31-37.

Espinosa, D., Llorente, J. & Morrone, J. J. (2006). Historical biogeographical patterns of the species of *Bursera* (Burseraceae) and their taxonomic implications. *J. Biogeogr, 33*, 1945-1958.

Espinosa, D. & Morrone, J. J. (1998). On the integration of track and cladistic methods for selecting and ranking areas for biodiversity conservation. *J. Comp. Biol., 3*, 171-175.

Espinosa, D., Morrone, J. J., Aguilar, C. & Llorente, J. (2000). Regionalización biogeográfica de México: Provincias bióticas. *In* Llorente J., González, E. & Papavero, N. (eds.). Biodiversidad, taxonomía y biogeografía de artrópodos de México: Hacia una síntesis de su conocimiento. Vol. II, *J. Universidad Nacional Autónoma de México,* Mexico, D.F., pp. 61-94.

García-Marmolejo, G., Escalante, T. & Morrone, J. J. (2008). Establecimiento de prioridades para la conservación de mamíferos terrestres neotropicales de México. *Mastozoología Neotropical, 15(1),* 41-65.

Graham, C. H., Ferrier, S. Huettman, F. Moritz, C. & Peterson, T. (2004). New developments in museum-based informatics and applications in biodiversity analysis. *Trends in Ecology and Evolution, 19(9)*, 497-503.

Grehan, J. R. (1989). Panbiogeography and conservation science in New Zealand. *New Zealand J. Zool., 16*, 731-748.

Grehan, J. R. (1993). Conservation biogeography and the biodiversity crisis: a global problem in space/time. *Biodivers. Letters, 1*, 134-140.

Grehan, J. R. (1995). Natural biogeographic patterns of biodiversity: the research imperative. *In* Herman, T. B., Bondrup-Nielsen, S., Martin, J. H., Willison, J. H. M. & Munro, N. W. P. (eds.). Ecosystem monitoring and protected areas. Proceedings of the 2nd international conference on science and the management of protected areas. Dalhouise University, Halifax, Nova Scotia, pp. 35-44.

Herrera, P., Delgadillo, C. Villaseñor, J. L. & Luna-Vega, I. (2008). Floristics and biogeography of the mosses of the state of Querétaro, México. *The Bryologist, 111*, 41-56.

Huidobro, L., Morrone, J. J., Villalobos, J. L. & Álvarez, F. (2006). Distributional patterns of freshwater taxa (fishes, crustaceans and plants) from the Mexican transition zone. *J. Biogeogr, 33*, 731-741.

Humphries, C. J. & Parenti, L. R. (1999). *Cladistic biogeography: interpreting patterns of plant and animal distributions.* 2nd ed. Oxford University Press, Oxford.

Kluge, A. G. (1988). Parsimony in vicariance biogeography: A quantitative method and a greater Antillean example. *Syst. Zool., 37*, 315-328.

Ladiges, P. Y., Kellermann, J. Nelson, G. Humphries, C. J. & Udovicic, F. (2005). Historical biogeography of Australian Rhamnaceae, tribe Pomaderreae. *J. Biogeogr., 32*, 1909-1919.

Ladiges, P. Y., Nelson, G. & Grimes, J. (1997). Subtree analysis, *Nothofagus* and pacific biogeography. *Cladistics, 13*, 125-130.

Luna-Vega, I. & Alcántara, O. (2002). Placing the Mexican cloud forests is a global context: a track analysis based on vascular plant genera. *Biogeographica, 78*, 1-14.

Luna-Vega, I., Alcántara, O., Espinosa, D. & Morrone, J. J. (1999). Historical relationships of the Mexican cloud forests: a preliminary vicariance model applying parsimony analysis of endemicity to vascular plant taxa. *J. Biogeogr., 26*, 1299-1305.

Luna-Vega, I., Alcántara, O., Espinosa, D. & Morrone, J. J. (2000). Track analysis and conservation priorities in the cloud forests of Hidalgo, Mexico. *Divers. Distrib., 6*, 137–143.

Luna-Vega, I., Alcántara, O., Espinosa, D. & Morrone, J. J. (2001). Biogeographical affinities among Neotropical cloud forests. *Plant Syst. Evol., 228*, 229-239.

Margules, C. R. & Pressey, R. L. (2000). Systematic Conservation Planning. *Nature, 405*, 243- 253.

Margules, C. R. & Sarkar, S. (2007). *Systematic Conservation Planning.* Cambridge University Press, Cambridge.

Martínez-Meyer, E. & Sánchez-Cordero, V. (2006). Uso de datos de colecciones mastozoológicas. In: Lorenzo, C., Espinoza, E., Briones, M. A. & Cervantes, Y. F. A. (eds.) Colecciones mastozoológicas de México. Mexico, D. F., AMMAC, pp. 177-186.

Méndez-Larios, I., Villaseñor, J. L., Lira, R., Morrone, J. J., Dávila-Aranda, P. & Ortiz, E. (2005). Toward the identification of a core zone in the Tehuacán-Cuicatlán Biosphere reserve, Mexico, based on parsimony analysis of endemicity of flowering plant species. *Interciencia, 30*, 267-274.

Morrone J. J. (1994). On the identification of areas of endemism. *Syst. Biol., 43*, 438-441.

Morrone, J. J. (1999a). Biodiversidad en el espacio: la importancia de los atlas biogeográficos. *Physis* (Buenos Aires) C, *55*, 47-48.

Morrone, J. J. (1999b). How can biogeography and cladistics interact for the selection of areas for biodiversity conservation?: A view from Andean weevils (Coleoptera: Curculionidae). *Biogeographica, 75*, 89-96.

Morrone, J. J. (2000). La importancia de los atlas biogeográficos para la conservación de la biodiversidad. *In* Martín-Piera, F., Morrone, J. J. & Melic, A. (eds.). Hacia un proyecto CYTED para el inventario y estimación de la diversidad entomológica en Iberoamérica: PrIBES, Monografías Tercer Milenio, no. 1. Zaragoza, pp. 69-78.

Morrone, J. J. (2001). *Biogeografía de América Latina y el Caribe*. Manuales y Tesis SEA. Vol. 3. Zaragoza. 148 pp.

Morrone, J. J. (2005). Hacia una síntesis biogeográfica de México. *Rev. Mex. Biodivers., 76*, 207-252.

Morrone, J. J. (2006). Biogeographic areas and transition zones of Latin America and the Caribbean Islands, based on panbiogeographic and cladistic analyses of the entomofauna. *Ann. Rev. Entomol., 51*, 467-494.

Morrone, J. J. (2007). Hacia una biogeografía evolutiva. *Rev. Chil. Hist. Nat., 80*, 509-520.

Morrone, J. J. & Crisci, J. V. (1992). Aplicación de métodos cladísticos y panbiogeográficos en la conservación de la diversidad biológica. *Evol. Biol.*, (Bogotá), *6*, 53-66.

Morrone, J. J. & Crisci, J.V. (1993). El retorno a la historia y la conservación de la diversidad biológica. *In* Goin, F. & Goñi, R. (eds.). Elementos de política ambiental. Cámara de diputados de la provincia de Buenos Aires, La Plata, pp. 361-365.

Morrone, J. J. & Espinosa, D. (1998). La relevancia de los atlas biogeográficos para la conservación de la biodiversidad mexicana. *Ciencia* (Mexico), *49,* 12-16.

Morrone, J. J., Espinosa, D., Aguilar, C. & Llorente, J. (1999). Preliminary classification of the Mexican biogeographic provinces: a parsimony analysis of endemicity based on plant, insect, and bird taxa. *Southwest. Nat., 44*, 507-514.

Morrone, J. J. & Gutiérrez, A. (2005). Do fleas (Insecta: Siphonaptera) parallel their mammal host diversification in the Mexican transition zone? *J. Biogeogr., 32*, 1315-1325.

Morrone, J. J. & Llorente, J. (eds.). (2006). *Componentes bióticos principales de la entomofauna mexicana.* Vols. I y II. Las Prensas de Ciencias, Universidad Nacional Autónoma de México. Mexico, D. F.

Morrone, J. J. & Márquez, J. (2003). Aproximación a un atlas biogeográfico mexicano: componentes bióticos principales y provincias biogeográficas. *In* Morrone, J. J. & Llorente, J. Una perspectiva latinoamericana de la biogeografía. CONABIO-Universidad Nacional Autónoma de México. Mexico, D.F., pp. 217-220.

Nelson, G. & Ladiges, P. Y. (1991). Three-area statement: standard assumptions for biogeographic analysis. *Syst. Zool., 40*, 470-485.

Nelson, G. & Ladiges, P. Y. (1992). TAS and TAX: MsDos computer programs for cladistics. New York and Melbourne.

Nelson, G. & Ladiges, P. Y. (1993). Missing data and three-item analysis. *Cladistics, 9*, 111-113.

Nelson, G. & Ladiges, P. Y. (1995). TAX: MsDOS computer programs for systematics. Published by the authors. New York and Melbourne.

Nelson, G. & Platnick, N. I. (1980). A vicariance approach to historical biogeography. *Bioscience, 30*, 339-343.

Nelson, G. & Platnick, N. I. (1981). *Systematics and biogeography: cladistics and vicariance.* Columbia University Press, New York. 567 pp.

Nihei, S. S. (2006). Misconceptions about parsimony analysis of endemicity. *J. Biogeogr., 33*, 2099-2106.

Prance, G. T. (1994). A comparison of the efficacy of higher taxa and species numbers in the assessment of biodiversity in the Neotropics. *Phil. Trans. R. Soc. Lond. B, 345*, 89-99.

Rodríguez-Tapia, G. & Escalante, T. (2006). Manejo e importancia de las bases de datos en colecciones biológicas. En: Lorenzo, C., Espinoza, E., Briones, M. A. & Cervantes, F. A. (eds.), Colecciones mastozoológicas de México. AMMAC, Mexico, D.F., pp. 133-150.

Rosen, B. R. (1988). From fossils to earth history: Applied historical biogeography. *In* Myers, A. A. & Giller, P. (eds.). *Analytical biogeography: An integrated approach to the study of animal and plant distributions.* Chapman and Hall. London, pp. 437-481.

Rosen, D. E. (1976). A vicariance model of Caribbean biogeography. *Syst. Zool., 24*, 431-464.

Salinas, J. L. (2003). Índices filogenéticos para la conservación: una discusión del método. *An. Inst. Biol., Ser. Zool., 74(1)*, 21-34.

Sarkar, S., Aggarwal, A., Garson, J., Margules, C. R. & Zeidler, J. (2002). Place prioritization for biodiversity content. *J. Biosci., 27(4)*, 339-346.

Sarkar, S. & Margules, C. (2002). Operationalizing biodiversity for conservation planning. *J. Biosci., 27(4)*, 299–308.

Sarkar, S., Pressey, R. L., Faith, D. P., Margules, C. R., Fuller, T., Stoms, D. M., Moffett, A., Wilson, K. A., Williams, K. J., Williams, P. H. & Andelman, S. (2006). Biodiversity conservation planning tools: present status and challenges for the future. *Annu. Rev. Environ. Resour., 31*, 123–159.

Szumik, A. C., Cuezzo, F., Goloboff, P. A. & Chalup, A.E. (2002). An optimality-criterion to determine areas of endemism. *Syst. Biol., 51*, 806-816.

Szumik, A. C. & Goloboff, P. A. (2004). Areas of endemism: an improved optimality criterion. *Syst. Biol., 53*, 968-977.

Thirgood, S. J. & Heath, M. F. (1994). Global patterns of endemism and the conservation of biodiversity. *In* Forey P. L., Humphries, C. J. & Vane-Wright, R. I. (eds.). *Systematics and conservation evaluation.* Clarendon Press, Oxford, pp. 207-227.

Torres, A. & Luna-Vega, I. (2006). Análisis de trazos para establecer áreas de conservación en la Faja Volcánica Transmexicana. *Interciencia, 31*, 849-855.

Vane-Wright, R., Humphries, C. & Williams, P. (1991). What to protect? Systematics and the agony of choice. *Biol. Conserv., 55*, 235-254.

Whittaker, R. J., Araújo, M. B., Jepson, P., Ladle, R. J., Watson, J. E. M. & Willis, K. J. (2005). Conservation biogeography: Assessment and prospect. *Divers. Distrib., 11*, 3-23.

Wiley, E. O. (1987). Methods in vicariance biogeography. *In* Hovenkamp, P., Gittenberger, E., Hennipman, E., DeJong, R., Roos, M. C., Sluys, R. & Zandee, M. (eds.). Systematics and evolution: a matter of diversity. University of Utrecht, Utrecht, pp. 283-306.

INDEX

I

J

N

O

T

U

urban areas, 175, 176, 177, 180, 185, 186, 187, 188, 192, 200
urban population, 185
urbanization, 175, 176, 177, 180, 181, 182, 184, 186, 187, 188, 189, 192, 198, 200
urbanized, 176, 184, 185, 186, 187, 188
utricle, 204, 210

V

Valencia, 175, 177, 178, 180, 181, 182, 183, 184, 185, 187, 188, 189, 190, 196, 197, 199, 200, 201
validation, 43, 221, 225
validity, 12
values, 16, 29, 35, 141, 143, 224, 225, 234
variability, 71, 96, 129, 132
variables, 23, 74, 91, 181, 218
variance, 74, 77, 85, 87
variation, 37, 59, 63, 65, 70, 71, 81, 82, 85, 86, 87, 91, 93, 94, 95, 97, 99, 101, 103, 106, 121, 130, 131, 135, 146, 150, 175, 184, 186, 187, 197, 210, 220
vector, 209
vegetation, 177, 199
Venezuela, 32, 60
vertebrates, 47, 54, 76, 138, 150, 189, 198
village, 189
visible, 210
visualization, 223
voice, 47
volcanic activity, 120

W

walking, 178
water, 44, 66, 69, 71, 76, 77, 78, 88, 96, 99, 103, 116, 191, 198, 206

waterfowl, 188, 209, 212
watershed, 122
wealth, 105, 119
web, 77, 144, 150
weedy, 185
West Africa, 105, 109, 110, 111, 119, 121, 123, 124, 126, 127, 128
West Indies, 34, 138, 139, 147, 150, 168, 169, 170, 171, 173, 174, 237
Western Europe, 203, 208
Western Hemisphere, 100
wetlands, 177, 186, 188, 189, 204, 205, 210
wildland, 177
wildlife, 199, 219, 225, 227
wind, 77
windows, 87
winter, 45, 77, 178, 182, 184, 185, 226
witnesses, 115
wood, 106, 115, 177, 188, 190
woodland, 132, 175, 179, 181, 185, 188
workers, 148
worms, 219

Y

Y-axis, 82
yield, 20, 172, 221

Z

Zambezi, 122
Zimbabwe, 108, 110
zoogeography, 33, 93
zooplankton, 93, 101